International Conference on Sustainable Development on Energy and Environment Protection 2017

IOP Conference Series: Earth and Environmental Science
Volume 86

Yichang, China
28 – 30 July 2017

ISBN: 978-1-5108-5109-2
ISSN: 1755-1307

Printed from e-media with permission by:

Curran Associates, Inc.
57 Morehouse Lane
Red Hook, NY 12571

Printed by Curran Associates, Inc. (2017)

For permission requests, please contact the Institute of Physics at the address below.

Institute of Physics
Dirac House, Temple Back
Bristol BS1 6BE UK

Phone: 44 1 17 929 7481
Fax: 44 1 17 920 0979

techtracking@iop.org

Additional copies of this publication are available from:

Curran Associates, Inc.
57 Morehouse Lane
Red Hook, NY 12571 USA
Phone: 845-758-0400
Fax: 845-758-2633
Email: curran@proceedings.com
Web: www.proceedings.com

Table of contents

Volume 86

2017 International Conference on Sustainable Development on Energy and Environment Protection

28–30 July 2017, Yichang, China

Accepted papers received: 29 September 2017
Published online: 2 November 2017

Preface

Papers

IOP Conf. Series: Earth and Environmental Science **86** (2017) 011001 doi:10.1088/1755-1315/86/1/011001

Preface

The 2017 International Conference on Sustainable Development on Energy and Environment Protection (SDEEP 2017, http://www.sdeep.org) was held during July 28-30, 2017 in Yichang, China, a beautiful city by the Three Gorge Dam. SDEEP 2017 covers a broad range of frontier topics related to its themes include but not limited to environmental science and ecology, terrestrial biology and ecology, aquatic biology and ecology, atmospheric chemistry, environmental analyses and monitoring, phytoremediation and ecosystem restoration, environmental microbiology/bio-based energy sources, assessment of risks and interactions of pollutants in the environment, conservation biology and sustainable agriculture, impact of chemicals/pollutants on human and animal health, new energy, oil and gas, etc., and serves as an optimal platform for academic communications converging specialists and scholars to exchange ideas and inspirations in the fields related to sustainable development and environment protection, including hot issues such as environmental science, energy development and climate change, aiming to promote the sustainable developments of human society, nature and earth. All papers included in the conference proceedings were peer reviewed according to IOP Publishing standards.

Great thanks to the conference Chairs and Committees, organizers, all the attendees and authors, and IOP Publishing and so on for their helps and supports, which made this conference a success. We hope the future conferences will continue to be an attractive platform for the academics and scholars in the fields to attend and contribute.

Professor Shihong Qin
Organizing Committee of SDEEP 2017

IOP Conf. Series: Earth and Environmental Science **86** (2017) 011002 doi:10.1088/1755-1315/86/1/011002

Peer review statement

All papers published in this volume of *IOP Conference Series: Earth and Environmental Science* have been peer reviewed through processes administered by the proceedings Editors. Reviews were conducted by expert referees to the professional and scientific standards expected of a proceedings journal published by IOP Publishing.

IOP Conf. Series: Earth and Environmental Science **86** (2017) 012001 doi:10.1088/1755-1315/86/1/012001

Remodeling and Reflection Historic District - Taking Qianmen Street as an example

Z WANG

Institute of Agricultural Information and Economics, Beijing Agriculture and Forestry Academy of Science, Beijing 100097.

E-mail:angelwang1009@163.com

Abstract. Qianmen area is no longer in those years of prosperity spectacular, there is a historical reason, a more important reason is its commercial positioning fuzzy, overall positioning error caused. This paper studies the Qianmen Street commercial development context, to extract the elements of Qianmen Street economy and cultural prosperity. We would like to see through the nature of the phenomenon and find Qianmen Street "soul", which stimulate the revival of the potential for the revival of the historic district of cultural, economic revival and morphological remodeling, providing development ideas and implementation strategies.

1.Introduction

Qianmen Street was founded 570 years ago, it is the only way forroyal ritual, hunting, patrol at Ming and Qing Dynasties, known as "Heaven Street". Sedimentary nearly 600 years, making Beijing Qianmen Street became the architectural culture, business culture, cultural hall, opera culture, folk culture precipitate profound characteristic historic district, is one of Beijing landmark. Today, however, Qianmen Street no longer have the same spectacular, so many people confused by its development. History Commercial Streets become ordinary street shops, tasteless attractions. By combing historical context of Qianmen Street, the author were compared for business forms and characteristics in different periods, to extract the Qianmen Street's economic and cultural boom operating mode. We would like to see through the appearance essence, Claim this lost Qianmen Street "soul" unleash the potential of Qianmen Street revival, provide ideas and implementation strategies for the development of the historic district of remodeling.

2.Overview of Qianmen Street

Qianmen Street is the most famous commercial street in Qianmen area. It is a traditional Beijing Commercial Street. Located in Beijing's central axis, starts at the northern end of Qianmen Door, reach the southern end of the bridge crossing. Jiajing twenty-nine years (1550), before establishing outside the city, this street was the Royal Road which Ming Royal to the Temple of Heaven, after outside the

IOP Conf. Series: Earth and Environmental Science **86** (2017) 012001 doi:10.1088/1755-1315/86/1/012001

city was established the street was the main north-south street. It was called as Zhengyang Street by the republic at the Ming and Qing Dynasty, commonly known as Qianmen Street. Since the Republic of China after the liberation, Beijing Qianmen Street has been one of the three most important commercial Old Town streets. It's officially named Qianmen Street in 1965.

Qianmen area is a commercial circle that the center is Qianmen Street, which includes both sides of the fish mouth, Dashanlan Street, Jewelry Street and other traditional city street bazaar. Many old and famous brands located here, it's one of the urban areas of Beijing's most historical and cultural charm.Literature References.

3.Causes of Qianmen Street boom

Qianmen Street in history to be Beijing's most bustling commercial street, the reasons are mainly following

3.1Special location

It was the only way for royal ritual, hunting, patrol at Ming and Qing Dynasties, known as "Heaven Street". Qianmen Street was also the only way for Beijing businessman. After the Ming Dynasty moved the capital to Beijing, Beijing's commercial center has moved to Qianmen Street. In addition, the expansion outside the city after the Ming Dynasty, the government building houses to attract resident stay in here, and also to attract businessmen operating. Wealthy businessmen in Zhejiang, Suzhou province and other places moved to here, brought prosperity to the region many passengers and goods, and promote the rapid development of all aspects of the Qianmen Area.

3.2 Old and famous shop or time-honored brand gathered

Historically, Qianmen Street is a gathering place of time-honored brands, these Old and famous are: Founded in the Qing Tongzhi-three years, as the first Chinese eat named "Quanjude"; founded in Kangxi Dynasty eight years, the country's most famous traditional Chinese medicine hall called "Tong ren-tang Pharmacy"; one of Beijing's oldest opera house, built in 1796, "Guangde Opera House Park," and so on, over twenty Old and famous shops, including the food, clothing, and other life forms.

4. The outstanding problems facing the Qianmen Street

2008 years ago, the street door reopened. Now, years later, it has experienced a brilliant century, but was decline of Qianmen Street, still did not revive the glory. Its ability to achieve revival, still faces the following outstanding issues:

4.1 Lack of cultural accumulation neither fish nor fowl.

Qianmen Street attractive is that the old Beijing's history and cultural accumulation however, reopened Qianmen Street, there are many international brands have entered the marketplace Qianmen Street's inherent cultural and historical accumulation has been greatly hurt. And also, the local business climate has been heavy damaged to. There are all kinds of incompatible and inconsistent between Qianmen Street and international brands. The original consumer groups are rapidly lost.

4.2 Loss of cultural traditions, a vague position.

Many folklorists believe that the transformation of Qianmen Street is demolished and sacrifice of alleys as the price, lost their original culture label. As the positioning fuzzy, poor management, this ancient street has been reduced to a "dull" attraction.

5. Ideas and suggestions for Qianmen Street revival and remodeling development

5.1The Collating of Qianmen Street's value, reasonable positioning its remodeling function.

Business location is a critical decision-making, this is about the historic district remodeling and development success. Qianmen area as commercial district has a rich tradition of historical value, cultural value, artistic value and commercial value, resource value and social value. Must fully understand the basis of these values, in order to objectively and reasonably given the overall positioning of the front door area.

I believe that the Qianmen Street's revival and remodeling can be positioned as follows: to protect and display the "Civil Culture" as the core, focused on "Beijing old" culture. traditional old, characteristic industry, folk customs as such Qianmen Street both pay their last respects, tourism, recreation, education, dining, shopping and other functions, and ultimately become a commercial, leisure, entertainment, experience is a comprehensive history of one of the commercial cultural areas.

5.2 Enhance the cultural connotation of Qianmen Street, making it a tourist destination for the through connection with "Forbidden City, Tiananmen Square and the front door."

Qianmen Street retains the full traditional pattern of streets, many streets, alleys still follow the traditional name of the Ming and Qing Dynasties, become a traditional memory of the city. Courtyards there are an unique urban fabric of traditional Beijing city, are very rich in geographical characteristics.

Qianmen Street as the geographic coordinates of the capital, occupys an excellent central location. Its wide streets scale reinforces the axis of the charm of the ancient city.

Focus on refining the cultural connotation Qianmen Street, depth excavation its special historical position, take advantage of its rich cultural heritage and other physical resources, and actively guide, training people's attention on Beijing folk customs and traditional style. Thereby forming a travel chain throughout "Forbidden City, Tiananmen Square and Qianmen Street", reproduce the old spectacular and reputation of Qianmen Street.

5.3 Based on tradition, reasonably and effectively development of Qianmen Street intangible cultural heritage.

Qianmen Street once the "Civil Culture" and "marketplace atmosphere", with rich human touch. This human touch and rich meaning of life that makes Qianmen Street reflect a spirit of place, so that the street is full of multi-color, multi-exotic atmosphere of public life, but also makes people produce their sense of identity and sense of belonging.

Remodeling Qianmen Street, the main way to make it re-energize is to "the continuation of the old city social life, heritage spirit of place." to form "Beijing school culture" typical and representative, it is necessary to organize the collection of information by mining its form, manner and other activities, and then re-creation, then, demonstrated.

Development of experiential tourism. By arrangement visitors to participate in commercial or production activities or the development of private goods and other hotel visitors or residents to exchange with the owner, to understand old Beijing customs, traditions and crafts. This makes many of the traditional crafts, folk customs are preserved not only in the exhibition, handed, carry forward, inheritance, but also enhance tourism in the process of entertainment, fun. The effect of making the same strain theme, bring tourists "an unique" cultural tourism experience.

Acknowledgement
This research was financially supported by “2017 Beijing Agricultural Science and Technology Project” and “Key Cultivation Ability of Beijing Academy of Agriculture and Forestry”.

References
[1] http://www.bj.xinhuanet.com/bjpd sdwII1 /2009—10/15/content—17956624.htm
[2] Jia Miao, Zhang Bin. Qianmen Area Historical Origin[J]. Architectural Creation,(2007)
[3] Zhou Nan.Qianmen Street: commercial renaissance Rye[J]. Chinese market,(2009)
[4] Zhou Chun-yan. Old Beijing, New Beijing --Urban Consumption nature of the business landmark refracted[J]. Sales and Marketing Manager, (2013)
[5] Selection paved rebirth road[J]. Investment Beijing, (2008)
[6] http://news.xinhuanet.c0m/shuhua/20l3-l0/15/c 125540405.htm

IOP Conf. Series: Earth and Environmental Science **86** (2017) 012002 doi:10.1088/1755-1315/86/1/012002

The Technological Mineralogical Research of Molybdenum in Skarn-type Ore of Huangshaping Polymetallic Mining Area, Hunan, China

W H Liu[1,2,*] and J Y Pan[3]

1. State Key Laboratory Breeding Base of Nuclear Resources and Environment, East China University of Technology, Nanchang, 330013, Jiangxi, China
2. Key Laboratory for Digital Land and Resources of Jiangxi Province, East China University of Technology, Nanchang, 330013, Jiangxi, China
3. School of Earth Sciences, East China University of Technology, 330030 Nanchang, Jiangxi, China

Corresponding author E-mail: liuwh@ecit.cn

Abstract. Huangshaping is one of the most important polymetallic deposits in the south of Hunan Province. Through field investigation, chemical analysis, observation under the optical microscope, energy spectrum analysis of the SEM and X-ray diffraction, the author made a technological mineralogical research of molybdenum on skarn-type ore, and the result shows that the ore containing molybdenum is mainly on the contact of the granite porphyry and the impure limestone in the lower carboniferous Shidengzi group. Besides molybdenum, the ore minerals contain scheelite, native bismuth, bismuthinite, magnetite and so on; and the gangue minerals are mainly andradite, fluorite and wollastonite. Part of the molybdenum exists in the scheelite in form of isomorphism, and there is an obvious negative correlation between MoO_3 and WO_3. The molybdenite granularity is mainly located in the 0.04~0.08mm area, which accounts for 29.5% of the total and is the finely disseminated ore. For samples of 70%, 90%, and 100% with the particle size of more than 200 meshes, the maximum recovery of the molybdenite are 75.15%, 86.45% and 91.25% respectively. So there will be a better use of molybdenum if we properly improve the grinded particle size of the comprehensive samples. As part of the molybdenum is distributed in the scheelite lattice, the actual recovery rate in this area may decline compared with the ideal value.

1 Preface

Huangshaping polymetallic deposit with a long history is one of the most representative deposits in the south of Hunan Province, and an important polymetallic origin of lead and zinc in our country [1-2]. In the early days, lead-zinc mine is the main mining object, but with the lead-zinc resource gradually decreasing, it is urgent to find new economical minerals to prolong the productive life. Tungsten molybdenum bismuth ore body produced from the acidic magma eroding in the carbonate rock becomes the research priority[3]. There are a lot of researches about Huangshanping deposit[4-7], but few about the systematical study of the minerals containing Mo. On the basis of principles and methods of process mineralogy[8-12], and with the help of microscope, scanning electron microscope and X-ray diffractometer to make a process mineralogical study of Mo-bearing minerals, the author provides a convincing ground for further research on the mineral smelting.

2 Overview of the Deposit Geology

Huangshaping polymetallic deposit is located in the collision convergence belt of Neocathaysian plate and Yangtze plate, and in the complex area of the north of the Nanling WE and Leiyang-Linwu NS tectonic belt [3-4]. The main stratums belong to the upper Devonian and lower Carboniferous, and the main ore-bearing stratum is the impure limestone belonging to the Shidengzi Formation of lower Carboniferous, secondly is the sandshale in the Ceshui Formation and the dolomite in the Xinmenqiao Formation. Affected by Yanshanian movement, The tectonic structure is very complicated with many folds and fractures inside. The magmatic rocks are mainly the Yanshanian hypabyssal - ultra shallow intrusions which contain the quartz porphyry granodiorite - porphyry, granite porphyry and granite. The deposit discussed in this paper is a skarn-type polymetallic one in the contact zone of the granite porphyry and the impure limestone in the Shidengzi Formation of lower Carboniferous. (Figure 1).

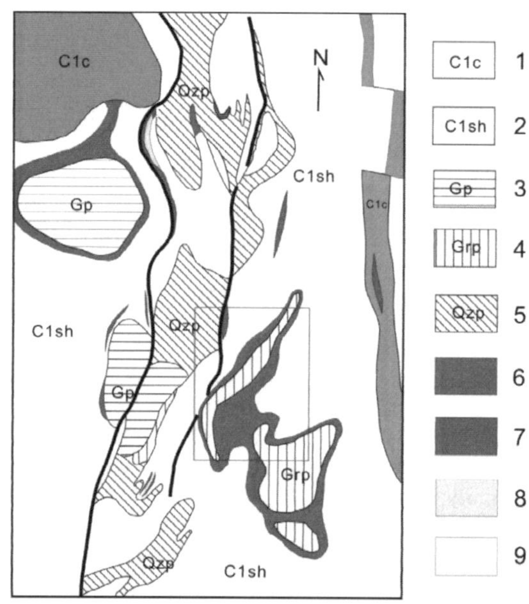

1- lower series of Carboniferous Ceshui Formation; 2- lower series of Carboniferous Shidengzi Formation; 3- Yanshanian granophyre; 4- Yanshanian granite porphyry; 5- Yanshanian quartz porphyry; 6- skarn rock; 7- pb-zn ore body; 8- copper ore body; 9- research scope

Figure1. 16 m middle geological map of Huangshaping deposit

3 Types of Ores

Table 1. Main Minerals and Their Content

Mineral	Granularity (mm)	Average Content (wt.%)	Mineral	Granularity (mm)	Average Content (wt.%)
Scheelite	0.01- 10	0.4	Magnetite	0.01- >10	3
Molybdenite	0.01- 10	0.3	Pyrite	0.01-0.2	1
Bismuthinite	0.005-0.07	0.05	Fluorite	0.01- >10	12
Native bismuth	0.005-0.1	0.06	Andradite	0.1 - >10	40
Chalcopyrite	0.005-0.1	0.01	Diopside	0.05- 5	5
Galena	0.005-0.1	0.04	Calcite	0.05- 5	8
Sphalerite	0.005-0.1	0.1	Quartz	0.1-0.5	5
Pyrrhotite	0.005-0.1	tiny	Epidote	0.05-0.5	8
White iron ore	0.01-0.1	0.1	Wollastonite	0.05-0.5	12
Cassiterite	0.001-0.01	tiny	Potash feldspar	0.05-0.1	5

Huangshaping Polymetallic deposits with molybdenum are mainly divided into three types. The first, also the main type is the skarn-type molybdenum polymetallic ore body in the contact zone, and its main output (figure 1) is 301 # granite porphyry. The output of single ore body with the highest industrial value goes up to 200 - 200 tons. The second type, a small-scale one, is the skarn-type tungsten molybdenum ore body in the outside of the contact zone. The third type is the vein polymetallic ore body in the fracture tectonic belt, a high grade one, but with low degree of thickness. The research of this study is mainly focused on the first type of ore body.

In order to find out the types and contents of minerals in molybdenum polymetallic ore of the deposit, the ore samples are observed by a polished-thin microscope, supplemented by electric energy spectrum quantitative analysis and X-ray diffraction analysis of typical samples. The statistics of the main minerals contents and the estimated secondary mineral contents based on chemical analysis are shown in Table 1. As listed in the table, the main ore minerals in the molybdenum polymetallic ore are molybdenite, scheelite, natural bismuth, bismuthinite and magnetite, and the ore also contains a small amount of pyrite, sphalerite, white iron ore, galena and chalcopyrite. Gangue minerals are mainly andradite, fluorite, wollastonite, and there are also a small amount of calcite, and epidote, etc.

According to the natural endowment and characteristics of element combination, molybdenum polymetallic ore in the deposit can be divided into the skarn-type containing molybdenum and diopside-magnetite skarn-type ore, and the latter accounts for the vast majority. The results of multielement composition analysis of ore samples (see Table 2.) are: the grade of molybdenum in skarn type ore are 0.073% to 0.073%, and the average is 0.485%; molybdenum grades in diopside magnetite skarn type ore, skarn type surrounding ore and comprehensive deputy samples in different particle sizes are 0.42%, 0.13%, 0.11% (80 - 200 meshes) and 0.22% (> 200 meshes) respectively. In addition, it can be seen that besides molybdenum, the ore sample also contains much tungsten and bismuth, and small amounts of lead, zinc, tin and other metal elements.

Table 2. Multielement Composition Analysis of the Ore

Number	Ore Type	Inspection Results									
		$Cu/10^{-6}$	$Pb/10^{-6}$	$Ag/10^{-6}$	$Zn/10^{-6}$	$Bi/10^{-6}$	$Au/10^{-9}$	$W/10^{-2}$	$Sn/10^{-2}$	$Mo/10^{-2}$	$S/10^{-2}$
200-36	Skarn-type ore	6.1	48.3	0.58	130	375	10.06	1.96	0.35	0.17	0.045
165-Mo	Skarn-type ore	4.5	56	0.1	267	51.2	12.44	0.13	0.023	0.57	0.37
GK54-176	Skarn-type ore	3.5	29.6	0.2	64	33.3	8.18	0.54	0.04	0.14	0.072
56-21-5	Skarn-type ore	4.4	47.9	0.23	82	85.2	9.3	0.33	0.14	1.47	0.88
GK100-11	Skarn-type ore	2.6	30.7	0.5	320	500	9.48	0.29	0.85	0.073	0.014
92-57	Diopside magnetite skarn ore	22	39.4	2.4	314	1565	5.72	0.29	0.26	0.42	0.75
56-171	Skarn surrounding rock	1.5	11.1	0.1	152	38	4.34	0.012	0.0084	0.13	0.08
heald+200	Comprehensive deputy samples(granularity 80-200 section)	204	281.4	1.53	1800	487	20.26	0.21	0.27	0.11	0.73
heald-200	Comprehensive deputy samples(granularity >200 section)	267.4	760	1.65	3700	699	20.8	0.37	0.18	0.22	0.91

4 Endowment State of Molybdenite

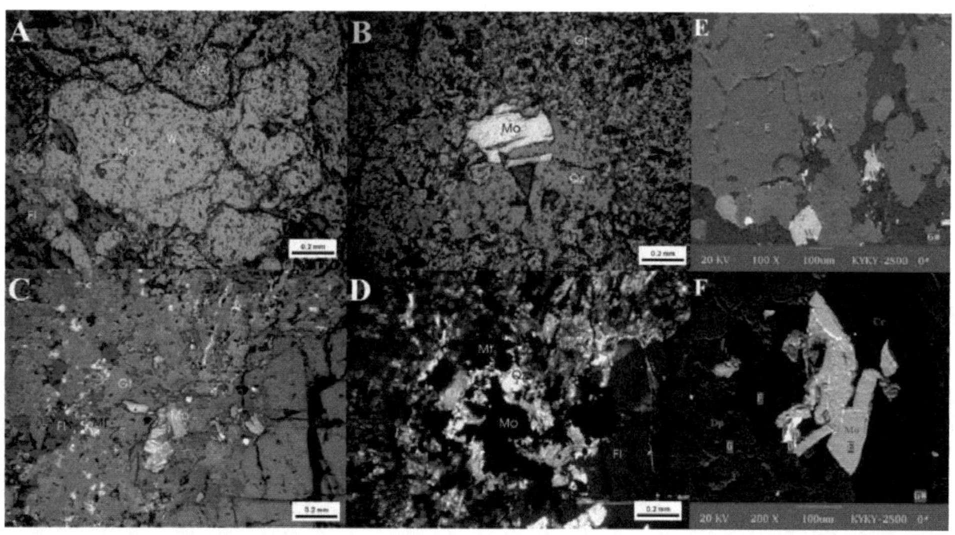

A. Symbiosis of molybdenite (Mo) and scheelite (W), produced in the garnet (Gt) and fluorite (Fl); B. Molybdenite (Mo) plate shaped distributed in the garnet (Gt) and quartz (Qz); C. Molybdenite (Mo) plate shaped distributed in the garnet (Gt) - magnetite (Mt) - fluorite (Fl); D. Molybdenite (Mo) distributed in diopside (Dp) - magnetite (Mt) - quartz (Qz) - fluorite (Fl); E. Symbiosis of molybdenite (Mo) and white tungsten (W), produced in magnetite (Mt) and garnet (Gt); F. Symbiosis of flake molybdenite (Mo), natural bismuth (Bi), and calcite (Cc), distributed in diopside (Dp);

Figure 2. The Endowment State of Huangshaping Molybdenite

From the field research, it can be seen the molybdenite is usually a vein aggregate filled in the crack of the skarn whit a width of 1-10 mm and a length of 10 cm. Figure 2 is the endowment state of the molybdenite samples observed under microscope and scanning electronic microscopy. The molybdenite in this area are 0.01-0.1 mm granules or patchy distributed in the skarn, part of which contain scheelite (Figure 2. a, E) and a small amount of natural bismuth (Figure 2. f) . The paragentic minerals are mainly magnetite, garnet, diopside, calcite, and fluorite, etc.

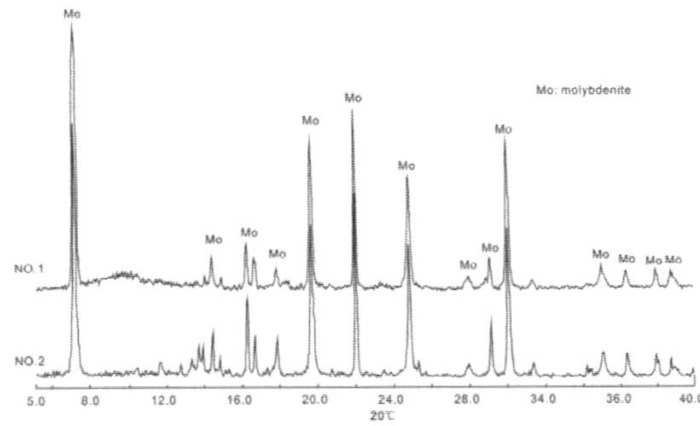

Figure 3. X-ray Diffraction Diagram of Huangshaping Molybdenite

The samples of molybdenite have been researched through X-ray diffraction, analyzed and tested by Changsha Research Institute of Mining and Metallurgy. The instrument used is Dmax / 2200 - gamma A10 X-ray diffractometer produced by Ricoh company of Japan and the test onditions are as follows: copper target for the X-ray tube, the pipe pressure is 50 kv, the pipe flow is 100 mA,4°/min, and automatic slit system. The test result is in Figure 3, showing the fact that there is no solid impurity component existing in the molybdenite, which is the feature of the 2H type. Previous studies have pointed out that molybdenite polymorphism is associated with Re contained, which can reflect different metallogenic temperatures [13], therefore it can be concluded the molybdenite in this area is formed at high temperatures. Former studies also proved that the mineralization temperature of skarn type molybdenum polymetallic ore is about 300-400 ℃ [14], which is consistent with the conclusion drawed in this article.

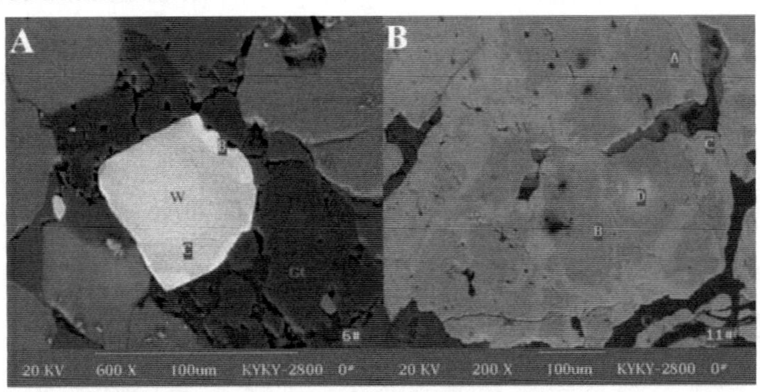

A. Scheelite (W) produced in magnetite (Mt) and garnet (Gt). It's darker with high content of molybdenum (C) and brighter with low content of molybdenum (B);
B. Scheelite (W) produced in magnetite (Mt), symbiosis with calcite (Cc). It's darker with high content of molybdenum (B) and brighter with low content of molybdenum (D)

Figure 4. SEM Image of Huangshaping's Molybdenum in Scheelite

During the spectroscopy quantitative analysis of the scheelite on this deposit, it was found that scheelite generally contains a certain amount of molybdenum oxide (Table 3), whose content is 0.64% -21.74% with an average content of 7.21%. The analysis under the scanning electron microscopy shows that scheelite with more molybdenum is much less bright than that with less molybdenum (Figure 4). Through cluster analysis of various components in scheelite, the author got the following results, as shown in Table 4, that related index between

WO_3 and MoO_3 in scheelite is -0.950, which is a negative correlation, showing that the content of WO_3 in scheelite decreases while the content of MoO_3 increases. Since 6+ ion valence of the tungsten molybdenum elements has similar atomic radius and the same charge, it's easy for the Mo^{6+} to get into the lattice of scheelite to replace W^{6+} in the process when the minerals separate out from the ore fluid, causing part of the molybdenum to exist in scheelite in the form of isomorph [15].

Table 3. Molybdenum Content in Scheelite of Huangshaping Deposit

number	WO_3	CaO	MoO_3	FeO	number	WO_3	CaO	MoO_3	FeO
5#-A	80.16	19.84			12#b-e	68.56	15.93	14.93	0.23
5#-H	80.64	19.35			12#b-f	79.92	16.16	3.91	
6#-a	77.11	16.3	6.58		12#b-g	83.8	14.29	1.53	
6#-b	81.78	16.48	1.74		12#b-h	76.29	10.54	12.7	0.2
6#-c	74.32	17.18	7.54		12#b-I	80.62	16.35	2.15	0.54
11#-a	63.23	18.13	18.63		12#b-j	80.72	16.22	3.05	
11#-b	67.34	17.45	15.19		12#b-k	83.17	15.92	0.64	0.27
11#-c	83.44	15.36	1.19		12#b-l	81.83	13.77	4.4	
11#-d	77.79	12.37	9.83		12#b-m	69.71	18.23	12.05	
11#-e	72.44	17.89	9.66		12#b-n	57.49	20.31	21.74	0.05
12#b-a	69.66	15.23	14.72	0.38	12#b-o	74.81	15.21	9.73	
12#b-b	83.27	14.47	2.25		12#b-p	79.87	14.39	4.95	0.78
12#b-d	82.14	16.41	1.07	0.37					

Table 4. Cluster Analysis of Various Components in Scheelite of Huangshaping Deposit

	WO_3	CaO	MoO_3	FeO
WO_3	1			
CaO	-0.413	1		
MoO_3	-0.950	0.111	1	
FeO	0.103	-0.230	-0.066	1

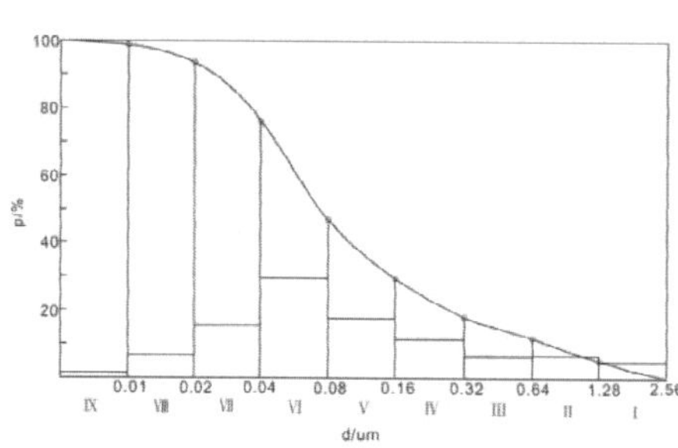

Figure 5. Distribution Features of Particle Content of Molybdenite in Huangshaping Deposit

5 Particle Feature Analysis of Molybdenite

Table 5. Measuring Results of Molybdenite's Particle Size

Sequence of Particle size	size range /mm	ratio size /d	particles /n	content ratio/n*d	content distribution/%	content distribution/%
I	2.56～1.28	256	3	768	5.06	5.06
II	1.28～0.64	128	8	1024	6.74	11.8
III	0.64～0.32	64	15	960	6.32	18.12
IV	0.32～0.16	32	54	1728	11.38	29.5
V	0.16～0.08	16	165	2640	17.39	46.89
VI	0.08～0.04	8	560	4480	29.5	76.39
VII	0.04～0.02	4	589	2356	15.52	91.91
VIII	0.02～0.01	2	513	1026	6.76	98.66
IX	0.01～0.005	1	203	203	1.34	100
Total			2110	15185	100	

The composition and distribution of the subject mineral's particles in ore have a determining impact on the grinding fineness and the process of mineral separation[16]. Due to the prominent optical character, the natural particle size of molybdenite calculated under the reflective light microscope was divided by a power of 2. For those particularly small particles (eg. particle size less than 50 microns), might be neglected, thus the results might miss the micro-fine fractions. Statistical results are shown in Table 5 and Figure 5. It shows that the particles are relatively small mainly between 0.01 ~ 0.08mm, the number is 1662, 78.77% of the total number 2110. The largest particle size ranges form 0.04 ~ 0.08mm, 29.5% of the total number; then followed by 0.08 ~ 0.16mm, 17.39% of the total. Molybdenite particle size is relatively less in coarse and microgranular area, over 200 mesh (0.074mm) , up to about 47% of the total, and is the fine grain disseminated.

Table 6 Determination of molybdenite's monomer dissociation degree

sample	franctions/ mesh	Monomer particles	Interlocked particles			fractions dissociation degree
			3/4	2/4	1/4	
Composite samples	80～200mesh	367	92	58	11	78.46%
	>200mesh	628	69	15	4	91.25%

The amounting granulation method was used in determinating the molybdenite's monomer dissociation degree of composite samples of the ore, which were classified into two fractions, i.e., 80 to 200 mesh level, and> 200 mesh level. According to the relative value of volume molybdenite occupied in the interlocked particle, the interlocked particle was divided into 3/4, 2/4 and 1/4 level, then the number of the monomer particle and interlocked particle was recoded under the microscope, and the results are shown in Table 6.

According to Table 6, when the composite samples are grinded to 80 to 200 mesh, the number of the monomer particles is 367, the interlocked particles is 161, most of which is of 3/4 level, and the monomer dissociation degree of the molybdenite is 78.46%; when more than 200 mesh, the number of the monomer particles is 628, the interlocked particles is 88, and most of them is also of 3/4 level, the monomer dissociation degree is of 91.25%. On the basis of the analysis in Table 2, the grade of molybdenum is 0.11% in the samples of 80 to 200 mesh, but 0.22% in > 200 mesh. Thus, the appropriately increased grinding degree of particles can raise the monomer dissociation degree of molybdenite and its grade as well. What's more, it's favorable to mineral recycle.

Therefore, the monomer dissociation degree of molybdenite is high in the composite sample of > 200 mesh, which is considered as the ideal recovery rate. When 70% of the composite sample is grinded to the particles of bigger than 200 mesh, the maximum ideal recovery rate (R70%) of molybdenum is: R70% = [0.22 × 0.7 / (0.11 × 0.3 +0.22 × 0.7)] × 0.9125 = 75.15%; when 90% is grinded to > 200 mesh, it's R90%: R90% = [0.22 × 0.9 / (0.11 × 0.1 +0.22 × 0.9)] × 0.9125 = 86.45%; when 100% is grinded to > 200 mesh, it's R100%: R100% = 91.25%. Therefore, the appropriately increased grinding degree of the composite sample can guarantee a better recovery prospects. Furthermore, since the inner part of the molybdenum mine is isomorphous existed in scheelite lattice (Table 3), this region's actual recovery rate of molybdenum will be reduced compared with the ideal one. In addition, the relations between the 2H multi-type and its flotation performance should be considered in the actual production process.

6 Conclusion

(1) Huangshaping ore body with molybdenum is mainly located on the contact belt of granite porphyry and lower Carboniferous Shidengzi Formation with pure limestone, which is a skarn type deposit. In addtion to the fine granular, there are also scheelite, natural bismuth and magnetite in the ore molybdenite, among which gangue minerals are mainly andradite, fluorite, wollastonite, etc.

(2) Part of the molybdenum mine exists in scheelite in the form of isomorphism, where is negative correlation between MoO_3 and WO_3.

(3) The molybdenite granularity is mainly distributed in the 0.04 ~ 0.08 mm mining area, accounting for 29.5% of the total. Over 200 mesh (0.074 mm) accounts for about 47% of the total, which is the finely – disseminated ore.

(4) When 70%, 90%, 100% of the mineral composite samples are grinded to > 200 mesh, the most ideal molybdenum recovery rate is 75.15%, 86.45% and 91.25% respectively, so appropriately improving the grinded particle size of comprehensive samples can increase the recovery rate of molybdenite. As part of molybdenum exists in scheelite in the form of isomorphism, the actual recovery rate of molybdenite in this area will be relatively reduced compared with the ideal value.

Acknowledgments

Foundation Item:Research supported by Scientific Research Foundation of ECUT(DHBK201409) and Open Fund of State Key Laboratory Breeding Base of Nuclear Resources and Environment(NRE1613).

IOP Conf. Series: Earth and Environmental Science **86** (2017) 012002 doi:10.1088/1755-1315/86/1/012002

References

[1] Xu Y M , Gong S Q, Jiang Y C, Lei Z H and Li Y S 2007 Hunan Huangshaping lead zinc mine of the deep analysis of the prospects for lead zinc mine resources *Geol Prospec* 43 pp 38-43

[2] Tong Q M 1986 The characteristics of metallogeny of the Huangshaping lead-zinc ore deposit of southern Hunan *Geol Rev* 32 pp 565-577

[3] He H Q, Wang J C and Jiang Y C 2010 Preliminary analysis on the geological characteristics and genesis of Fe-W-Mo-Bi(-Sn) polymetallic deposit in the southern Huangshaping lead-zinc mine district,Hunan *Miner Exploration* 1 pp 323-333

[4] Wei D G, He L, Pang B C, Hu Y H, Yang X F, Cao X C and An Z H 2007 Joints characteristics and their geological significance of the Huangshaping Pb-Zn ore field in Guiyang County,Hunan province *Miner Resour Geology* 21 pp 555-559

[5] Li Y Z 1986 On the genesis of the Huangshaping Pb-Zn deposit,Guiyang,Hunan in special reference to fluid inclusion studies of its minerals *Acta Mineral Sin* 6 pp 77-85

[6] Deng S F 1997 Zoning regularity of mineral assemblages in the Huangshaping lead-zinc deposit *Miner Resour Geology* 11 pp 314-318

[7] Zhu E J, Wang J G and Qiu Y M 1995 The distribution regularity and occurrence of the associated silver in the Huangshaping Pb-Zn ore field,Hunan province *Geological Exploration for Non-ferrous Metals* 4 pp 89-95

[8] Zhou L G 2002 *technological mineralogy* (Beijing: Metallurgical Industry Press) pp 202-267

[9] Ding J and Zhang S G 2011 The tin technological mineralogical research of Skarn-type multi-metallic ore in Huangshaping *Acta Mineral Sin* 31 pp 80-87

[10] Peng M S, Liu X W, Liu Y and Yang Z J 2012 The main advances of process mineralogy in China in the last decade *Bull Mineral, Petrol Geochem* 31 pp 210-217

[11] Luo X P, Chen S H, Wang F H and Miao J C 2010 Process mineralogy research on deep ore of Qixiashan lead-zinc-silver mine in Nanjing *Met Min* (6) pp 90-95

[12] Cao Y J, Gui X H, Ma Z L,Yu X X, Chen X D and Zhang X P 2009 Process mineralogy of copper-nickel sulphide flotation by a cyclonic-static micro-bubble flotation column *Min Sci Techno* 19 pp 784-787

[13] Han Y W 1988 Molybdenite polytypes and the mechanism of polytype formation *Earth Sci: J China U Geosci* 13 pp 385-394

[14] Tian X F and Gong S Q 2007 Future prospect of tungsten and molybdenum resources in Huangshaping lead and zinc mine *J U South China(Sci Techno)* 21 pp 110-114

[15] Han Y W, Ma Z D, Zhang H F, Zhang B R, Li F L, Gao S and Bao Z Y 2003 *geochemistry* (Beijing: Geological Publishing House) pp 78-79

[16] Liao Q, Feng Q M, Ou L M, Lu Y P, Zhang G F and Cao J H 2011 Process mineralogy for a low grade nickel ore in Jinchuan *J Mineralogy Petrology* 31 pp 5-10

IOP Conf. Series: Earth and Environmental Science **86** (2017) 012003 doi:10.1088/1755-1315/86/1/012003

Land Degradation is The Instinctive Source of Poverty in Rural Areas of Pakistan

L L Qiu[12], M A Koondhar[3], Y Y Liu[3], W Z Zeng[12*]

[1] College of Economics, Sichuan Agricultural University, Huimin RoadNo.211, Wenjiang District, Chengdu, 611130 China
[2] Southwest Center for Poverty Alleviation and Development Research, Sichuan Agricultural University, Huimin RoadNo.211, Wenjiang District, Chengdu, 611130 China
[3] College of Management, Sichuan Agricultural University, Huimin RoadNo.211, Wenjiang District, Chengdu, 611130 China

*Corresponding author's email: **zwz2011@foxmail.com**

Abstract. This review paper focused on the correlation between land degradation and poverty. Pakistan is an agricultural country and agriculture is the backbone of Pakistan`s economy. For the rapid growth of population food security should be under guarantee as well as the food production. In that farmers overused agrarian inputs, such as fertilizer, pesticide and water, environment and farmers were affected from the perspective of contamination and disease increase respectively. Due to over-exploitation of fertilizer and irrigation, ground water was contaminated, soil fertility weakening,salinity increasing and waterlogged. Consequently, soil was hard to be cultivated. In Pakistan 70% of people live in rural areas who are directly or indirectly involved in agriculture. As a result of land degradation farmers can not gain much benefit from agricultural activities and they are also unable to feed their children. Many of them became criminals, therefore, poverty deepened day after day. In order to alleviate poverty, Pakistan government should subsidize farmers on environmentally friendly inputs and; government should also open agricultural training schools to engage farmers in modern methods of cultivation, and provide modern technologies with subsidy rate. When the farmers are aware of how to increase the fertility of soil by employing modern methods, they can gain higher production, and obvious higher production is critical for living a better life and reducing poverty.

1. Introduction

Pakistan is a predominantdry-land country with the majority of land arid or semi-arid. Rapid growth of population depends a lot upon dry-land areasas well as the support from the agriculture sector [1]. In Asia there are many other developing countries affected by land degradation, but Pakistan is seriously affected by land degradation and desertification due to untenable management of land use practices, rapidly increasing demand of renewable and non-renewable natural resources[2]. What's more, enormous threats of environment were created in Pakistan, such as dry land degradation, reducing soil fertility, increasing floods, loss of biodiversity, increasing desertification, decreasing land productivity, soil erosion, water erosion, increasing salinization, loss of vegetation plants, increasing overgrazing, etc. In most of the rural areas what's commonly found is drought as a result of water scarcity, desertification owing to mismanagement of land planning and reducing production, which are main causes of the deepening poverty in Pakistan.

Land in Pakistan covers79.61 million hectares, from which 59.47 million hectares were recorded through the survey of Pakistani government. Nearly 21.85 million hectares of land was utilized for agricultural activities, of which 4.04 million hectares is forestland and 9 million hectares is cultivable wetland[3]. Approximately 90 percent of land is located in the river deltas of Sindh and Punjab provinces. Furthermore, 85 percent of the land was irrigated, and the remaining 5 percent was not irrigated which was denoted by Barani (Rainfed) area of agriculture in northern KPK and southern Punjab of Pakistan[4][5][6]. There are totally 140 million hectares of agricultural land, some 43 percent of which faced degradation problems. Moreover, among the degraded land, 23% was seriously affected, 45% moderately influenced andthe remaining 32% was mildly degraded.

Pakistan`s economy is primarily dependent on agriculture, which accounts for 21.9% of GDP [7]. To achieve a higher level of production, farmers carried out over-exploited application of inputs. However, as a result of overusing inputs salinity increased, leading to the decline of soil fertility and increase of water logging[8]. Some facts show in Sindh and Punjab land degradation issues are very serious due to poor planning and management. Some rural households rented or sold their land to bricks factories and engaged in non-agricultural activities, such as grazing, pastureand destroy vegetation which can readily lead to land degradation. Overgrazing also has correlations with land degradation, especially in Indus Delta River wherealmost all farmers ranch livestock whichare reared in intensively cultivated irrigated plains by stall-fed crop residues and forages [9].

The relationship between land degradation and poverty is instinctive [10], whether poverty was induced bythe mismanagement of farmers or the consistent relationship of population with land which is used for productive purposes [11]. Here is an extreme dearth of studies seeking to rigorously test these relationships, however, the lack of appropriate data leads to the paucity. Reliance has to be placed on studies from which the relationships can be inferred. Most of the rural areas are exposed to the following problem: behaviors of small-scale farmers and land degradation, which denotes that small-scale farmers have been the primary agents responsible for land degradation activities [12].The market and institutional failure are the prime factors for farmers to adopt non-sustainable practices. The price regulation on agricultural products in Pakistan didnot provide incentives for the poorsmall-scale farmers to conserve their land [13]. In some cases this has led to the decrease of resources for maximum output. Whereas,on the other hand, evidence was found that small-scale farmers were willing to give up short-term income gains even under price and famine pressure to pursue long-term sustainable management strategies [14]. The existence or non-existence of secure land tenure systems might explain the contradiction about small-scale farmers' behaviors. Secure land tenure was the primary reasonwhy poor farmers cultivated their land excessively to exhaustion[15][16][17]. The main objective of this study is to make plans to strengthen the awareness of land resource conservation, to sensitize the policy makers and practice partnership with due authorization of successful planning.

2. Effect of exploitations of agricultural practices on rural households

Changes in agricultural practices have primary effects on land degradation, which will consequently influence the income level ofrural households[18]. Such changes have taken place worldwidethrough adopting green revolution technologies. Agricultural change can also occur where green revolution technologies have not beenadopted yet. In the case of the latter, the primary effects on the environment are generally stated to be in the form of desertification, deforestation, watershed degradation, soil erosion and soil fertility decline. The secondary effects can be droughts and floods. These environmental effects can be transformed into specific effects at the household level. And theymay take the form of impoverishment or productivity decline, migration-related health stress, vector borne and airborne diseases, communicable disease, chronic food insecurity, seasonal malnutrition and famines. Under the circumstance of the green revolution technology, potential environmental degradation can result from every element of the technology package which can be estimated by direct and indirect use of technology elements. For example, over-irrigation can lead to aggravated underground water quality and increasing salinization which are harmful for humans health and agricultural productivity. Inappropriate overuse of pesticide has venomous effects on households. Over-application of fertilizercan reduce the soil fertility and nitrate leach down and mix with underground water. As for households, these environmental activities may triggerdiseases like diarrhea, cholera, typhoid, malaria,schistosomiasis,onchocerciasis, poisoning and diseases of the circulatory system in infants. The secondary effects of the utilization of such technology can be crowding, sanitation deficiency, diet change and vector control (through inappropriate pesticideuse), which leads to communicable diseases, nutritional diseases, poisoning, etc. The effectsabove denote a reduction in welfare, which ismeasured under the conventional consumption-based method. Still, it is important to include the non-income measures of poverty likeanthropological measurements in assessing the poverty status.Other paragraphs are indented.

3. Land degradation and poverty

Over the last few decades awareness and anxiety about land degradation have increasedon the global scale, which was shared by people from different nations, cultures, religions and social classes. Poverty was defined as a condition of being poor [19]. Landdegradation was influenced by human activities such as deforestation, overgrazing, destroying herbs and vegetationand changing land cover[20].

Considering the relationship between poverty and land degradation has great significance.Regional overgrazing canseriously destruct the grazing land. Forestry, soil, air pollution and water pollution have significant relationship with national overgrazing [21]. The carrying capacity of the natural environmental resources has been decreasing because as the poor became too impoverished they consumed the resources very quickly[22]. Poor people tend to overuse renewable and non-renewable resources, in that they are in bad living conditions. They need to consume natural resources for the survival, what's more, to meet their basic requirements, such as demand for firewood, agricultural production (Maize), water and herbal plants for medicine.

3.1. Morbidity, mortality and rural households

Land degradation glitches obviously led to the ratio increase of morbidity and mortality through contamination, dearth of drinking water and decreasing productivity of farm crops due to over-exploitation of fertilizer, pesticides and misuse of natural resources. Therefore, here we discusshow morbidity and mortality affected livelihood strategies and strengthened dependency on indigenous environmental resources. Generally, the minor nonetheless growing body of literature provides evidence for significant relationships between household morbidity, mortality and natural environmentmediated through livelihood coping strategies. The basic assumption of this developing form is that natural resources have significant mitigating effects when households are involved in morbidity and mortality through providing additional sources of income and substantial goods that might otherwise be purchased. Natural resource is in urgent demand during the period of drought, famine and natural disasters[23]. Morbidity and mortality willexertheavier burden on family income under the condition that most of the rural households in Pakistan have already lived in margin [24].

Overall, mortality will be the principal cause of the rising household vulnerability if the adult household members decease[25][26]. In recent years, AIDS was a serious challenge for Pakistan which heavily impacted household livelihoods. Haddad and Gillespiebluntly addressed AIDS striped folks, households and community resources[27]. Morality of adults relating to AIDS can influence the householdcapitalin all forms. Humanoid wealth is missing as a result of the sank efficiency [28], the diversion of healthy laborers and the possible injury of remittances from employed refugee family members [29]. Into the bargain, inter-generational information transfer also suffers [30].

Overall, mortality will be the principal cause of the rising household vulnerability if the adult household members decease[31][32]. In recent years, AIDS was a serious challenge for Pakistan which heavily impacted household livelihoods. Haddad and Gillespie bluntly addressed AIDS striped folks, households and community resources[33]. Morality of adults relating to AIDS can influence the householdcapitalin all forms. Humanoid wealth is missing as a result of the sank efficiency [34], the diversion of healthy laborers [35] and the possible injury of remittances from employed refugee family members [36]. Into the bargain, inter-generational information transfer also suffers [37].

There is a lot of health-related literature on impacts of land cover changes subsequently after frontier relocation, and many current studies concentrate on rising malaria occurrence [38][39][40]. The case above can deliver influences to micro-level research on mutual relationships between frontier relocation and rising malaria occurrence. Land changes due to household migration are responsible for environmental transformation which led to household morbidity and mortality that affect natural resources use (Based on literature review).

4. Influence of land degradation on rural households

Rural households have access to degraded land that is complex for health caring and incomeearning. However,land degradation will influence rural households directly and indirectly.For example, it will bring about lower yield because of the fertility deficiency, which is a kind of direct impact.Moreoverit can indirectly affect the physical condition of laborers due to the poisonous food harvested on the degraded land. In the situation where the labor force is weak due to malnutrition, laborers tend toengage inlow-profit activities like fuel wood collection [41].

Agriculture is a sector which greatly depends upon environmental condition for sustainable productivity. Natural resources degradation and poverty indexes were defined in the study that those who browbeat natural resources for the growth of country, such as farmers, were facing whims of nature like disease, drought, spates, inferior agricultural productivity efficiency and poor socioeconomic condition, which are all signs showing the status of poverty [42]. They still are operating less input owing less output domains with low productive efficiency rises poverty index. Furthermore they will live in poverty trap till actual measures of reintegration, renewing and up-keeping of natural resources are taken. Anadditional cause may be that the sufferers have not realized the severity yet. Although the development and research institutes including public and private are working for the natural resources renewal and protection,situations are still worrying. Farmer is thereal guardian of land resources which is serious affected by poverty and productivity consequences, nevertheless is slightly referred to during land preparation and use of applications inputs. Agriculture has been sound recognized as aneffective tool of poverty reduction by way of leading to the overall growth in agro-based industries of the world. Farmer's active involvement due to case studies shows that many technological involvement were abundant through growers after the removal of official input/backup, the productivity of natural resources that the poor manage, the deterioration tortuously connect with poverty, population and environmental [43].

5. Conclusion

Pakistan is a predominantdry-land country. Rapid growth of population depends upon dry-land areas and the agriculture sector supports their livelihoods. Pakistan`s economy is primarily dependent on Agriculture, which mainly consists of production of wheat, cotton, rice and sugarcane and other cash crops. To achieve higher production farmers over-exploited inputs,which led to salinity increase, soil fertility decline and water logging. As a result of land degradation, production diminished. In some cases, farmers became criminals due to the hunger of their children. They can not bear the hunger of children so they may take the illegal actions and commit a crime. Due to the overuse of agrarian inputs, such as fertilizer and pesticide, which tend to leave components of poisons in food, people who eat them are likely to suffer from diseases. With regard to policies, we find that since the environmental impacts of mortality and morbidity are generally passed on tothe household poverty, policies that improve livelihoods, natural resource management and public health as well as mitigating risks or providing insurance for vulnerable families need to be made. Specific natural resource conservation interventions ought to be taken to encourage the establishment of low-input enterprises based on the sustainable production and use of non-timber forest products, agroforestry efforts, and production of wild indigenous products for meeting dietary needs.

IOP Conf. Series: Earth and Environmental Science **86** (2017) 012003 doi:10.1088/1755-1315/86/1/012003

References

[1]Government of Pakistan, 2006. Economic survey of Pakistan 2006-2007, Ministry of Finance, Islamabad, Pakistan.
[2]Zia M S, Muhmood T, Baig M B, et al. Land and environmental degradation and its amelioration for sustainable agriculture in Pakistan[J]. Sci Vision, 2004, 9: 1-2.
[3]Government of Pakistan, 2004. Economic survey of Pakistan 2006-2007, Ministry of finance, Islamabad, Pakistan.
[4]WHO U. World Bank (2009)[J]. State of the world's vaccines and immunization, 2013.
[5]Food and Agriculture Organization, 1997. Aquastat: Pakistan. FAO, Rome. Available at: http://www.fao.org/nr/water/aquastat/ countries/pakistan/index.stm.
[6]Muhammad D, Khattak R A. Wheat yield and chemical composition as influenced by integrated use of gypsum, pressmud and FYM in saline-sodic soil[J]. Journal of the Chemical Society of Pakistan, 2011, 33(1): 82-89.
[7]Government of Pakistan, 2015.Economic survey of Pakistan 2006-2007, Ministry of Finance, Islamabad, Pakistan.
[8]Niazi T. Land tenure, land use, and land degradation: A case for sustainable development in Pakistan[J]. The Journal of Environment & Development, 2003, 12(3): 275-294.
[9]Muhammad D, Khattak R A. Wheat yield and chemical composition as influenced by integrated use of gypsum, pressmud and FYM in saline-sodic soil[J]. Journal of the Chemical Society of Pakistan, 2011, 33(1): 82-89.
[10]Grepperud S. Poverty, land degradation and climatic uncertainty[J]. Oxford Economic Papers, 1997, 49(4): 586-608.
[11]Scherr S J, Yadav S. Food agriculture and the environment discussion paper 14[J]. International Food Policy Research Institute, 1995.
[12]Southgate D D W. economics of land degradation in the Third World[J]. 1988.
[13]Pagiola S. Environmental and natural resource degradation in intensive agriculture in Bangladesh[M]. Washington DC: World Bank, 1995.
[14]Mittermeier R A, Myers N, Thomsen J B, et al. Biodiversity hotspots and major tropical wilderness areas: approaches to setting conservation priorities[J]. Conservation biology, 1998, 12(3): 516-520.
[15]Southgate D D W. economics of land degradation in the Third World[J]. 1988.
[16]Mink S D, Mundial B. Poverty, population, and the environment[M]. Washington, DC: World Bank, 1993.
[17]Repetto R C, Magrath W, Wells M, et al. Wasting assets: natural resources in the national income accounts[R]. Washington, DC: World Resources Institute, 1989.
[18]Vittor A Y, Gilman R H, Tielsch J, et al. The effect of deforestation on the human-biting rate of Anopheles darlingi, the primary vector of falciparum malaria in the Peruvian Amazon[J]. The American Journal of Tropical Medicine and Hygiene, 2006, 74(1): 3-11.
[19]Walsh, T, 2005. Causes of Poverty. [Internet] 2006 Jan 10th, 09:21 UTC. Available from: http://www.gdrc.org/icm/poverty-causes.html.
[20]Anon, 2007. Definition of poverty (Online). 2007 Feb 23rd, 11:08 UTC. Available at:http://www.google.co.za/url?sa=X&start=0&oi=define&q=http://wordnet.princetonedu/perl/webwn%3Fs%3Dpoverty&usg=W9VMjHxmt2dtS9Q6i1C3dXliaiI.
[21]Mittermeier R A, Myers N, Thomsen J B, et al. Biodiversity hotspots and major tropical wilderness areas: approaches to setting conservation priorities[J]. Conservation biology, 1998, 12(3): 516-520.
[22]Mittermeier R A, Myers N, Thomsen J B, et al. Biodiversity hotspots and major tropical wilderness areas: approaches to setting conservation priorities[J]. Conservation biology, 1998, 12(3): 516-520.
[23]McSweeney K, Arps S. A" demographic turnaround": the rapid growth of the indigenous populations in Lowland Latin America[J]. Latin American Research Review, 2005, 40(1): 3-29.

IOP Conf. Series: Earth and Environmental Science **86** (2017) 012003 doi:10.1088/1755-1315/86/1/012003

[24]De Waal A, Tumushabe J. HIV/AIDS and food security in Africa[J]. A report for DFID, 2003.
[25]White J, Robinson E. HIV/AIDS and rural livelihoods in sub-Saharan Africa (NRI Policy Series 6)[M]. Natural Resources Institute, 2000.
[26]Yamano T, Jayne T S. Measuring the impacts of working-age adult mortality on small-scale farm households in Kenya[J]. World Development, 2004, 32(1): 91-119.
[27]Haddad L, Gillespie S. Effective food and nutrition policy responses to HIV/AIDS: what we know and what we need to know[J]. Journal of International development, 2001, 13(4): 487-511.
[28]Barnett T, Blaikie P. AIDS in Africa: its present and future impact[M]. Belhaven Press, 1992.
[29]Rugalema G, Khanye V. Mainstreaming HIV/AIDS in the education systems in sub-Saharan Africa: Some preliminary insights[J]. Perspectives in Education, 2002, 20(2).
[30]Haddad L, Gillespie S. Effective food and nutrition policy responses to HIV/AIDS: what we know and what we need to know[J]. Journal of International development, 2001, 13(4): 487-511.
[31]White J, Robinson E. HIV/AIDS and rural livelihoods in sub-Saharan Africa (NRI Policy Series 6)[M]. Natural Resources Institute, 2000.
[32]Yamano T, Jayne T S. Measuring the impacts of working-age adult mortality on small-scale farm households in Kenya[J]. World Development, 2004, 32(1): 91-119.
[33]Haddad L, Gillespie S. Effective food and nutrition policy responses to HIV/AIDS: what we know and what we need to know[J]. Journal of International development, 2001, 13(4): 487-511.
[34]Barnett T, Blaikie P. AIDS in Africa: its present and future impact[M]. Belhaven Press, 1992.
[35]Hunter L M, Twine W, Johnson A. The natural resource context of the African HIV/AIDS pandemic[J]. Institute of Behavioral Science Working Paper, 2006.
[36]Rugalema G, Khanye V. Mainstreaming HIV/AIDS in the education systems in sub-Saharan Africa: Some preliminary insights[J]. Perspectives in Education, 2002, 20(2).
[37]Haddad L, Gillespie S. Effective food and nutrition policy responses to HIV/AIDS: what we know and what we need to know[J]. Journal of International development, 2001, 13(4): 487-511.
[38]Barbieri A F, Soares-Filho B S. Population and land use effects on malaria prevalence in the southern Brazilian Amazon[J]. Human Ecology, 2005, 33(6): 847-874.
[39]de Castro M C, Monte-Mor R L, Sawyer D O, et al. Malaria risk on the Amazon frontier[J]. Proceedings of the National Academy of Sciences of the United States of America, 2006, 103(7): 2452-2457.
[40]Vittor A Y, Gilman R H, Tielsch J, et al. The effect of deforestation on the human-biting rate of Anopheles darlingi, the primary vector of falciparum malaria in the Peruvian Amazon[J]. The American Journal of Tropical Medicine and Hygiene, 2006, 74(1): 3-11.
[41]Kumar S K, Hotchkiss D. Consequences of deforestation for women's time allocation, agricultural production, and nutrition in hill areas of Nepal[M]. Intl Food Policy Res Inst, 1988.
[42]Scadding G K, Durham S R, Mirakian R, et al. BSACI guidelines for the management of allergic and non- allergic rhinitis[J]. Clinical & Experimental Allergy, 2008, 38(1): 19-42.
[43]Mink S D, Mundial B. Poverty, population, and the environment[M]. Washington, DC: World Bank, 1993.

IOP Conf. Series: Earth and Environmental Science **86** (2017) 012004 doi:10.1088/1755-1315/86/1/012004

The Study of Ecological Service Value of Farmland Ecosystem in the Beijing-Tianjin-Hebei Region

C H Jiang[1,a], G Y Li [2,b], H Q Li[3] and M Li[4]

[1.] Institute of Agricultural Information and Economics BAAFS, Key Laboratory of urban Agriculture Ministry of Agriculture, Beijing 100097，China
[2]National Geomatics Center of China, Beijing, China
[3] Liang Shan Bureau of land and resources,Shan Dong Province，China
[4]Liang Shan County Water Conservancy Bureau, Shan Dong Province，China
Email: [a]caujiangch@126.com,[b]lihonglei1979@163.com

Abstract: The coordinated development of Beijing-Tianjin-Hebei Province has become major national strategy of the CPC central committee and the state council under the new historical conditions. The farmland of Hebei Province has made great contribution to food supply for Beijing and Tianjin in a long time. Hebei Province has played a more important role in the ecological function especially.The geographic grid method was adopted in this study to account the ecosystem service value of farmland in Beijing-Tianjin-Hebei region. The result showed that total farmland ecosystem service value in 2015 was 110.3 x 109 yuan and the proportion of ecosystem service value in Hebei Province was the highest which reached up to 92.56%. The results of the evaluation of ecological services could provide the basis for determining the ecological compensation standards for farmland in Beijing-Tianjin-Hebei region.

1. Introduction

As the publication of 'The Plan for The Collaborative Development of Beijing-Tianjin-Hebei Province' and 'The Coordinated Development for Modern Agriculture Planning(2016-2020) of Beijing-Tianjin-Hebei Province', the coordinated development of Beijing-Tianjin-Hebei Province has become major national strategy of the CPC central committee and the state council under the new historical conditions.

There is the phenomenon of the struggle of industrial and agricultural land, urban and rural land in the Beijing-Tianjin-Hebei Province region with the high-speed economic development, the agricultural overall value of the Beijing-Tianjin-Hebei Province region was not fully reflected in the value of the national economy accounting system.Agricultural products output were accounted only.

As one of three ecological systems of global land,farmland ecosystem has provided important role in local and national grain security,job security for the farmers,rural natural habitat,also with a certain maintenance of biodiversity, air purification, water conservation and the function of regulating climate, etc[1]. The farmers directly got few agricultural economic value and service value as farmland in contribution to the huge ecological service value.The theory and practice are all looking forward to the establishment and implementation of ecological compensation echanism in Beijing-Tianjin-Hebei Province region.

Administrative geographic units at various levels of farmland ecosystem service function value and evaluation method has been unified in Beijing-Tianjin-Hebei Province region. Based on this, the total value of farmland ecosystem were calculated so as to provide the basis for determining reasonable compensation standard for the farmland.

2. Ecosystem Services Classification and Standard Value Determination

The scientific ecosystem services classification and the standard value of value are the prerequisite for evaluation and application.

2.1. Ecosystem services classification

Ecosystem services came from the ecological system of logistics, energy flow and information flow[2]. At first, complex structure and process should be decomposed into limited number of functions which should be able to produce the profits including resources supply, environmental control, cultural entertainment and etc.

Costanza and others divided ecosystem services into 17 categories and identified the standard value of services early.The primary function of ecosystem services could be divided into four categories which were the support service,supply service, adjust service and culture service according to the Millennium Ecosystem Service Assessment(MA). 11 kinds of the secondary function had been divided further according to the research results of experts like Xie G D[3].

Supply services were divided three secondary types which included food production, raw material production and water supply. Adjustment services were divided 4 secondary types which included gas regulation, climate regulation, purification environment and hydrological regulation. Support services were divided three secondary type which included soil conservation, nutrient recycling and maintenance of biodiversity. Cultural service mainly included a secondary type of aesthetic landscape service.

According to the relation between supply and demand of the ecosystem and the human society,the value of ecosystem services were divided into market value and non-market value by Costanza and etc[4].Market value was the value of ecosystem services which could be reflected or achieved in market mechanisms, including food and raw materials.The non-market value played a very important role for human beings without the value of ecosystem services that are reflected in the market mechanism.

2.2 Equivalent Factor of Ecological Service Value

Per unit area equivalent factor table of ecosystem service valuereflects the various ecosystems and the ecological service function of the national average service value in the year. The reasonable equivalent factor table is the premise and foundation which could manifest district space difference and years time of the value of ecosystem services[5].This researchrevised Chinese secondary ecosystem services value equivalent factor tablemainly according to ecosystem services value evaluation result by Xie G D.

The net profit of grain production per unit area about farmland ecosystem was a standard equivalent factor of ecosystem services value according to processing methods by Xie G D. The grain yield of the farmland ecosystem was calculated based on the main products of rice, wheat and corn.To make the farmland ecosystem service value more accurate measurement results,The equivalent factor should been modified because equivalent factor was used for the national grain yield data[6].

The following formula is used.

$$Y = \frac{x_1}{x_2} \times Z$$

Y stands for the equivalent factor of field ecosystem services value in the study area. x_1 stands for the average grain yield in the study area, x_2 stands for the national average grain yield, Z stands for the unit area equivalent factor of farmland ecological service valuein China's land ecosystem.

The economic value of the standard ecosystem Eco-service value equivalent factor reflects the potential ability of the ecosystem to contribute relative to the ecological service[7].Classification of ecosystem services and equivalent factor tables were adopted and adjusted by "production method". The natural food production of cultivated land per unit area in Beijing-Tianjin-Hebei Province region was taken to adjust the economic value and the equivalent factor of the Beijing-Tianjin-Hebei Province region were determined.

Grain yields per unit area and food prices in 2015 had been consulted to revise the equivalent standard ecosystem services value factors which was 3651.8 yuan/hm2 of Beijing-Tianjin-Hebei region. As a special agricultural ecosystem,greenhouse ecosystem was not considered in this study.

3 Methods of Ecological Service Value Accounting for Farmland Ecosystem

The geographic grid method were adopted to achieve the spatial expression of the ecological service value in this study[7]. The vector data was handled by geographic grid and each grid cell was a regional ecological service value of internal statistics unit cell.The value of each type of ecological service in each grid was calculated and the statistical results was the value of the ecological services provided by all the coverage types in the grid[8].

$$V_i = \sum_{j=1}^{12} A_{ij} * V_{ij}$$

V_i stands for the value of ecological services in the gird i; A_{ij} stands for the area of type j in the grid i; V_{ij} stands for the ecological service value of the type j in the grid i.

4. General Situation in the Study Area

The region of Beijing-Tianjin-Hebei Province covers an area of 216,000 square kilometers with a total population of 10865,000. It is the core of the political, economic and cultural development of our country area and urbanization level is high, which has become the third most dynamic city cluster in our country after the Yangtze river delta and the pearl river delta. It is the highest level of economic development in north China area[10].

The region of Beijing-Tianjin-Hebei Province involves Beijing and Tianjin municipalities directly under the central government and 11 cities in Hebei Province which included Shijiazhuang, Tangshan, Qinhuangdao, Baoding, Cangzhou, Langfang, Handan, Zhangjiakou, Hengshui, Xingtai.

In this study six districts were amalgamated into Beijing downtown which included Dongcheng , Xicheng , Haidian , Chaoyang, Fengtai and Shijingshan. Ten districts were amalgamated into Tianjin urban area which included Six zones within,Dongli , Xiqing ,Jinnan and Beichen .

Farmland area of Beijing in 2015 was 1.375 million hectares, accounting for 8.2 percent of the land area and 1.93 percent of the area in the Beijing-Tianjin-Hebei region.

Farmland area of Tianjin in 2015 was 389,000 hectares, accounting for 32.6 percent of the land area and 5.45 percent of the area in the Beijing-Tianjin-Hebei region. Hebei province has 660.6 million hectares of farmland, accounting for 35.0% of the land area and 92.62% of the total area of the Beijing-Tianjin-Hebei area.

5 Results and Discussion

5.1 Results

5.1.1 Ecological Service Value of Farmland Ecosystem in the Beijing-Tianjin-Hebei Region

The total farmland ecosystem service value of Beijing-Tianjin-Hebei Region in 2015 was 110.3 x 109 yuan. The proportion of total farmland ecosystem service value in Beijing was 2.18% and Tianjin proportion was 5.26%. The highest proportion in Hebei Province was as high as 92.56%. The value of ecological services provided by farmland in the Beijing-Tianjin-Hebei region is enormous which was 15464.4 yuan per hectare.

5.1.2 Total Value of Different Services in the Beijing-Tianjin-Hebei region

The value of farmland regulation service in the Beijing-Tianjin-Hebei region was calculated as high as 39.45x 109 yuan, accounting for 35.8 percent of the total value of farmland ecological services. The second was support service and supply service, accounting for 31.8% and 30.9% of the total value of farmland ecological services. Cultural service was the least valuable, accounting for only 1.5% of the total value of farmland ecological services.

5.2 Discussion

When the ecological compensation policies is formulated, we should realize that the reality of ecological compensation ability is hard to meet in the ecosystem service[11] .Because of the influence of the economic development gap,the data is often beyond the jurisdiction of the government's fiscal capacity.Therefore, it is only as a reference and theoretical reference for ecological compensation when determining the standard of lateral financial transfer in practice.

Acknowledgments

This work was financially supported by Beijing Academy of Agriculture and Forestry Youth Fund(QNJJ201618).

Reference

[1] Wu D Q, Liu J, He T L, et al. Profit and loss analysis on ecosystem services value based on land use change in Yellow River Delta[J].Transactions of the Chinese of Society of Agriculture Engineering,2009,25(8):256-261.

[2] Costanza R, D Arge R, de Groot R, et al. The value of the world's ecosystem services and natural capital [J].Nature,1997,387(6230):253-260.

[3] Ouyang Z Y, Wang R S, Zhao J Z. Ecosystem services and their economic valuation[J]. Chinese Journal of Applied Ecology,1999,10(5):635-640.

[4] Li W H. Functional Valuation of Ecosystem Services:Theory, Method and Application[M].Beijing:China Renmin University Press,208.

[5] Xie G D, Lu C X, Leng Y F, et al. Ecological assets valuation of the Tibetan Plateau[J].Journal of Natural Resources,2003,18(2):189-196.

[6] Xie G D, Zhen L, Lu C X, et al. Expert knowledge based valuation method of ecosystem services in China[J].Journal of Natural Resources,2008,23(9):911-919.

[7] Yu Y N,Peng S L.Review on evaluation of ecosystem service[J]. Ecology and Environment Sciences,2010,19(9):2246-2252.

[8] Yang Z Y, Yang H Y, Guo Z X. A literature review on evaluation of agricultural ecosystem services [J].Chinese Journal of Eco-Agriculture,2009,17(5):1045-1050.

[9] Zheng D F,Zang Z,Sun C Z. An improved ecosystem service value model and application in ecological economic evaluation [J].Resources Science,2014,36(3):584-593.

[10] Hebei Provincal Government. The industry development plan of the area surrounding Beijing and Tianjin in Hebei Province(2010-2015)[R].Jizheng[2010] No.89.

[11] Zhang M, Cui H L, Liang C H, et al. Regional eco-compensation based on ecosystem service assessment-A case study of Liaoning Province [J]. Territory and Natural Resources Study,2014(2):53-55.

IOP Conf. Series: Earth and Environmental Science **86** (2017) 012005 doi:10.1088/1755-1315/86/1/012005

Study on The Application of Composed TiO_2-diatomite in The Removal of Phenol in Water

S Liu[1] and J Li[1]

[1] School of Environmental Science and Engineering, Southern China University of Science and Engineering, Shenzhen, 518055, China

Abstract. As an environmentally friendly pollution control technology, TiO2 photocatalytic technology has a broad prospect in the field of environmental protection. In this paper, composed nano-TiO2–diatomite were prepared by depositing TiO2 nanoparticles on the surface of diatomite microparticles. The nano-TiO2/diatomite composed photocatalyst is used to remove phenol in water in a specific designed reaction box under 4 different operation factors such as different reaction time, different pollutant concentration, different UV light powers and different amount of catalytic powder. The experimental results indicate that the phenol removal percentages are influenced by the reaction time most significantly, the second is the phenol concentration, the next one is the photocatalyst amount and the UV light powers' effect is quite limited. Tthe degradation of phenol typically slows down at the reaction time about 30 or 60 minutes. Besides that, the phenol removal kinetic removal rates were also investigated.

1. Introduction

Global drinking water pollution has brought serious challenges to human health. WHO survey showed that 80% of the world's disease and 50% of children's deaths are related to poor drinking water quality [1]. With the aggravation of water pollution, the conventional water treatment process has exposed a lot of limitations. For example, this cannot effectively remove trace organic contaminants, inorganic contaminants, pipe network and water storage tanks attached microorganisms . At present, domestic and foreign has developedTiO2adsorption technology, photocatalysis technology, advanced treatment technology of membrane and enhanced coagulation and filtration of traditional processing technology to improve and ensure the quality of drinking water [2].

Since the early development of this technology in the 1970s, TiO2 constitutes the archetypical photocatalyst due to its relatively high efficiency, low cost and availability [3]. Photocatalytic technology has the advantages of no twice pollution, low energy consumption, fast response, simple operation and high efficiency, so it has gradually become a research hotspot in the advanced treatment of drinking water. Thiruvenkatachari et al. (2008)[4] and Fan et al.(2001)[5] found out that TiO2 photocatalytic oxidation can effectively remove the humic substances in drinking water. Humus is the main precursor to produce toxic by-products (DBPs) in drinking water disinfection, and it is also the main factor leading to the increase of the mutagenic activity of drinking water. TiO2 photocatalytic technology can remove organic pollutants such as dyes, pesticides, phenols, hydrocarbons and surfactants in water, and can make the halogen atoms (X), sulfur atoms, phosphorus Atoms and

nitrogen atoms into X-, SO42-, PO43- and NO3- and other inorganic salts, to reduce or even completely eliminate its original harm [6]. Liu et al. (2009) [7] used the nano-TiO2/Diatomite compound photocatalyst to degrade rhodamine B dye wastewater in photochemical reactor. The best technical conditions are concentration of rhodamine B solution 10mg/L, ultraviolet light 300W, the compound photocatalyst amount used 1g/L, the pH 5.8, reaction time 20min. Under these conditions the rate of photodegradation of rhodamine B may reach as high as 97.80%. TiO2 photocatalytictechnology in the pharmaceutical wastewater, pesticide wastewater, oil-containing wastewater, paper waste water and other fields have been applied [8]. Cheng Cangcang et al. [9] fixed TiO2 photocatalyst film on the inner wall of stainless steel reactor, and the introduction of Fenton reagent in Wuhan, a pharmaceutical factory for pharmaceutical wastewater treatment. The results showed that the removal percentage of CODCr was 92.3% and the content of nitrobenzene compounds decreased from 8.05 mg / L to 0.41 mg / L. Zhang Haiyan et al. [10] prepared nano-TiO2 semiconductor photocatalyst, investigated the catalyst crystal structure, particle size, dosage, pH and the photocatalytic degradation of Fe3+ or H2O2 coexisting wastewater containing oil. The results show that TiO2 has a high photocatalytic activity of oil degradation. When it is coexisted with Fe3 + or H2O2, the removal percentage of oil can be increased by 5% ~ 16% under the same illumination time. Degradation properties of phenol using nano-TiO2 as photocatalyst in aqueous solution were investigated. In Wang et al.'s work as well [11], the effect of annealing temperature and ionic modification on the degradation was studied. The results showed that photocatalyst nano-TiO2 material doped with Fe3+ was prepared quickly by sol-gel process and was used as photocatalyst to degrade phenol solution of 100mg/L under UV irradiation for 3 hours. The phenol degradation percentage was 94.18%. SP Kunjalukkal et al. [12] in 2014 found that the photocatalytic reaction follow a pseudo-first-order rate law and the composite catalyst crystallized at 500°C shows the highest catalytic activity toward RhB degradation. It was observed that diatomite substrate significantly enhanced the photodegradation efficiency of TiO2 due to higher surface area, mesoporous structure and uniform deposition of the small sized TiO2 particles.
From the above summary, we can see that the investigation of composed nano-TiO2-diatomite material in the application of phenol degradation in water is still quite limited. The different influence reaction factors and kinetic removal rate were not yet explored in depth. With this in mind, we did some preliminary experimental work.

2. Experiments

2.1. The preparation of the composed nano-TiO2 and diatomite

In this paper, according to the method proposed by Akpan in 2009 [13] and Shi enbin in 2014 [14], we used the modified bought P25 nano-TiO2 (40% by weight) loaded onto the porous diatomite (60% by weight) and dried. Then the composed powder particles were sieved to the size larger than 45 μ m (figure 1).

Moreover, the reasons why we preferred to use the composed nano-TiO2 and diatomite material instead of only nano-TiO2 is the composed material should have higher effective efficiency than the pure nano-TiO2 material in removing pollutants because it has diatomite which has porous structure. The diatomite was responsible for the porous matrix so that this kind of materials can have relatively huge surface area covered by P25 nano-TiO2.

IOP Conf. Series: Earth and Environmental Science **86** (2017) 012005 doi:10.1088/1755-1315/86/1/012005

Figure 1. The composed P25 Nano-TiO2 and diatomite material after sieving

2.2. Design of the UVC reaction box

Different from other researcher' s work, most of them applied the method by directly inserting the high power UV lights into the liquid to make reaction happen more efficiently. But this method can be limited to use in lab and the UV lights、catalytic materials have relatively shorter life time. In this paper, we used the designed reaction box to carry out the experiments to avoid the UV lights contact with liquid. In this box, there are maximum 4 UV lights (power 55W) can be stalled at the top ceiling which can be shown in figure 2. From above description, the UV light tube power in this paper can be changed to 55W (1 light tube installed), 110W (2 light tubes installed), 165W (3 light tubes installed) and 220W (4 light tubes installed). In figure2, we also can see that during the experimental process, the glass ware can be settled on the platform in the UVC reaction box and the top of water level in glass wares is about 5 cm from the UV tubes. During the experiments, when the UV lights should be on we could close the UVC box and press the ON/OFF button as figure 3 showed. In this way, the UV lights can make reaction happen as expected even not contact with the liquid, and the harm caused by the UV lights could be reduced to minimum.

Figure 2. The inside of the Reaction box(UVC)

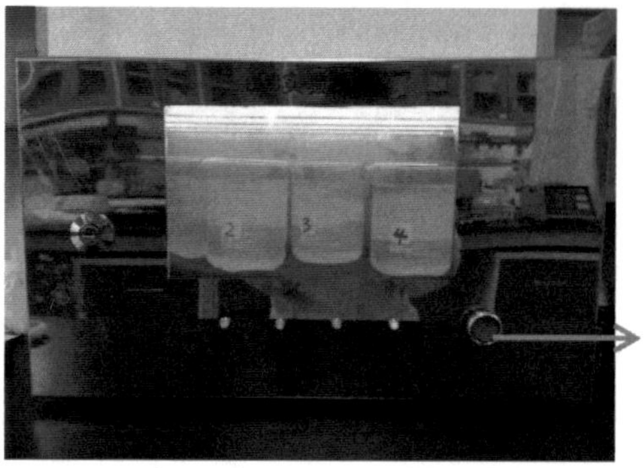

Figure 3.UVC box with UV lights on

2.3. The preparation of water sample

The water samples we used in the experiments are the purified water polluted by phenol (C6H5OH, Sigma-Aldrich, 99%, Shanghai, China) and were prepared by 0.005mg/L, 0.01mg/L, 0.02mg/L (phenol concentration limit in drinking water in the New National Standard is 0.002mg/L). The catalytic powder amount in water samples were chosen to 0.05g/L or 0.2g/L.

2.4. Experimental Process

Based on the 4 different main influence factors to effect phenol removal efficiency, we got amount of samples. The 4 different factors are:

(1) Water samples with different initial phenol concentration, Co, 0.005mg/L, 0.01mg/L, 0.02mg/L;
(2) Different catalytic powder amount, 0.05g/L or 0.2g/L;
(3) Different UV lights tube powers, 110 W and 220W;
(4) Reaction time, 15 mins, 30 mins, 45 mins, 60 mins, 90 mins, 120 mins.

The water samples obtained from experiments under the above different factors were collected out the UVC box and then filtrated by the 22 μ m membrane as figure 4. As described before in this paper, the catalytic powder chosen in this paper were sieved to > 45 μ m so the catalytic powder can be removed all. The residue phenol concentration in water samples after filtrated from catalytic powders were measured by the UV Visible spectrophotometer UV 1700 in the following figure 5.

Figure 4. Filtration membrane (22 μm)

Figure 5.UV Visible Spectrophotometer UV1700

The standard calibration line of the UV Visible spectrophotometer UV 1700 were obtained as well in figure 6 so from figure 6, we can see that the analysis results of the UV visible Spectrophotometer 1700 in this paper were reliable.

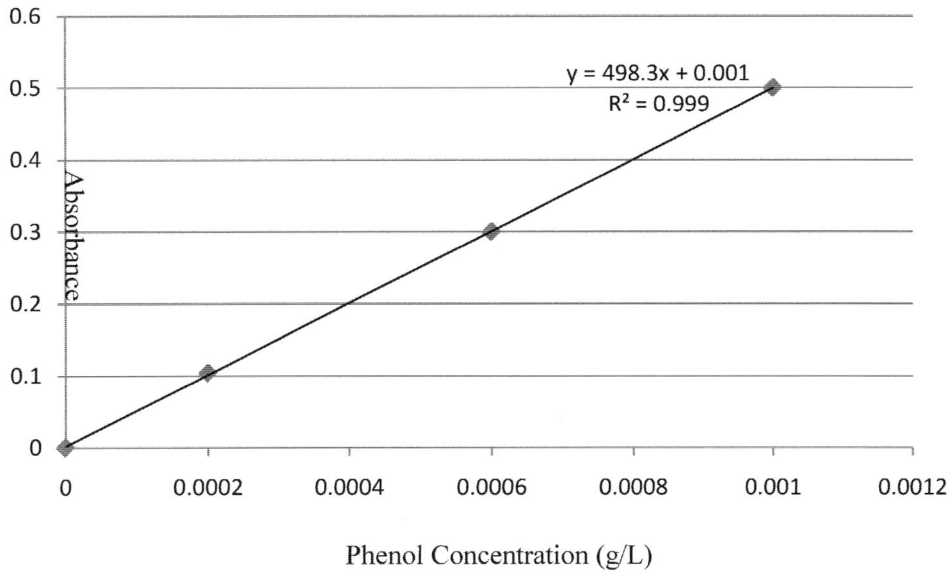

Figure 6. The standard calibration line of the UV Visible spectrophotometer UV 1700

3. Results and Discussion

3.1. The effect of reaction time on removal percentage and kinetic rates

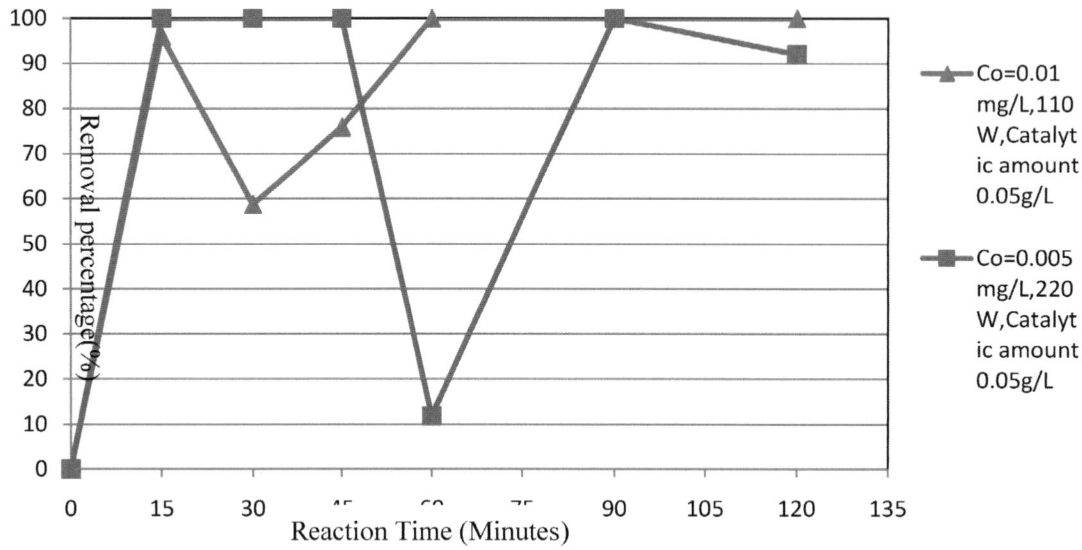

Figure 7. The removal percentage at different time (minutes)

In figure 7, the x-axis is the reaction time (unit: minutes) and y-axis is the phenol or other pollutants removal percentage which can be calculated as: (Ct-Co)/Co (unit: %).

From abovefigure 7, we can see that for different operation conditions such as different UV light powers or water samples with different initial phenol concentrations, the phenol removalpercentages all appeared a declined trend at the reaction time about 30-60 minutes. The reason why the degradation removal percentages have declined is because that as reaction going on, the composed nanto-TiO2 and diatomite materials absorption ability achieved quite relatively high and the desorption of phenol happened. Then the TiO2 photocatalytic reaction happened again and the phenol removal percentage became stable.

The phenol degradation typical declined is one of the most innovative findings in this paper. Based on this result, it gives us a hint that when we use composed nano-TiO2 and diatomite catalytic powder to remove phenol or some other organic pollutants in water or air, we should consider about the removal percentage slowed down.

IOP Conf. Series: Earth and Environmental Science **86** (2017) 012005 doi:10.1088/1755-1315/86/1/012005

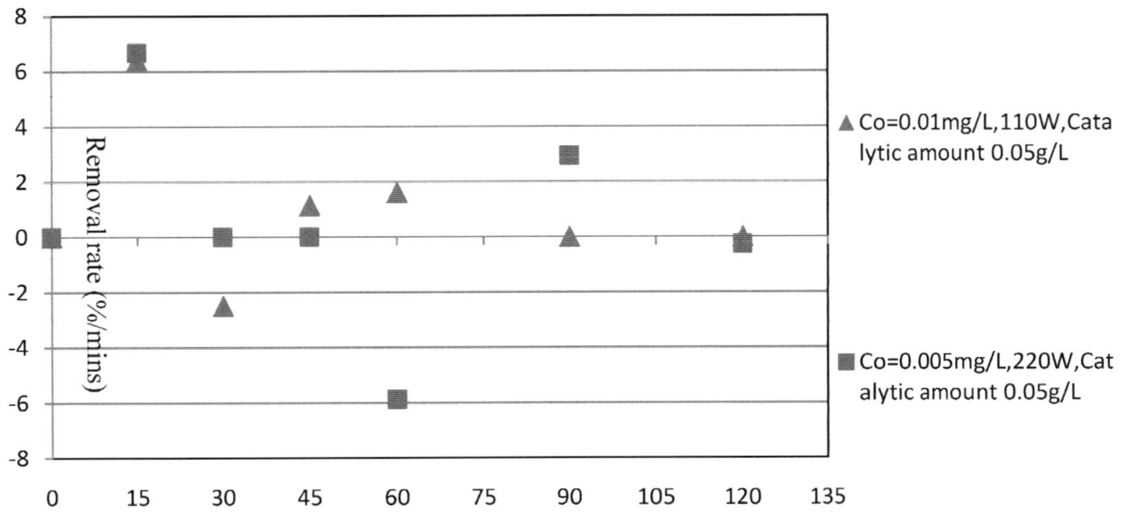

Figure 8. The effect of reaction time on removal kinetic rate

To investigate the reaction kinetic removal rate, we got figure 8. The x-axis is reaction time which has the unit as minutes. Y-axis is the removal kinetic rate which can be calculated as removal percentage/time lapse and the unit can be expressed as %/minutes. From figure 8, it can be seen that the kinetic removal rate appeared wave shape. At reaction time about 30 minutes, the kinetic removal rate decreased and then the removal rate appeared increased. We tried to find the relationship between reaction time (minutes) and removal rate (%/minutes), but from figure 9, it can be seen that even the 2 first order lines describing the relationship between reaction time (minutes) and removal rate (%/minutes) appeared quite close but both the correlation coefficients R square appeared quite low, so the 2 lines are not reliable.

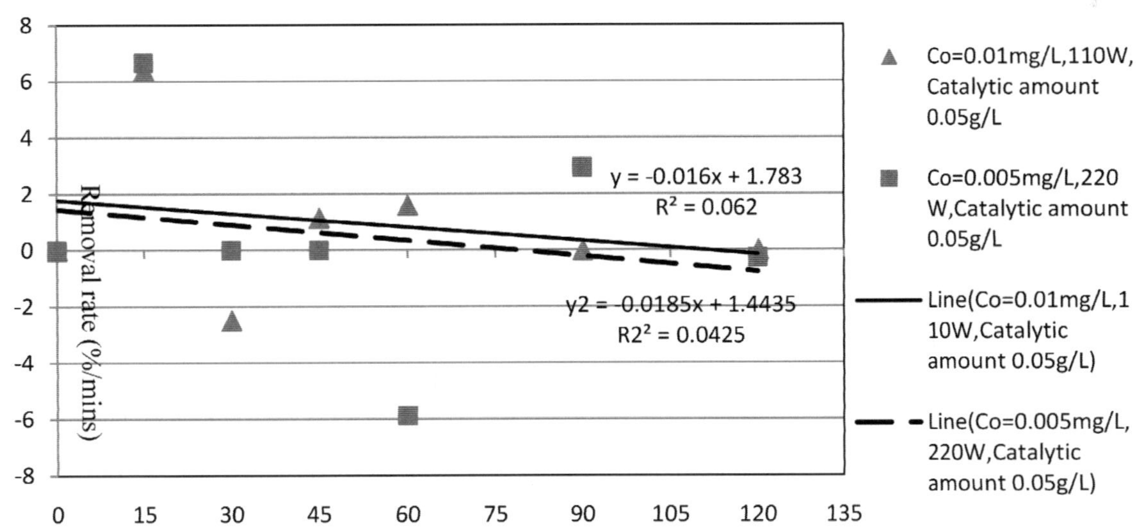

Figure 9. The effect of reaction time on removal kinetic rate with the correlation coefficients R square

IOP Conf. Series: Earth and Environmental Science **86** (2017) 012005 doi:10.1088/1755-1315/86/1/012005

3.2 The effect of water samples with different initial phenol concentrations on removal percentages and kinetic rates

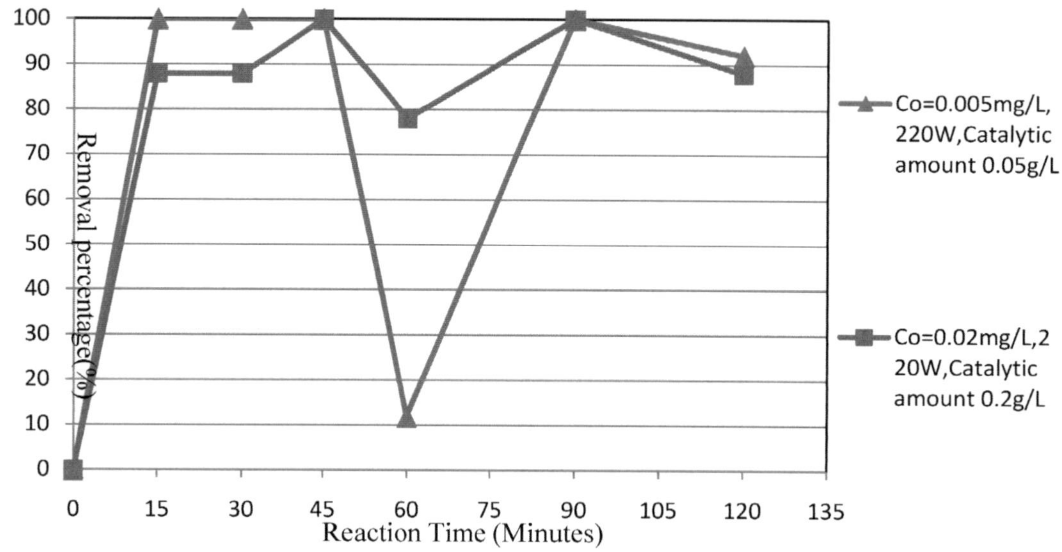

Figure 10. The effect of water samples with different initial phenol concentrations on removal percentages

From figure 10, we can see that for different water samples (different C0), the phenol remove percentage decreased as the C0 increased under the same experimental condition. It means that the photocatalytic technology in our paper was effective in the removal of trace concentration of phenol in water.

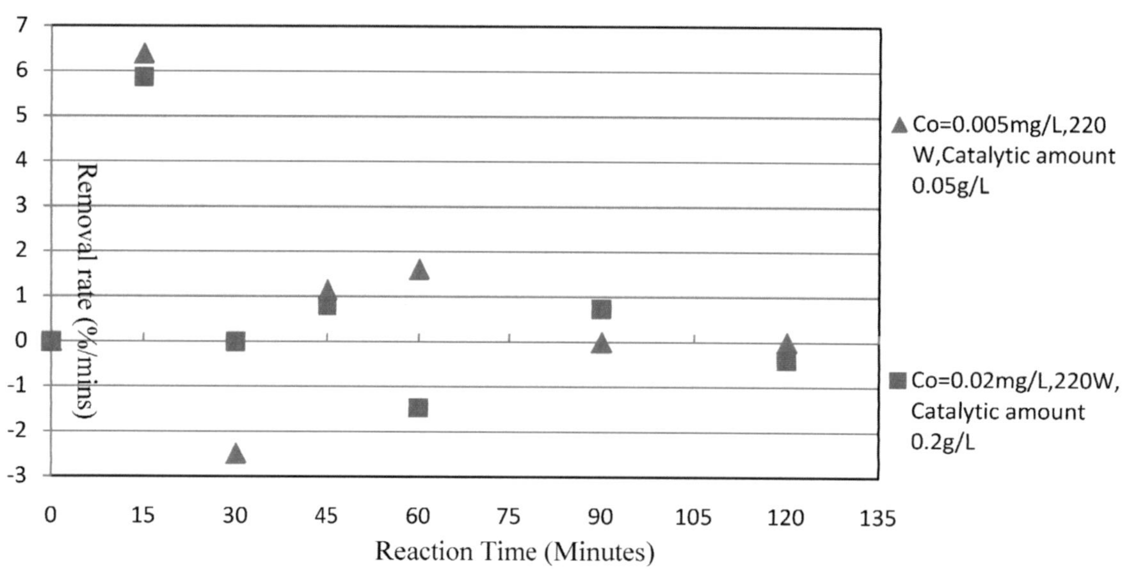

Figure 11. The effect of water samples with different initial phenol concentrations on removal kinetic rate

Similar with figure 8, from figure 11 it can be seen that the kinetic removal rate still appeared wave shape as reaction time passed. For water samples with different initial phenol concentrations, the lower initial phenol concentration, the change rate higher.

3.3 The effect of UV light powers on removal percentages and kinetic rates

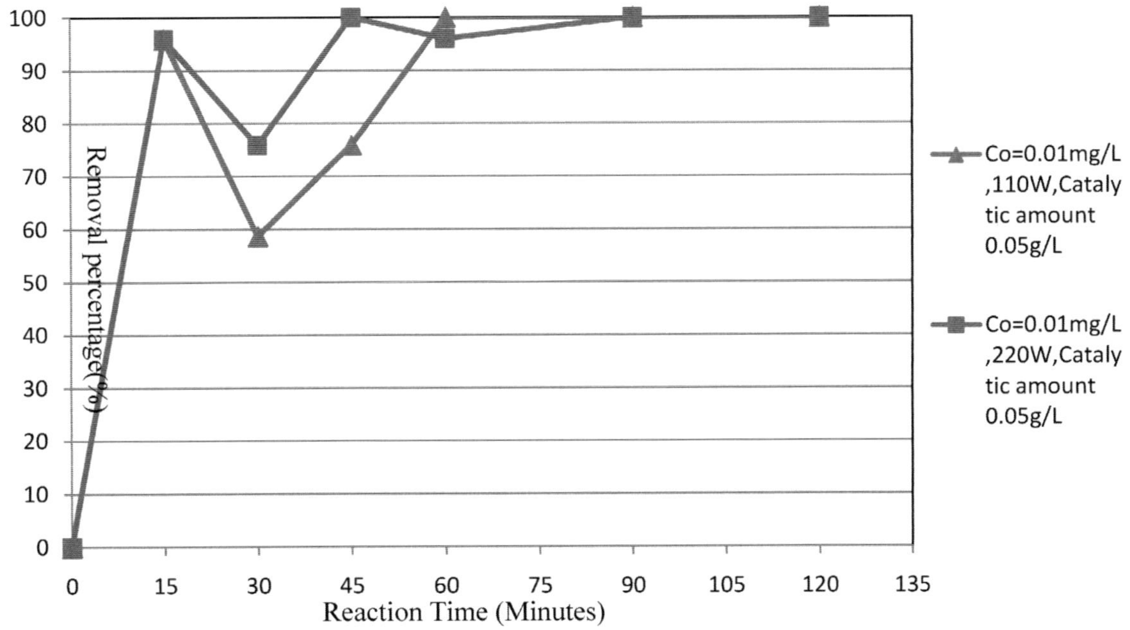

Figure 12. The effect of UV light powers on removal percentage

From figure 12, we can see that the effect of UV light powers on removal percentages appeared more effective at the reaction time before 60 minutes. The higher UV light power, the higher phenol remove efficiency we got.

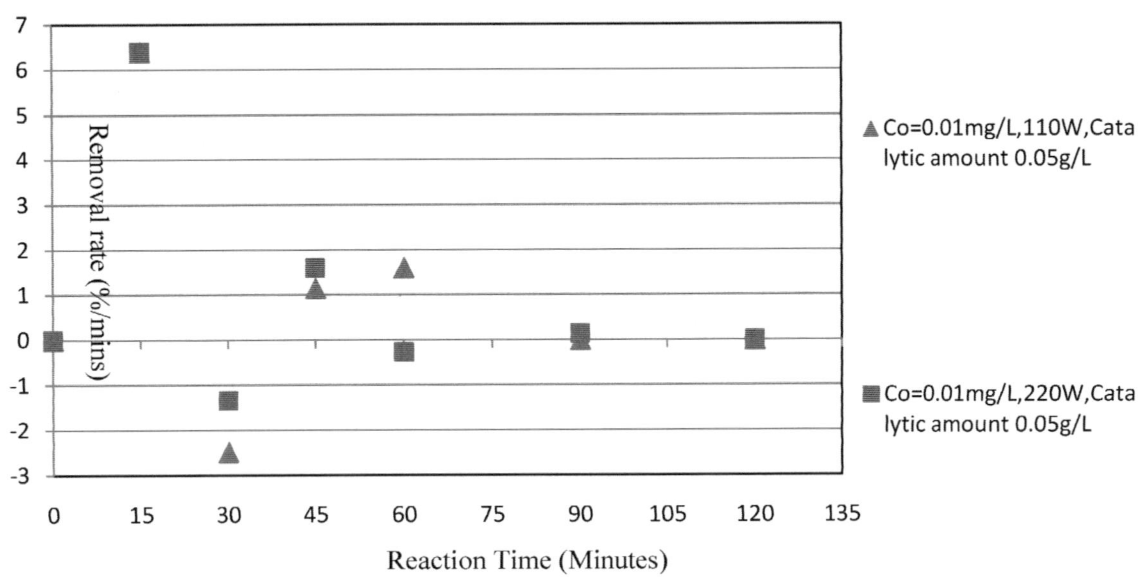

Figure 13. The effect of UV light power on removal kinetic rate

IOP Conf. Series: Earth and Environmental Science **86** (2017) 012005 doi:10.1088/1755-1315/86/1/012005

Again, the removal kinetic rate appeared wave shape and the UV light power has quite limited effect on the removal kinetic rate in figure 13.

3.4 The effect of different catalytic powder amounts on removal percentages and kinetic rates

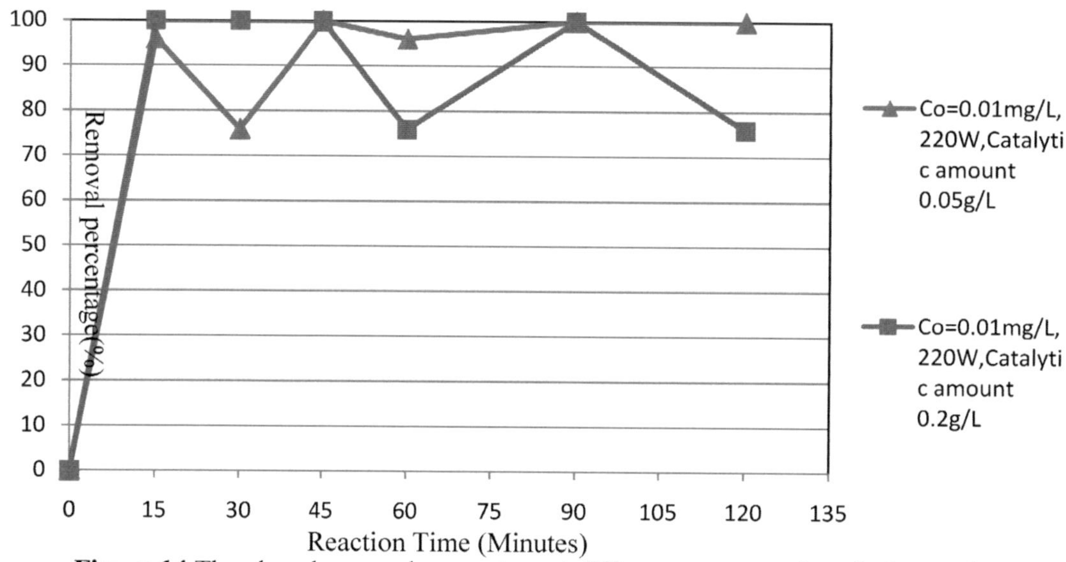

Figure 14. The phenol removal percentage at different amounts of catalytic powder

In figure 14, we get 2 different lines about removal percentages under the same UV tube powers and same C0 0.01mg/L but different catalytic powder amounts 0.05g/L, 0.2g/L. It can be seen from figure 14 that at the beginning stage of reaction, the more catalytic powder amount, the higher removal percentage. But the removal percentage increased not so much compared with the increase of catalytic amount from 0.05g/L to 0.2g/L. And as reaction going on, the removal percentage decreased a little bit for the operation condition 220W, C0=0.01mg/L, catalytic amount 0.2g/L. So the effect of the catalytic powder amount can be summarized not significant. Very little amount of catalytic powder can achieve quite good removal results.

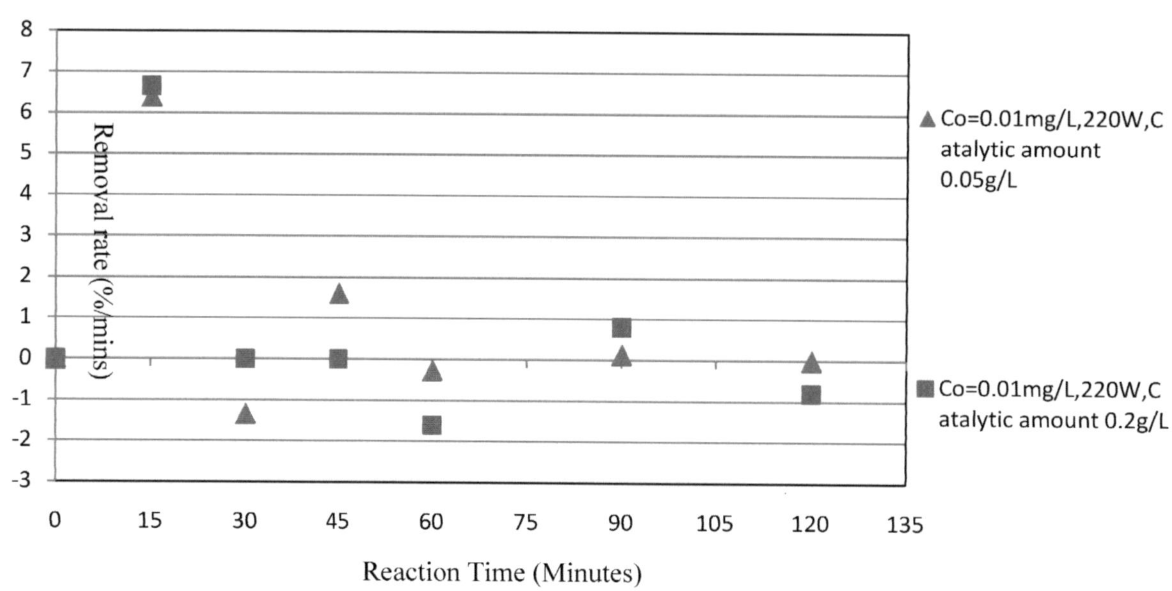

Figure 15. The phenol removal kinetic rate at different amounts of catalytic powder

From figure 15, we can see that the phenol removal kinetic rate still appeared wave shape but the effect of different catalytic amount on removal kinetic rate can not be ignored. The lower catalytic powder amount, the kinetic removal rate change more significantly.

Besides the above experimental data, blank experiments were done as well. The phenol concentration in water samples under UV without catalytic powder, or with catalytic powder without UV both appeared not effected and proved that the actual efficiency of composed nano-TiO2 diatomite as catalytic materials.

In this study, we firstly investigated the research background and research status of the topic; then on the basis of theoretical knowledge, we constructed composed nano-TiO2 and diatomite material as catalytic powder in the water treatment; through the analysis of specific experiments and experimental results of the system, we verify the feasibility and effectiveness of the treatment system. It proves that the research is fruitful and meaningful.

4. Conclusions

The conclusions can be summarized as: (1) The phenol removal percentages are influenced by the reaction time most significantly, the second is the phenol concentration, the next one is the photocatalyst amount and the UV light powers' effect is quite limited. (2) The degradation of phenol typically slows down at the reaction time about 30 or 60 minutes. Because the type of reaction kinetics which depends on pollutant concentration. (3) The higher UV light power, the higher phenol removal percentage at the beginning of the reaction stage. But after time passed, lower UV tube power can achieve the 100% removal percentage. But for kinetic removal rate, the UV light power has limited effect. (4) The more catalytic powder amount, the higher removal percentage at the beginning of the reaction time. But the removal percentage increased not so much compared with the increase of catalytic amount. But also it appeared that the removal percentage can achieve very high at about 1 hour and later. So the effect of the catalytic powder amount can be summarized not significant. Very little amount of catalytic powder can achieve quite good removal results. But the effect of different catalytic amount on removal kinetic rate can not be ignored. The lower catalytic powder amount, the kinetic removal rate change more significantly.

Acknowledgments
This work was supported by Shenzhen Municipal Science and Technology Innovation Committee through project Application of TiO2 in the Advanced Treatment of Drinking Water in the New National Standard (NO. JCYJ20150831141958719).

5. References

[1] WuC, MaurerC, WangY, XueSand DavisD L1999. "Water Pollution and Human Health in China,"*Environmental Health Perspectives*, 107(4):251-6
[2] LiuX, ShenLand HuY2016. "Preparation of TiO2-Graphene Composite by a Two-Step Solvothermal Method and its Adsorption-Photocatalysis Property,"*Water Air and Soil Pollution*,227(5)
[3] HernándezalonsoM D, FrensoF, Suárez Sand CoronadoJ M 2009. "Development of alternative photocatalysts to TiO2: Challenges and opportunities,"*Energy & Environmental Science*, 2 (12) :1231-1257
[4] Thiruvenkatachari R, Vigneswaran S and Moon S2008. "A review on UV/TiO2 photocatalytic oxidation process," *Korean Journal of Chemical Engineering*, 25(1):64-72
[5] FanCand SunY2001. "Effect of Calcium and Magnesium on the Photocatalytic Oxidation of Humic Acid in Water over TiO2," *The Chinese Journal of Process Engineering*
[6] Zang D and Yang Y 2009. "Preparation 8 application of nano-titaniaphotocatalyst,"*Tianjin Chemical Industry*, 23(5): 1-4

[7] Liu Y, Zheng S, Du G, ShuFand Chen J 2009. "Photocatalytic Degradation property of Nano-TiO2/Diatomite for Rodamine B Dye wastewater,"*International Journal of Modern Physics B*, 23:1683

[8] Lu J and Chen H 2009. "Advances in studying on modification of nanoscale titanium dioxide photocatalyzer and its application in the treatment of waste water,"*Journal of Chongqing University of Arts and Sciences*, 28(3): 74-78.

[9] Cheng C, Xiao Z, Hu D and Zhang J 2001. "Study on Pharmaceutical Wastewater Treatment by UV/TiO2-Fenton Agent,"*Research of Environmental Science,* 14(2): 33-35.

[10] Zhang H, Wang B and Chen Ying 2003. "A Study of Oilfield Sewage Oxidation through Photocatalysis,"*Chemical Industry and Engineering Progress*, (1): 67-70

[11] Wang J, Huang D, Liu W and Pan Q and Min Y 2014. "Degradation of Phenol by Fe3+ Doped Nano-TiO2 Photocatalyst,"*Advanced Materials Research*, 859 (2):361-364

[12] Kunjalukkal S P, Pal S, Licciulli A and Haq E U 2014. "Nanocrystalline TiO2–Diatomite Composite Catalysts: Effect of crystallization on the Photocatalytic Degradation of Rhodamine B,"*Applied Catalysis A General*, 485 (966) :157-162

[13] Akpan U G and Hameed B H 2009. "Parameters affecting the photocatalytic degradation of dyes using TiO2-based photocatalysts: a review,"*Journal of Hazardous Materials*, 170 (2-3) :520

[14] Shi E, Zhu H and Chen X 2014. "Preparation of diatomaceous with modified nano-TiO2 loaded onto diatomite and its performance study,"*New Building Materials*, 41(12): 84-88

Experimental Investigation of the Reactivity of the Copper-based Oxygen Carrier Used in Chemical Looping Air Separation

K Wang, Q B Yu*, Q Qin and J L Liu

School of Metallurgy, Northeastern University, Shenyang, Liaoning 110004, China

Abstract. Chemical looping air separation is a novel oxygen production technology based on the reduction and oxidation cycle of the oxygen carrier. In this work, reactivity tests of the reduction and oxidation reactions with CuO supported on SiO_2, TiO_2, ZrO_2, and Al_2O_3 were performed on STA409PC thermogravimetricanalyzer under different temperatures and particle sizes. The results show that SiO_2, TiO_2, ZrO_2 are the suitable binders for CuO. The reaction rates have the tendency of increasing as particle sizes decrease, but the influence is not apparent. The binders of SiO_2, ZrO_2 and TiO_2 combined with CuO show high reactivity and conversion ratio. The reaction rates increase greatly as the temperature increases.

1. Introduction

CO_2 is the main cause of global climate change and the CO_2 concentration in the atmosphere has increased greatly during last decades. Several measures have been proposed to reduce the CO_2 emission and oxy-fuel combustion is one of the effective measures [1, 2]. In oxy-fuel combustion, high purity of oxygen is needed [3-5]. However, the shortcomings and high price of the current air separation technologies restrict its application in industry widely[3-5].

Chemical looping air separation (CLAS) is anewair separation technology for oxygen production. The average specific power of the CLAS process is just 26 % of an advanced cryogenic air separation system[6]. The integration of CLAS and oxy-fuel combustionpower plant has been investigatedby Moghtaderi*et al*. The results indicated that for acceptable process kinetics CLAS can potentially run at much loweroperating costs compared to an advanced cryogenic distillation system for oxy-fuel thermal power plant [5].

The CLAS system contains two interconnected reactors (oxidation and reduction reactors). Oxygen carrier is the bond of the two reactors. In the oxidation reactor, the oxygen carrier captures oxygen from air and in the reduction reactor,the oxygen carrier releases oxygen to carrier gas.The carrier gas used in the reduction reactor can be steam or flue gas to produce pure O_2 or oxygen enriched gases.

The key issue for large-scale application of CLAS is the capability of releasing oxygen of oxygen carriers under suitable temperature and oxygen concentration. In addition, some other properties are also needed, such as sufficient oxygen transport capacity,high reactivityand mechanical strength, chemical stability [7-9]. Ni-, Cu-, Co-, Fe-and Mn-oxides were the common oxygen carriers in chemical looping technologies [10, 11]. Besides the oxides, NiO and Fe_2O_3cannotrelease oxygen even when the temperature are higher than 1600K.CuO, Mn_2O_3 and Co_3O_4 could evolve gaseous oxygen at suitable temperatures [12-14]. Among the three oxides, copper oxide has the highest oxygen transport capacity and has been studied widely. Gayán*et al*. [15] tested the oxidation-reduction reactivityand the stability of CuO supported on $MgAl_2O_4$ or ZrO_2in abatch fluidized-bed reactor. The oxygen carriers were identified as suitable materials.Adánez-Rubio*et al*. [16, 17] evaluated thestability of

$CuO/MgAl_2O_4$oxygen carrier under different temperatures. They foundthat both reduction and oxidation reaction rateswere stable with cycles.Mattisson*et al.* [18] investigated the reactivity of $CuO/ZrO_2$with petroleum cokein a small quartz laboratory fluidized bed reactor. In this work, SiO_2, TiO_2, ZrO_2, and Al_2O_3 as binders were prepareoxygen carriers [19]. Reactivitytests were performed onSTA409PC under different temperatures and particle sizes.

2. Experimental

2.1. Materials

The materials used in the process were copper oxide combined with binders and these materials were prepared by mechanical mixing method. SiO_2, TiO_2, ZrO_2, and Al_2O_3powders were used as binders. The preparation process could be seen from our previous studies [21]. Three particle sizes of oxygen carrier were obtained: 76~150μm, 150~200μm and 200~315μm.

The phases of oxygen carriers preparedwere analyzed using XRD.The XRD patterns were shown in Figure 1.

When ZrO_2, TiO_2 and SiO_2 were used, the phases in the oxygen carriers prepared were mainly CuO and the corresponding binders. When Al_2O_3 was used, $CuAlO_2$ were formed. The formation of $CuAlO_2$ followed the Eqs. (1) and (2).

$$CuO + Al_2O_3 = CuAl_2O_4 (1)$$

$$4CuAl_2O_4 = 4CuAlO_2 + 2Al_2O_3 + O_2(g) \quad (2)$$

Unlike Cu_2O, the phase of $CuAlO_2$ cannot be oxidized to $CuAl_2O_4$ in the oxidation process [22]. The phases of Al_2O_3 and $CuAlO_2$ all were binders in the Cu/Al oxygen carrier.

The BET surface area of the particleswas evaluated by N_2-adsorption and the values were shown in Figure 2. The BET values of the oxygen carriers prepared by the four binders all increased with the binder adding ratio increasing. The effect of binder could increase the anti-sintering property of CuO. Among the four binders, the BET values of Cu/Al oxygen carriers were smallest. The probable reason was that the formation of $CuAlO_2$ increased the compactness of oxygen carrier.

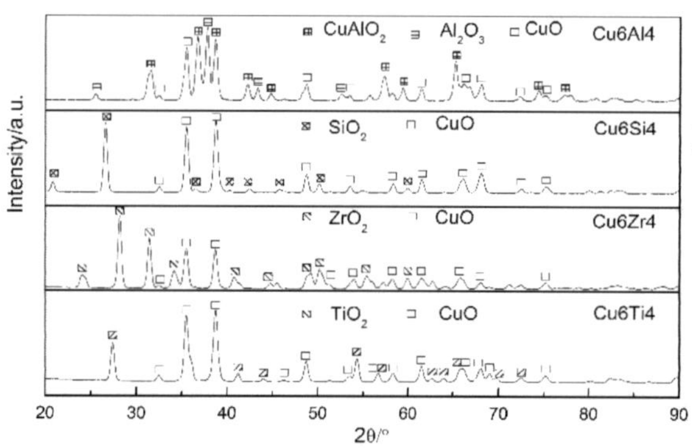

Figure 1 The XRD patterns of oxygen carriers

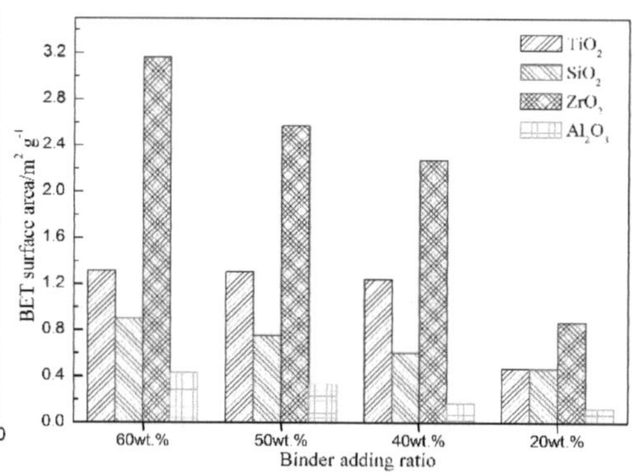

Figure 2 The BET values of oxygen carriers

2.2. Experimental setup

Experimental tests were performed on the STA409PC. For reduction, the sample was heated to 700℃ with heating rate of 30℃/min and kept at 700℃ for 30min in air atmosphere to make the sample oxidized fully. Then the sample was heated tothe required reduction temperatures. At these points, air was stopped andnitrogen was turned on. The reductionprocess was maintained for 30min to complete the full decomposition of oxygen carrier. For oxidation, the sample was cooled to the requiredoxidation temperatures in nitrogen atmosphere. Then nitrogen was stopped and air was turned on. Theoxidation process was maintained for 30min to complete the full oxidation of oxygen carrier. The mass of sample at each experiment was 10mg.

The conversion ratio (α) can be calculated according to Eqs. (3) and (4).

For reduction: $$\alpha_{red} = \frac{m_{ox} - m}{m_{ox} - m_{red}} \quad (3)$$

For oxidation: $$\alpha_{oxi} = 1 - \frac{m_{ox} - m}{m_{ox} - m_{red}} \quad (4)$$

Where m, m_{ox}, $m_{red}$are the masses of the sample at each time, fully oxidized and fully reduced, mg.

3. Results and discussion

3.1. Effect of binder

Figure 3 shows the conversion ratio with time using different binders. The oxygen carriers with ZrO_2, TiO_2 and SiO_2 supports had high reduction and oxidation conversion ratios. However, the oxygen carrier with Al_2O_3 supporthad conversion ratio only about 55%. The reason was that the spinel of $CuAlO_2$was formed during the preparation. Based on the maximum oxygen transfer capacity, ZrO_2, TiO_2 and SiO_2 were the suitable binders for CuO. The reduction times of Cu/Zr and Cu/Ti oxygen carriers were small and the reduction rates were fast. The reduction rate of Cu/Si oxygen carrier was slow. The possible reason was that the BET values of Cu/Si oxygen carriers were small. Compared to the reduction rates, the oxidation rates of Cu/Zr, Cu/Si and Cu/Ti oxygen carriers were all high.

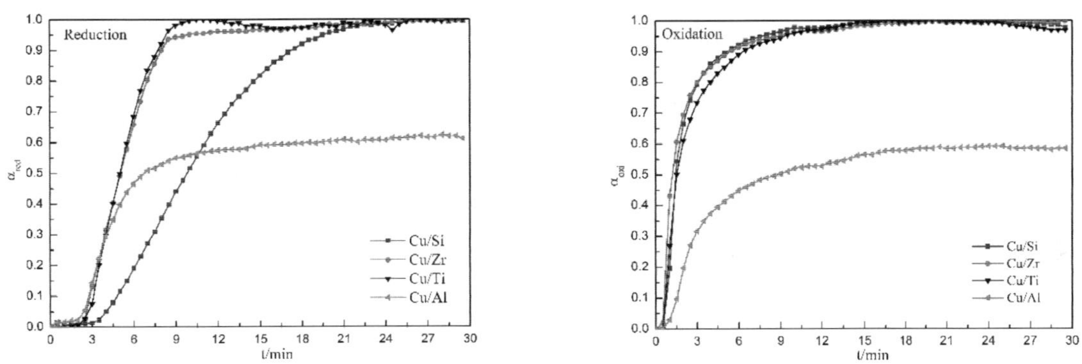

Figure 3The conversion ratio with time using different binders

3.2. Effect of temperature

Figure 4 shows the conversion ratio with time under different reaction temperatures. The reduction processes were divided into two stages: incubation and rapid oxygen releasing stages. At the beginning of the reduction, the atmosphere in the reactor was air. With the nitrogen inletting, the oxygen concentration in the reactor decreased. When the concentration in the reactor was the equilibrium oxygen partial concentration, the reduction began. The equilibrium oxygen partial concentrations for 900, 950, 1000℃ were 1.5%, 4.5% and 12.4% respectively. Thus, the times for incubation stage decreased with reduction temperature increasing. After the incubation stage, the oxygen releasing rates increased greatly. For oxidation, the equilibrium oxygen partial concentrations for 600, 650 and 700℃ all were very low (lower than 0.01%). The oxygen concentration in the air (21.0%) was much higher than the equilibrium oxygen partial concentrations of the oxidation reactions. Thus the incubation times for oxidation were short. Moreover, the oxidation rates increased with the oxidation temperatures increasing. Under the oxidation temperatures investigated, high temperatures could promote the oxidation reactions [23].

3.3. Effect of particle size

Figure 5 shows the conversion ratio with time under different particle sizes. For reduction, time for overall conversion increased with the particle size increasing. During the reduction process, the oxygen released should diffuse from the inside of the oxygen carrier to outside. However, the effect of particle size on the reduction rate is not obvious and the reduction rates for the three sizes of oxygen carriers were all very high. The phenomenon was important for the fluidizationoperation in the CLAS technology. For oxidation, the curves of the three conditions were almost overlapping and the effort of particle size was little. The reason was that the oxygen concentration in the air (0.21) / equilibrium oxygen partial concentration (5.8×10^{-5}) at 700℃was so high that the effect of particle size could beignored. The oxidation rates were high and only about three minutes were needed to oxidize the Cu_2O to CuO.

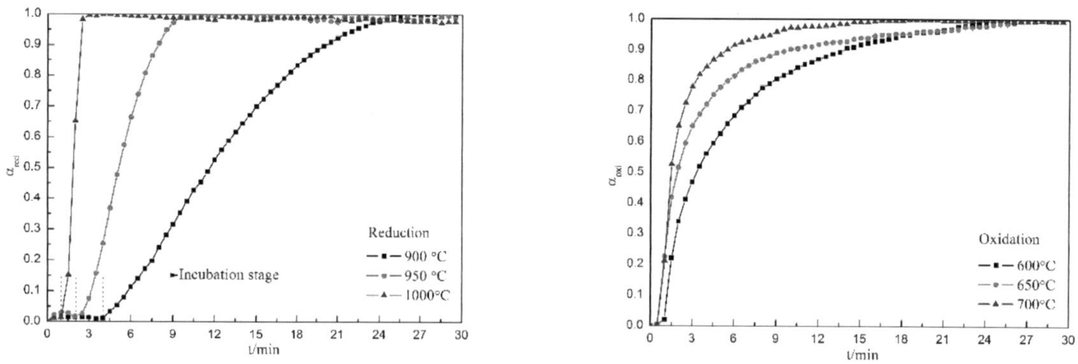

Figure 4The conversion ratio with time using different reaction temperatures

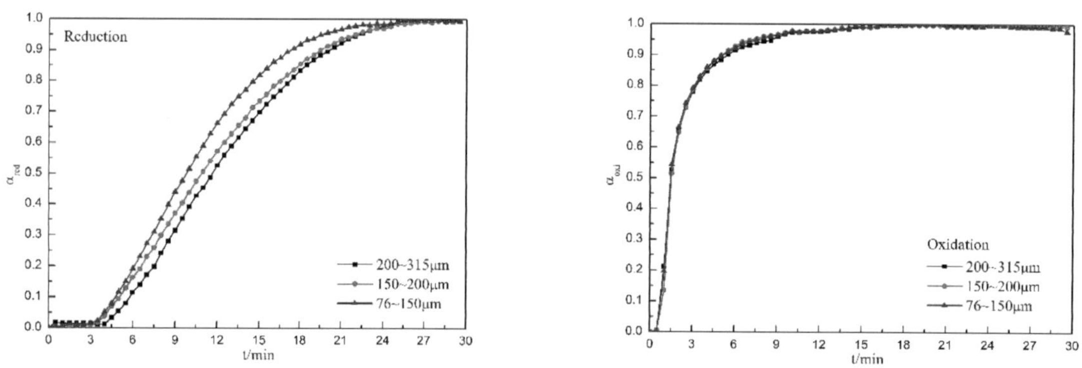

Figure 5The conversion ratio with time using different particle sizes

4. Conclusions

TGA tests showed that the binders of SiO_2, ZrO_2 and TiO_2 combined with CuO show high reactivity and conversion ratio.The reaction rates had the tendency of increasing as particle size decreased in the reduction, but the influence was not apparent. The reaction rates increase greatly as the temperature increases.

Acknowledgments

This research was financially supported by National Natural Science Fund of China (51604078, 51576035),Fundamental Research Funds for the Central Universities (N162504012), Post-Doctoral Science Foundation (2017M610185, 20170101).

References

[1] B. Erlach, M. Schmidt, G. Tsatsaronis,2011. Comparison of carbon capture IGCC with pre-combustion decarbonisationand with chemical-looping combustion, Energy 36: 3804-3815.
[2] T. Walla, Y. Liu, C.Sperob, L. Elliotta, S.Kharea, R.Rathnama,F.Zeenathala, B.Moghtaderia, B.Buhred, C. Shenge, R.Guptaf,T. Yamadac, K. Makinoc, J. Yu, 2009. An overview on oxyfuelcoal combustion—State of the artresearch and technology development, Chem. Eng. Res. Des. 87: 1003-1016.
[3] T. Burdyny, H.Struchtrup,2010. Hybrid membrane/cryogenic separation of oxygen from air for use in the oxy-fuel process,Energ.35: 1884-1897.
[4] A.R. Smith, 2001. A review of air separation technologies and their integration with energy conversion processes. Fuel. Process.Technlol. 70: 115-134.
[5] K. Shah, B. Moghtaderi, T. Wall, 2013. Effect offlue gas impurities on the performanceof a chemical looping based air separation process for oxy-fuel combustion,Fuel 103: 932–942.
[6] B.Moghtaderi, 2010. Application of chemical looping concept for air separation at high temperatures, Energ. Fuel 24: 190–198.
[7] A. Fossdala, E. Bakkenb, B.A. Øyea, C. Schøninga, I. Kausa, T. Mokkelbosta, Y. Larring, 2011. Study of inexpensive oxygen carriers for chemical looping combustion, Int.J.Greenh.Gas. Con. 5: 483-488.
[8] A. Dueso, M. Ortiz, A. Abad, F. García-Labiano, L. F. de Diego,P.Gayán, J. Adánez, 2012, Reduction and oxidation kinetics of nickel-based oxygen-carriers forchemical-looping combustion and chemical-looping reforming, Chem. Eng. J. 18: 142-154.
[9] K. S. Kang, C. H. Kim, K. K. Bae, W. C. Cho,S. H. Kim,C. S. Park,2010. Oxygen-carrier selection and thermal analysisof the chemical-looping process for hydrogen production, Int. J. Hydrogen Energ.35: 12246-12254.
[10] H.Gu, L. Shen, J. Xiao, S. Zhang, T. Song, 2011. Chemical Looping Combustion of Biomass/Coal with Natural Iron Ore as OxygenCarrier in a Continuous Reactor, Energ.Fuel 25: 446-455.
[11] J. Adanez, A. Abad, F. Garcia-Labiano, P.Gayan, L. F. de Diego, 2012. Progress in Chemical-Looping Combustion and Reforming technologies, Prog. Energ. Combust. 38: 215-282.
[12] K. Wang, Q. Yu, Q. Qin, 2013. The thermodynamic method for selecting oxygen carriers used for chemical looping air separation, J. therm. Anal. Calorim. 112: 747–753.
[13] M. Johansso, T. mattisson, A. Lyngfelt, 2006. Investigation of Mn_3O_4With Stabilized ZrO_2 for Chemical-Looping Combustion, Chem. Eng. Res. Des. 84: 807–818.
[14] A. Shulmana,, E.Cleverstam, T.Mattisson, A. Lyngfelt, 2011.Chemical-looping with oxygen uncoupling using Mn/Mg-based oxygen carriers-oxygen release and reactivity with methane, Fuel 90: 941-950.
[15] Abad, I.Adánez-Rubio, P. Gayán, F.García-Labiano, L. F. de Diego, J.Adánez, 2012. Demonstration of chemical-looping with oxygen uncoupling (CLOU) process in a1.5 kW_{th} continuously operating unit using a Cu-based oxygen-carrier, Int. J.Greenh.Gas. Con. 6: 189-200.
[16] Adánez-Rubio, P. Gayán, A. Abad, L. F. de Diego, F. García-Labiano,J. Adánez, 2012. Evaluation of a spray-dried $CuO/MgAl_2O_4$oxygen carrier for the chemical looping with oxygen uncoupling process, Energ. Fuel 26:3069-3081.
[17] Adánez-Rubio, P. Gayán, F. García-Labiano, L. F.de Diego, J. Adánez, A. Abad, 2011, Development of CuO-based oxygen-carrier materials suitable for Chemical-Looping with Oxygen Uncoupling (CLOU) process, Energ. Procedia 4: 417-424.
[18] T. Mattisson, H. Leion, A.Lyngfelt, 2009. Chemical-looping with oxygen uncoupling using CuO/ZrO_2 with petroleum coke, Fuel 88: 683-690.
[19] N. Ding, Y. Zheng, C. Luo, Q. Wu, P. Fu, C. Zheng, 2011. Development and performance of binder-supported $CaSO_4$ oxygen carriers forchemical looping combustion, Chem. Eng. J. 171: 1018-1026.
[20] K. Wang, Q. Yu, Q. Qin, Z. Zuo, T. Wu, 2016. Evaluation of Cu-based oxygen carrier for chemical looping air separation in a fixed-bed reactor, Chem. Eng. J. 287: 292–301.

[21] K. Wang, Q. Yu, L. Hou, Z. Zuo, Q. Qin, 2016. Redox performance of Cu-based oxygen carrier used in chemical looping air separation combined oxy-fuel combustion technology, Appl. Therm.Energ. 98: 440–448.

[22] E. Karimi, H.R. Forutan, M. Saidi, M.R. Rahimpour, A. Shariati, 2014. Experimental studyof chemical-looping reforming in afixed -bed reactor: performance investigationof different oxygen carriers on Al_2O_3 and TiO_2 support, Energ. Fuel 28: 2811–2820.

[23] K. Wang, Q. Yu, W. Duan, Q. Qin, H, Xie, 2014, The adaptability of Cu/Zr oxides as oxygen carrier usedfor chemical looping air separation (CLAS), J. Therm. Anal. Calorim. 115: 1163–1172.

Sustainable Development Strategies of Biomass Energy in Beijing

H Z Zhang[1] and B R Huang[2*]

[1] Institute of Agricultural Information and Economics, Beijing Academy of Agriculture and Forestry Sciences, Beijing, 100097, China.
[2] Institutes of Science and Development, Chinese Academy of Sciences, Beijing, 100190, China.

Abstract. The development of biomass energy industry can effectively improve the rural environment and alleviate the shortage of living energy in rural areas, especially in mountain areas. In order to make clear the current situation of biomass energy industry development in Beijing, this paper analyzed the status of biomass resources and biomass energy utilization and discussed the factors hindering the development of biomass energy industry in Beijing. Based on the analysis, suggestions for promoting sustainable development of Biomass Energy Industry in Beijing are put forward.

1. Introduction

Renewable energy has become an important strategic solution for addressing climate change and energy crisis while achieving sustainable development [1]. Biomass energy, as one of humanity's earliest sources of energy, can be produced from raw materials such as agricultural waste, manure, municipal waste, plant material, sewage, green waste or food waste. Countries around the world are becoming increasingly concerned about the development of biogas industry [2].

In China, alongside the rise in awareness of the importance of biogas, the government has realized that biogas is the one of the best substitutes for fossil energy and traditional biomass energy, and thus, setting up the biogas system should be an inevitable strategic option for the sustainable development of rural economy and society [3]. The biogas system in rural China is composed of two patterns, that is, the decentralized household-based biogas digesters for scattered rural households and the centralized biogas plants for large-scaled biogas production [4]. Most of the decentralized and centralized systems have been supported by the Chinese government. Until to the end of 2010, 40 million household biogas digesters and more than 50 thousand medium and large-size biogas projects have been built. The total tank volume reached 7.15 million m3 and annual output of 917 million m3 of biogas [5].

In Beijing, a large number of household biogas digesters were constructed in rural areas in 1980s. By the late 1990s, medium and large biogas projects were began to implement, which is a key project of renewable energy development program in "Eleventh Five-Year Plans" period in Beijing. Most medium and large biogas projects in Beijing were constructed in 2006 to 2009. The plan of 'New countryside construction in Beijing' which was implemented from 2006 to 2012 was the main driving

IOP Conf. Series: Earth and Environmental Science **86** (2017) 012007 doi:10.1088/1755-1315/86/1/012007

force of the construction of large biogas projects. There were less than 10 large biogas projects in 2000, while more than 100 projects until the end of "Eleventh Five-Year Plans" period. At the same time, some of the existing large biogas projects were expanded. Biogas projects and straw gasification projects together constructed the renewable energy system in Beijing rural areas. In the "Twelfth Five-Year Plans" period, medium and large-size biogas projects provide high-quality, clean, renewable energy to farmers. Life convenience of local residents has been greatly improved. Meanwhile, biogas projects significantly improved the structure of energy consumption and utilization and played an increasingly important role in the development of circular economy and ecological environment construction in rural countryside in Beijing.

2. Biomass resources in Beijing

The total output of biomass resources in Beijing is nearly 30 million tons per year (Table 1). It is including 4.39 million tons of agricultural and forestry biomass, 4 million tons of garden waste, 7.42 million tons of livestock manure and 12.18 million tons of sewage produced by intensive farming and nearly two million tons of other types of biomass resources. The main ways of biomass utilization in Beijing are biogas projects (straw gasification projects, methane fermentation projects and the utility of biomass fuel by household biomass gasify stove), biomass briquette, biomass power generation, bio-ethanol and bio-diesel, cooking firewood and others. There are a large amount of biomass resources in Beijing, but only a small amount of biomass has been used.

Table 1 Biomass resources and the utilization in Beijing

	Amount of resources (million tons)	Utilization rate	Utilization patterns
Straw that can be collected	2.26	70%	Most of the straw return to field directly, part is used as raw material for biogas production and a small part is used for cooking energy directly.
Firewood	2.13		Which is used for cooking energy directly
Livestock manure	7.42	70%	Most is used as raw material for organic fertilizer and biogas production.
Sewage produced by intensive farming	12.18	50%	Most returns directly to orchard and garden and part is used to produce biogas.
garden waste	4.00	<4.5%	Used as substrate for mushroom cultivation and raw material of biogas production
Rural domestic waste	0.65		There is hardly any use of rural domestic waste
Others	1.00		
Total	29.64		

3. Development of biomass energy projects in Beijing

Biomass can be converted into bio-energy through three ways: physical conversion, biological

IOP Conf. Series: Earth and Environmental Science **86** (2017) 012007 doi:10.1088/1755-1315/86/1/012007

conversion and chemical conversion. The common technologies for the production of bio-energy from biomass are combustion, solidification, fermentation, liquefaction and pyrolysis gasification. To make clear the status of utilization and development of biomass energy, an investigation was conducted by Beijing Municipal Bureau of Agriculture in 2014. Results showed that straw gasification system and marsh gas system for centralized village cooking gas supply, biomass solid fuel and household biomass gasification furnace were the mayor ways of biomass energy utilization in Beijing.

Large and medium scale biogas project is the most important way of biomass energy utilization in Beijing. There are more than 200 biogas projects constructed in Beijing. But only 128 biogas projects are still running normally at the end of 2014 due to a variety of reasons (Table 2). In those 128 projects, 15 projects have the daily biogas production capacity of more than 500 m^3. Biogas supplied by these 128 stations can meet the needs of 30 thousand households in rural Beijing.

Table 2 Construction and operation of biogas projects in Beijing

District	Methane fermentation projects		Straw gasification projects	
	Constructed (unit)	Running (unit)	Constructed (unit)	Running (unit)
Fangshan	25	15	27	10
Shunyi	13	11	10	5
Daxing	13	11	12	3
Yanqing	10	3	41	5
Tongzhou	10	2	9	0
Pinggu	5	3	5	0
Changping	5	2	6	4
Huairou	5	0	21	7
Miyun	4	2	5	0
Chaoyang	3	2	0	0
Haidian	0	0	1	1
Total	93	51	137	35

Combustion of biomass solid fuel by biomass gasification furnace is another important pattern of biomass energy utilization in Beijing. There are 13 Large and medium biomass solid fuel production plants constructed in Beijing with annual output of biomass solid fuel for each plant between 5 thousand and 10 thousand tons. Almost 50 thousand household biomass gasification furnaces are used in mountain areas in Beijing now for cooking and winter heating. Despite the relatively high cost of using biomass solid fuels, it can greatly reduce the amount of coal usage and alleviate the difficulty of energy usage in daily life in mountain areas.

4. Factors hindering the development of biomass energy industry

The utilization of efficient energy conversion from biomass has changed the traditional way of life in rural areas. It not only reduces the air pollution caused by the direct burning of straw, protects the ecological environment, but also improves the quality of life of rural residents. However, due to the technical level, equipment performance and other factors, there are still many problems in the utilization of biomass energy.

Firstly, in spite of the large output of biomass raw materials, it is difficult to collect the biomass in the rural areas, especially in mountain area. And generally, the cost of collecting biomass is too high to reduce its economic value. More than twenty biogas projects have been suspended because of inadequate of raw materials in Beijing.

Secondly, the application of biomass energy conversion technology and equipment needs to be improved. Large amount of tar can be produced in the process of straw pyrolysis in straw gasification centralized biogas supply project. If not handled properly, it will lead to serious environmental pollution. Half of the straw gasification projects have been suspended because of the tar treatment problem. The biomass cooking stove, biomass heating furnace and mixing furnace used in rural Beijing are not designed according to actual needs. It is hard to add biomass solid fuel to the stove and furnace. And the design of smoke exhaust system of stove and furnace is not reasonable too. A large amount of smoke and tar are produced when fuel burns.

Thirdly, faulty operation and management and high operation costs hinder the development of biogas industry in Beijing. Duo to the high operation cost, large number of fermented liquid and residues and poor work condition, most biogas stations can't be self-financing due to the restrictions on the scale of production. Half of the biogas stationmasters are not optimistic about the prospects for the development of biogas projects.

5. Suggestions for promoting sustainable development of Biomass energy Industry in Beijing

Under the constraint of development strategy of Beijing, the amount of available biomass resources are limited. Although biogas and biomass fuel are widely welcomed by residents in rural area, the total production of biomass energy in Beijing is difficult to increase in large scale, affected by insufficient supplies of raw materials. Thus, the future development of biomass energy industry is to improve production efficiency of existing projects. As to the biogas industry, production efficiency and resource utilization efficiency should be improved. The application of straw fermentation technology and multiple raw material fermentation technology can effectively promote the development of biogas industry. Taking the full advantage of existing biomass resources, the production of the biomass solid fuel industry could be moderately increased within policy permit. And the efficiency of biomass energy utilization should be promoted though technological innovation.

References:

[1] Chen L.H., Cong R.G., Shu B.R. A sustainable biogas model in China: The case study of Beijing Deqingyuan biogas project. Renewable and Sustainable Energy Reviews, 2017, 78: 773-779.

[2] Gosens J., Lu Y.L., He G.Z., et al., Sustainability effects of household-scale biogas in rural China, Energy Policy, 2013, 54, 273-287.

[3] He Y., Xu Y., Pang Y., Tian H., Wu R. A regulatory policy to promote renewable energy consumption in China: review and future evolutionary path. Renew Energy, 2016, 89: 695-705.

[4] Jiang X., Sommer S.G., Christensen K.V. A review of the biogas industry in China. Energy Policy, 2011, 39: 6073-6081.

[5] Jiang D., Zhuang D., Fu J., Huang Y., Wen K. Bioenergy potential from crop residues in China: availability and distribution. Renew Sustain Energy Rev, 2012, 16: 1377-1382.

An Accurate Co-registration Method for Airborne Repeat-pass InSAR

X T Dong [1,2], Y H Zhao [1,3], X J Yue [1,3] and C M Han [1,3]

[1]Microwave System Engineering Division, Institute of Remote Sensing and Digital Earth, Chinese Academy of Sciences, Beijing 100094, China

[2]University of Chinese Academy of Sciences, Beijing 100049, China

[3]Department of Airborne remote sensing, Sanya Institute of Remote Sensing, Sanya 572029, China

Abstract. Interferometric Synthetic Aperture Radar (InSAR) technology plays a significant role in topographic mapping and surface deformation detection. Comparing with spaceborne repeat-pass InSAR, airborne repeat-pass InSAR solves the problems of long revisit time and low-resolution images. Due to the advantages of flexible, accurate, and fast obtaining abundant information, airborne repeat-pass InSAR is significant in deformation monitoring of shallow ground. In order to getting precise ground elevation information and interferometric coherence of deformation monitoring from master and slave images, accurate co-registration must be promised. Because of side looking, repeat observing path and long baseline, there are very different initial slant ranges and flight heights between repeat flight paths. The differences of initial slant ranges and flight height lead to the pixels, located identical coordinates on master and slave images, correspond to different size of ground resolution cells. The mismatching phenomenon performs very obvious on the long slant range parts of master image and slave image. In order to resolving the different sizes of pixels and getting accurate co-registration results, a new method is proposed based on Range-Doppler (RD) imaging model. VV-Polarization C-band airborne repeat-pass InSAR images were used in experiment. The experiment result shows that the proposed method leads to superior co-registration accuracy.

1. Introduction

Synthetic Aperture Radar (SAR) [1,2,3], the main data source of Interferometric synthetic aperture radar (InSAR), has been a hot research issue in the fields of microwave imaging, opening numerous research areas and fields of applications. Comparing with spaceborne SAR, airborne SAR largely overcomes the problems of long revisit time and low-resolution images. InSAR is a promising recent technique for the generation of digital elevation models and the measurement of ground surface deformations. The co-registration is the first step in InSAR data processing, which is defined that the creation of the homologous points coordinate mapping and two images achieve the best space match. Airborne repeat-pass InSAR the same area's two or multiple images at different time or different angles. The differences of initial slant ranges and flight height lead to the pixels, located identical coordinates on master and slave images, correspond to different size of ground resolution cells. The reality lead to obvious mismatching of master and slave images.

Many works has been done on SAR/InSAR image co-registration. An accurate co-registration method of spaceborne repeat-pass SAR interferometry based on matrix transform is proposed in [4]. A three processing procedures method is proposed to resolve the problem of the offset amount between images changeing significantly in whole scene [5]. [6] presents a robust method which introduces the coarse-to-fine strategy, relaxation optimization technology and Maximum Spectrum Method to obtain the dense, reliable and accurate conjugate points for the registration of two single looking complex (SLC) SAR images. An adapted anisotropic Gaussian scale-invariant feature transform (AAG-SIFT) method is proposed in [7] to find feature matches for SAR image registration.

In the following sections, a novel method of accurate co-registration of airborne repeat-pass InSAR based on Range-Doppler (RD) imaging model [8] is proposed. RD imaging model with analytic geometry and physical significance is based on three function: Rrange Equation, Doppler Equation and Earth Ellipsoid Equation, conforming to the image formation theory of SAR[9]. RD imaging model is mainly used to calculate the correction values of initial slant ranges of overlapping swath and heights of the first and second flying paths.

2. Method

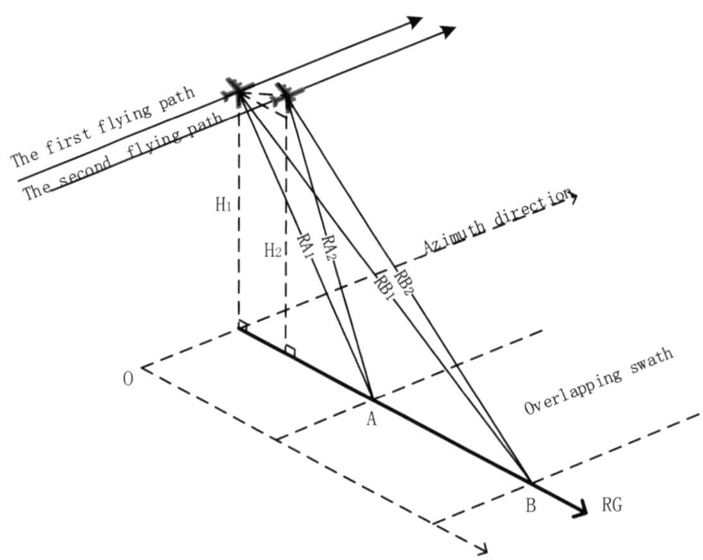

Figure 1. The geometric imaging model of airborne repeat-pass InSAR

As is shown in Figure 1, RA_1 and RB_1 are initial slant ranges of overlapping swath. RA_2 and RB_2 are maximum slant range of overlapping swath. H_1 and H_2 are respectively heights of the first and second flying paths. The pixel numbers of twice measurements are obviously different.

$$\frac{RB_2 - RA_2}{r} \neq \frac{RB_1 - RA_1}{r} \tag{1}$$

r is the range resolution.

RD imaging model is used to achieving that the pixels located identical coordinates on master and slave images correspond one-to-one. The main steps are

- Calculate the correction values of initial slant ranges of overlapping swath and heights of the first and second flying paths, using errors equation of RD imaging model, based on Position and Orientation System (POS) data and the location information of ground control points.
- Update the values of initial slant ranges of overlapping swath and heights of the first and second flying paths.
- Select one image as the master image and another as the slave image. According to geometrical relationship, only calculate the ground locations of pixels on master image.

- Calculate the coordinates corresponding to the ground locations on the slave images.
- Resample the slave image.

3. Experiment

VV-Polarization C-band airborne repeat-pass InSAR images were used in experiment. The platform acquiring these images operates in side-looking mode over Sanya of Hainan province in November, 2014. The basic information of SAR original echo shows in Table 1. The flight height of the master image is about 4459.97m, which of the slave image is 4446.57m. The initial slant range is 5644.84m. The velocity of the master image is 105.36m/s, which of the slave image is 103.43m/s. The baseline is 38.99m. The average height of object area is about 30m. Central incident angle of the image pair is about to 52°. The azimuth resolution of the image pair is 0.16m. The range resolution of the image pair is 0.2m.

Table 1. The basic of information of SAR original echo

Parameter	**Value**
Doppler central frequency	5.4e9
Bandwidth	5e8
Pulse width	2.0e-5
Pulse repetition frequency	1.25e3
Sampling rate	7.5e8
Antenna aperture	0.75
Doppler central frequency	5.4e9
Bandwidth	5e8
Baseline	38.99

Figure 2. The long slant range part of master image

Figure 3. Interferogram of original image pair

Figure 2 only shows the long slant range part of the master image. The image size is: 10600 pixels in range direction and 9380 pixels in azimuth direction. Figure 3 shows the interferogram calculated from figure 2 and the slave image with identical coordinates and size. Three to five coordinate offsets along range direction lead to unsatisfactory result. The processing result of our image registration approach shows in Figure 4 and Figure 5. The new initial slant ranges of master and slave image are respectively 5636.79m and 5639.965m. The new flight heights are respectively 4446.38m and 4434.51m. The results show that the proposed method leads to superior co-registration accuracy.

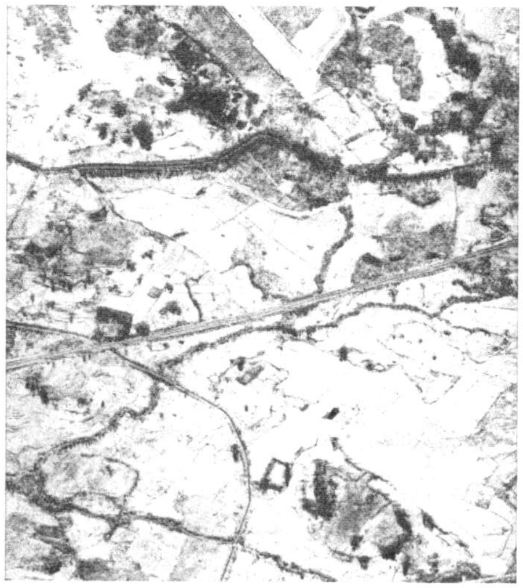

Figure 4. Coherence coefficient map of new image pair

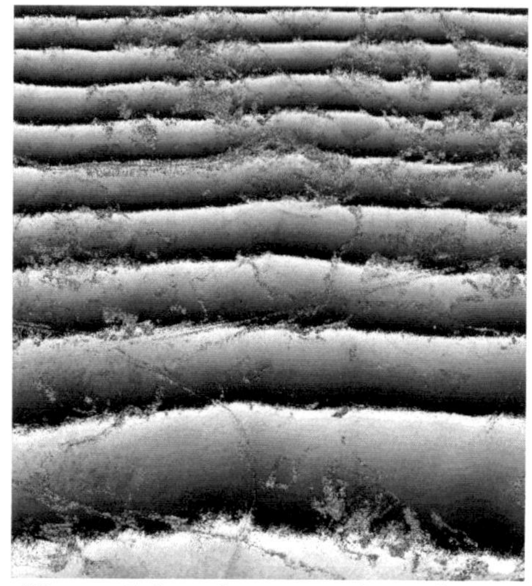

Figure 5. Interferogram of new image pair

Conclusion

In this paper, a new co-registration method based on RD imaging model is presented to the problem of pixels, located identical coordinates on master and slave images, corresponding to different size of ground resolution cells. Quality coherence coefficient map and interferogram were got from new method, which indicate that the proposed method performs superior co-registration accuracy to airborne repeat-pass InSAR images.

Acknowledgments

The authors would like to thank the anonymous reviewers for their valuable comments and suggestions. This work was supported by the key research and development plan of Hainan province under Grant NO. SY16ZY02132.

References

[1] Raney R K, Runge H, Bamler R, Cumming I G and Wong F H 1994 Precision SAR processing using chirp scaling *IEEE Trans.* GRS 32 (4) pp 786–799
[2] Carrara W G, Goodman R S, and Majewski R M 1995 Spotlight Synthetic Aperture Radar: Signal Processing Algorithms *Artech House Boston*
[3] Cumming I G and Wong F H 2007 Digital processing of Synthetic Aperture Radar Data: Algorithms and Implementation *Electronic Industry Press Beijing*
[4] Zou B, Hao L and Bao X 2005 An accurate co-registration method of spaceborne repeat-pass InSAR based on matrix transformation *IGARSS IEEE* pp 4560-4563
[5] Wang P, Li Z F, Suo Z Y, Liu Y Y 2009 An Image Co-registration Method for Wide-swath and High-resolution Spaceborne InSAR *2nd Asian-Pacific Conference on Synthetic Aperture Radar IEEE* pp 1031-1033
[6] Wu Y D, Zhang X Z and Ming Y 2009 Automatic and Reliable Registration for SAR Complex Image *Proceedings of SPIE* 7494(1) pp 0277-786X
[7] Wang F, You H and Fu X Y 2015 Adapted Anisotropic Gaussian SIFT Matching Strategy for SAR Registration *IEEE GEOSCIENCE AND REMOTE SENSING LETTERS*, 12(1) pp 1545-598X
[8] Curlander J C 1982 Location of Spaceborne SAR Imagery *IEEE Trans. on Geoscience and Remote Sensing* 20(3) pp 359-364
[9] Jihnsen H, Lauknes L and Guneriussen T 1995 Geocoding of fast-delivery ESR-1 SAR image mode product using DEM dat *International Journal of Remote sensing* 1(11) pp 1957-1968

Reliability Evaluation of Jacket-type Offshore Platforms Subjected to Wind, Wave, and Current Loads

Z F Song

School of Civil Engineering, North China University of Technology, Beijing 100041, China

Abstract: The aim of this study is to evaluate the reliability and sensitivity of jacket-type offshore platforms under wind, wave, and current loads. The reliability of the platforms are comprehensively determined based on the deformation and energy of the jacket platform. The Young's Modulus; the yield strength; the strain hardening rate; the diameter and thickness of members; the water depth; the wave height and period; the wave velocities on the surface, middle, and base of the platform; the drag coefficient; the inertial coefficient; and the horizontal concentrated scaled loads are defined in the study as random variables. The sensitivity of the performance functions to the above-mentioned variables is determined, and the variables that most influence the performance of the platform are identified. Main conclusions are drawn as follows: the increasing mean and decreasing standard deviation of wall thickness in Type-1, Type-4, and Type-2 jacket members can effectively reduce the deformation and total strain energy of platform structures. For jacket structures in the plastic state, increasing the mean and decreasing the standard deviation of yield strength is a reasonable method for reducing deformation. Maximum deformation and strain energy are highly correlated whether the load level is high or low, and the correlation of these indicators with the variables considered gradually increases with the load level.

Preface

An offshore platform possesses many features of a complicated structure, a large bulk, a high construction cost, and is located in the complex and harsh marine environment. The structure of an offshore platform is constantly subject to the forces of wind, waves, current, sea ice, and tides. Adverse factors, such as environmental corrosion, adhesion of marine creatures, the washing away of subsoil, basic dynamic softening, and material aging can weaken the structural members and the overall resistance of a platform, affecting its service, safety, and durability [1]. There have been a number of offshore platform accidents both domestically and overseas (see Fig. 1) that have resulted in significant economic losses and adverse social impact. For example, in the winter of 1964, two new drilling platforms located in the United States at Cook Inlet, Alaska, were severely damaged by sea ice. After hurricanes Hilda and Betsy, 22 of around 1000 platforms in the Gulf of Mexico collapsed or were otherwise damaged, becoming non-operational. After Hurricane Andrew in 1992, 19 of around 3850 platforms in the Gulf of Mexico were damaged and non-operational. On December 18, 2011, a Russian oil platform sank after impact with sea ice. The reliability of offshore platforms has been an important subject upon which both domestic and overseas researchers and engineers are focused.

Many domestic and overseas researchers have studied the reliability of offshore platform structures. Frieze [2] compared different regulations dictating the reliability level of offshore platform jacket structures in procedures API RP 2A-LRFD and SNAME T&R Bulletin 5-5A, and conducted an in-depth discussion of their reasoning. Zhu [3] discussed the main problems of reliability analysis of offshore platform jacket structures, and provided a general analysis procedure for the safety factor method. Deng [4] presented a reliability estimation method for offshore platform jacket structures, taking into account punching failure and buckling failure of members. This method can determine the main failure modes of the structural system of the platform. Salau [5] derived a time-dependent reliability equation for offshore platform jacket structures, and accurately ascertained the reliability of corroded jacket members. Jin [6] considered the coupling effect between offshore platform jacket members, piles, and soil, and evaluated the jacket reliability under extreme load. Hezarjaribi [7] combined the nonlinear pushover method with the wave load incremental method to evaluate the nonlinear response of the offshore platform jacket-pile-soil system under the action of waves through sensitivity analysis. Leira [8] studied the reliability of the offshore platform field-related response.

Fig.1 Collapse of offshore platform

However, previous studies have been limited to the reliability of the platform structural members and jacket systems, and do not take into account the sensitivity and correlation of jacket behavior to different loads. Therefore, the reliability and sensitivity of offshore platform jacket systems under the action of wind, wave, and current loads were selected as the object of this study, with the intent of comprehensively evaluating platform reliability from the perspective of maximum deformation and total strain energy in the platform jacket members. The sensitivity of the jacket structure deformation and strain energy to variable physical parameters, geometric parameters, and wind, wave, and current loads were studied, and the most influential variable is identified. Finally, the correlation between performance functions is discussed.

1. Reliability theory

1.1 Latin Hypercube sampling method

In order to accurately measure the reliability of offshore platform jacket structures under the action of wind, wave, and current, the Monte Carlo method is used for reliability computation in this study. The Latin Hypercube sampling method can efficiently generate multivariate samples and greatly reduce the number of simulations without compromising accuracy.

The mean is estimated as

$$Y = \overline{Y} = \frac{1}{N}\sum_{i=1}^{N} y_i \tag{1}$$

The mth value is estimated as

$$Y = \frac{1}{N}\sum_{i=1}^{N} (y_i)^m \tag{2}$$

The cumulative distribution function (CDF) is estimated as

$$F_Y(y) = \frac{y_i \leq y}{N} \tag{3}$$

1.2 Sensitivity

A sensitivity coefficient represents the relative importance of a random variable to the failure probability, which satisfies the following relation:

$$\alpha_1^2 + \alpha_2^2 + \cdots + \alpha_n^2 = 1 \tag{4}$$

where $\alpha_i (i = 1, 2, \cdots, n)$ represents the sensitivity coefficient of the ith random variable to the failure probability, which indicates the relative contribution of each random variable to the failure probability.

The sensitivity coefficient to the reliability index is expressed as

$$\frac{\partial \beta}{\partial u_i} = \frac{\partial}{\partial u_i}\sqrt{u_1^2 + u_2^2 + \cdots + u_n^2} = \frac{u_i}{\beta} = \alpha_i \tag{5}$$

where u_i represents the random variable in standard normal space.

The sensitivity coefficient to the failure probability P_f is expressed as

$$\frac{\partial P_f}{\partial u_i} = \frac{\partial}{\partial u_i}\Phi(-\beta) = \phi(-\beta)\frac{\partial \beta}{\partial u_i} \tag{6}$$

where $\Phi(-\beta)$ represents a standard normal distribution function, and $\phi(-\beta)$ represents a standard normal density function.

1.3 Correlation between performance functions in structural system

Performance functions Z_i and Z_j correspond to failure modes i and j, respectively, expressed here as Z_i, Z_j. The performance functions contain only two statistically independent variables, R and S. The standard errors of the mean for R and S are m_R, m_S and σ_R, σ_S, respectively. The performance functions Z_i, Z_j can be respectively expressed as:

$$Z_i = a_i R - b_i S \tag{7}$$

$$Z_j = a_j R - b_j S \tag{8}$$

The correlation coefficient of Z_i, Z_j can be expressed as:

$$\rho_{Z_i,Z_j} = \frac{COV(Z_i, Z_j)}{\sigma_{Z_i}\sigma_{Z_j}} = \frac{a_i a_j \sigma_R^2 + b_i b_j \sigma_S^2}{\sigma_{Z_i}\sigma_{Z_j}} \tag{9}$$

If the performance function is nonlinear, it can be expanded by a Taylor series at the check point X^*, the approximate value of the correlation coefficient is calculated by a simple equation (assuming the basic variables are not related), and the correlation coefficient of Z_i, Z_j is expressed as:

$$\rho_{Z_i,Z_j} = \frac{COV(Z_i, Z_j)}{\sigma_{Z_i}\sigma_{Z_j}} = \frac{\sum_k \left(\frac{\partial Z_i}{\partial x_k'}\right)_{X^*}\left(\frac{\partial Z_j}{\partial x_k'}\right)_{X^*}}{\sqrt{\sum_k \left(\frac{\partial Z_i}{\partial x_k'}\right)_{X^*}^2}\sqrt{\sum_k \left(\frac{\partial Z_j}{\partial x_k'}\right)_{X^*}^2}} \tag{10}$$

2. Computation of wave force

The computation of wave force is generally divided into the computation of small scale wave force and the computation of large scale wave force, based on the ratio of the structural scale to the wavelength (D/L) where D is the characteristic length of the object (such as diameter for a cylinder), and L indicates the wavelength. When D/L ≤ 0.2, the object under consideration is considered to be small scale. When D/L > 0.2, the object under consideration is considered to be large scale, for which the free surface effect and the relative scale effect, namely the diffraction effect, should be considered.

For the calculation of wave force on small-scale slender columns, the famous Morison Equation is widely used in engineering design [9]:

$$f_{wy} = \frac{1}{2} C_D \rho_w \left(u - \frac{\partial y}{\partial t} \right) \left| u - \frac{\partial y}{\partial t} \right| + C_M \rho_w \frac{\pi D^2}{4} \left(\frac{\partial u}{\partial t} - \frac{\partial^2 y}{\partial t^2} \right) \tag{11}$$

where: f_{wyd} is the wave force per unit length vertically acting on the pipe column in kN/m; ρ is the seawater density in kg/m^3; D is the pipe column diameter in m; u is the horizontal velocity of water particle on axis of pipe column in m/s; C_D is the resistance coefficient; C_M is the inertia force coefficient [10].

As Stokes wave theory can accurately describe the actual wave motion, standards for classification of marine platforms like ABS and DNV also suggest using three-order or five-order Stokes wave theory to conduct strength checks and structural design for marine structures. The Stokes wave theory can also be used for small-sized jacket parts [11].

This study used the ANSYS software, which can accommodate four different wave theories: the micro-amplitude wave theory with deep fading and empirical correction, the Airy wave theory, the five-order Stokes wave theory, and the stream function wave theory. If the five-order Stokes wave theory is selected, only wave height, period, and phase angle are required as inputs. If structural static analysis is performed, before inputting the phase angle, the phase angle of the maximum wave current coupling force must be determined, and this should be taken as the parameter for static analysis.

3. Reliability and sensitivity analysis of offshore platform jackets

3.1 Model parameters

The parameters of an offshore platform jacket structure were taken from Reference [10]. The platform consisted of an upper platform and a lower jacket structure (see Fig. 2), with the jacket bottom fixed by a pile foundation. The upper platform supports the frame and the deck, with external dimensions of 30 m × 20 m, and the height from the water surface to the top deck was 15 m. The majority of the lower jacket structure consisted of 4 main ducts using slim tubes for support, forming a spatial tower frame structure. The pile foundation was driven into the seabed soil through the main ducts. The size of main duct in the jacket structure above the mud line was Φ1.2 × 0.05 m; the size of main duct in the jacket structure below the mud line was Φ1.2 × 0.05 m; the size of main pipe used in the deck was Φ0.78 × 0.0381 m; the size of inter-story horizontal bracing in the jacket system was Φ0.78 × 0.0381 m; the size of diagonal braces in the jacket system was Φ0.508 × 0.0254 m; the section area of the deck beams was 0.16 m^2, with a height of 0.4 m and a width of 0.4 m; the deck was 0.025 m thick, 30 m long, and 20 m wide; the water depth was 45 m, with a wind velocity of 43.6 m/s, an effective wave height of 14.8 m, an effective wave period of 10.8 s, a surface flow rate of 2.35 m/s, a central flow rate of 1.96 m/s, and a bottom flow rate of 1.60 m/s. In order to simplify the computation, the influence of pile and soil on the behavior of the jacket structure was not considered, and the lower part of the jacket was assigned a fixed constraint.

The entire jacket platform model used four element types: PIPE20, PIPE59, BEAM4 and SHELL43 (see Fig. 2). The PIPE20 element was used for the main vertical supporting members of the deck framework above the water surface, as well as for members below the mud line. Because the hydrodynamic effect and the wave-current interaction can be considered in the PIPE59 element, it was used for jacket members below water surface but above mud line. The BEAM4 element was used for deck level frame beams, and the SHELL43 element used to model the horizontal deck. The steel material used was D36, with an elastic modulus of 2.0×10^{11} Pa, a Poisson's ratio of 0.3, a density of 7850 kg/m^3, a yield strength of 360 MPa, and a strain hardening rate of 0.02. The bi-linear elastic-plastic hardening model was adopted for the steel.

When the five-order Stokes wave theory is used, the value of the phase angle φ has a significant impact on the wave load acting on the structure. Before starting the static analysis, it was necessary to determine the phase angle φ from 0° to 360° at which the wave current coupling force is at its maximum. The authors determined that the phase angle corresponding to the maximum wave current coupling force was 46° [10]. Based on the characteristics of the jacket-type offshore platform, Jinping Ou proposed the whole advancement method [1], an approximate method for limit analysis that is essentially a pushover analysis method, which treats the platform structure as resisting the horizontal force as a whole to calculate the bearing capacity of entire structure. Based on this, for the convenience of computation, the wind loads were equally concentrated at the eight nodes where platform and jacket structure intersect, and were amplified by different multiples to study the elastic-plastic mechanical properties of the jacket structure under the action of the three different applied horizontal loads.

From the point of view of deformation and energy, the reliability and sensitivity of the jacket platform under wind, wave, and current loads were studied. The jacket physical parameters (elastic modulus, yield strength, and strain hardening rate), geometric parameters (pipe diameter and wall thickness) for four classes of sections (A–D), the water depth, the wave height and period, the surface, central, and bottom flow rate, the drag coefficient, the inertia force coefficient, and the amplified wind load were defined as random variables. The statistical characteristics of each random variable are shown in Table 1. See references [12] and [13] for parameter selection rules.

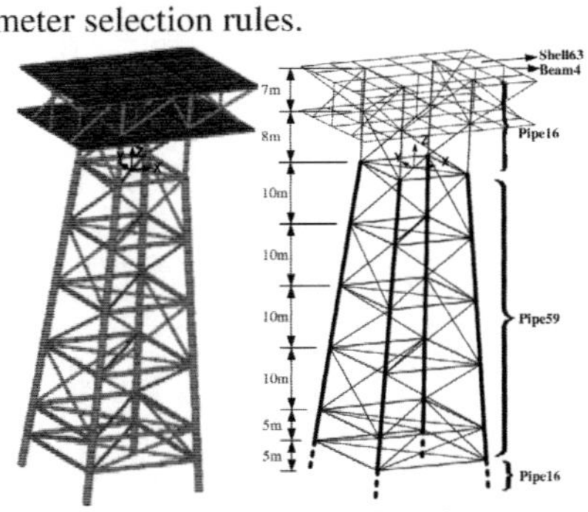

(a) Whole model (b) Unit types
Fig.2 Model of offshore platform

Table 1 Statistical parameters of random variables

Random variable	Description	Probability distribution	Mean value	Standard error	Random variable	Description	Probability distribution	Mean value	Standard error
DW/m	Water depth	Normal	45	4.5	D1	Class A diameter /m	Truncated Gauss	1.2	0.06
HW/m	Wave height	Normal	14.8	0.74	T1	Class A wall thickness/m	Truncated Gauss	0.05	0.0025
TW/s	Wave period	Normal	10.8	0.54	D2	Class B diameter /m	Truncated Gauss	1.2	0.06
VS/(m/s)	Surface flow rate	Normal	2.35	0.047	T2	Class B wall thickness/m	Truncated Gauss	0.05	0.0025
VM/(m/s)	Central flow rate	Normal	1.96	0.0392	D3	Class C diameter /m	Truncated Gauss	0.78	0.039
VB/(m/s)	Bottom flow rate	Normal	1.60	0.032	T3	Class C wall diameter /m	Truncated Gauss	0.0381	0.00191
CD	Resistance	Normal	0.70	0.014	D4	Class D	Truncated	0.508	0.0254

	coefficient					diameter /m	Gauss		
CE	Inertia force coefficient	Normal	2.0	0.04	T4	Class D wall thickness/m	Truncated Gauss	0.0254	0.00127
E/MPa	Elastic modulus	Log normal	2.06×10^5	1.03×10^4		Load level 1/kN		1400	140
f_y/MPa	Yield strength	Normal	360	18	LOAD	Load level 2/kN	Normal	2500	250
b	Strain hardening rate	Normal	0.02	0.001		Load level 3/kN		4000	400

3.2. Reliability analysis

First, the reliability of offshore platform jacket structures under the action of the three different horizontal loads considered was studied. Defining the maximum deformation and the total strain energy as performance functions, the CDF curves of probability for the maximum deformation and the total strain energy under three load levels are shown in Fig. 3 and Fig. 4, respectively. As can be seen from Fig. 3a, for Load Level 1, the average maximum deformation was 0.125 m, with a standard error of 0.0142 m; the probability for deformation beyond a threshold of 0.12 m was 63.23 %, and the probability for deformation beyond a threshold of 0.15 m was 4.34 %. With the increase in load level, the mean and the standard error of the maximum deformation also increased. According to Fig. 3b, for Load Level 2, the mean of maximum deformation reached 0.223 m with a standard error of 0.026 m, and according to Fig. 3c for Load Level 3, the mean of maximum deformation was 0.377 m with a standard error of 0.054 m.

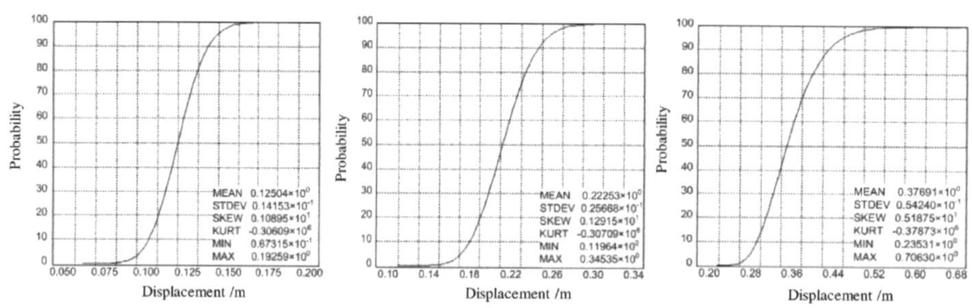

(a) Load level 1 (b) Load level 2 (c) Load level 3

Fig.3 CDF curves of maximum deformation

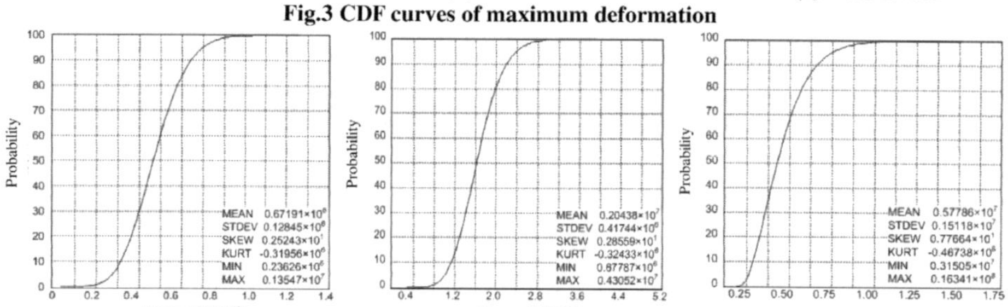

(a) Load level 1 (b) Load level 2 (c) Load level 3

Fig.4 CDF curves of total strain energy

Fig. 4 shows the CDF curve of probability for the total strain energy. As shown in Fig. 4a, for Load Level 1, which is relatively light, the mean of total strain energy was only 6.72×10^5 J, with a standard error of 1.29×10^5 J. As the load level increased, a number of members in the jacket entered into the plastic state, and the total strain energy increased as well. From Fig. 4b, it can be seen that when Load Level 2 was applied, the total strain reached 2.04×10^6 J, a two-fold increase compared to Load Level 1. As can be seen from Fig. 4c, for Load Level 3, the jacket has been significantly deformed and a large number of members have entered into the plastic state, with the mean of total strain energy reaching 5.78×10^6 J, which is a two-fold increase over Load Level 2.

3.3 Sensitivity and correlation analysis

In this section, the sensitivity of the maximum deformation and total strain energy to each random variable was analyzed by applying different load levels to the jacket platform. All calculation results are shown in Table 2 to Table 4, while Fig. 5 and Fig. 6 show the pie diagrams illustrating the sensitivity of the maximum deformation and total strain energy to each random variable under the three load levels applied. From Table 2 and Fig. 5a, it can be seen that for Load Level 1, the maximum deformation is mostly sensitive to changes in horizontal concentrated loads, which exhibit a sensitivity coefficient of 0.855—a positive value, indicating that the maximum deformation increases as the value of the horizontal concentrated load increases. The sensitivity coefficient for the elastic modulus was -0.424—a negative value, indicating that the maximum deformation decreases as the elastic modulus increases. The sensitivity coefficients for the wall thickness of Class A, D, and B members were -0.168, -0.144, and -0.027, respectively, indicating that an increase in mean wall thickness and reduction of their standard error can effectively reduce jacket deformation.

The sensitivity coefficients for the diameter of Class D, A, and C members were -0.019, -0.012, and -0.009, respectively, indicating that an increase in mean external diameter and reduction in their standard error can also reduce the jacket deformation to some extent. The sensitivity coefficients for water depth, wave height, wave period, central flow velocity, bottom flow velocity, resistance coefficient, inertia force coefficient, and yield strength were all so low that they can be neglected. With increasing load levels, there were some members entering the plastic state, and thus the sensitivity of structure to changes in yield strength increases. As displayed in Table 3 and Fig. 5b, for Load Level 2 the sensitivity of the maximum deformation to changes in yield strength increased from -0.001 to -0.010, while the sensitivity of other random variables remained unchanged. It can be seen from Table 4 and Fig. 5c that for Load Level 3, there are more members entering the plastic state, and the sensitivity of the maximum deformation to changes in yield strength was increased to -0.098. Therefore, it can be seen that for jackets in the plastic state, increasing the mean yield strength and decreasing its standard error can effectively reduce the maximum deformation.

According to the information in Table 2 to Table 4 and Fig. 6, for the three load levels, the total strain energy indicates the highest sensitivity to changes in the horizontal concentrated load, followed by changes in the elastic modulus. The data suggests similar high sensitivity to the wall thickness of the Class D, A, and B members, which is similar to the sensitivities of maximum deformation. In case of a low load level, the sensitivity of total strain energy to the yield strength can be ignored. As the load level increases, the sensitivity coefficient of the yield strength increases from -0.002, first to -0.012, then to -0.092, which is caused by the gradual entering of the members into the plastic state.

Table 2 Sensitivity of maximum deformation and total strain energy to each random variable (Load Level 1)

Performance function	DW	HW	TW	VS	VM	VB	CD	CE	E	fy
D	0.003	-0.001	0.002	-0.008	-0.001	-0.001	0.001	-0.001	-0.424	-0.001
E	0.002	-0.002	0.002	-0.008	-0.002	-0.001	0.000	-0.002	-0.247	-0.002

Performance function	B	D1	T1	D2	T2	D3	T3	D4	T4	LOAD
D	-0.003	-0.012	-0.168	-0.006	-0.025	-0.009	-0.008	-0.019	-0.144	0.855
E	-0.004	0.005	-0.082	-0.005	-0.016	-0.007	0.001	-0.012	-0.094	0.952

Table 3 Sensitivity of maximum deformation and total strain energy to each random variable (Load Level 2)

Performance function	DW	HW	TW	VS	VM	VB	CD	CE	E	fy
D	0.003	-0.001	0.002	-0.008	-0.001	-0.001	0.001	-0.001	-0.416	-0.010
E	0.002	-0.002	0.002	-0.008	-0.002	-0.001	0.000	-0.002	-0.231	-0.012

Performance function	B	D1	T1	D2	T2	D3	T3	D4	T4	LOAD
D	-0.003	-0.011	-0.165	-0.006	-0.027	-0.009	-0.006	-0.018	-0.143	0.860
E	-0.004	-0.005	-0.076	-0.005	-0.018	-0.007	-0.000	-0.012	-0.096	0.957

IOP Conf. Series: Earth and Environmental Science **86** (2017) 012009 doi:10.1088/1755-1315/86/1/012009

Table 4 Sensitivity of maximum deformation and total strain energy to each random variable (Load Level 3)

Performance function	DW	HW	TW	VS	VM	VB	CD	CE	E	fy
D	0.002	-0.001	0.002	-0.008	-0.001	-0.001	0.001	-0.002	-0.340	-0.098
E	0.001	-0.002	0.002	-0.008	-0.001	-0.001	0.000	-0.003	-0.189	-0.092
Performance function	**B**	**D1**	**T1**	**D2**	**T2**	**D3**	**T3**	**D4**	**T4**	**LOAD**
D	-0.003	-0.010	-0.132	-0.012	-0.054	-0.005	-0.007	-0.016	-0.122	0.899
E	-0.003	-0.000	-0.060	-0.003	-0.044	-0.004	-0.003	-0.011	-0.083	0.964

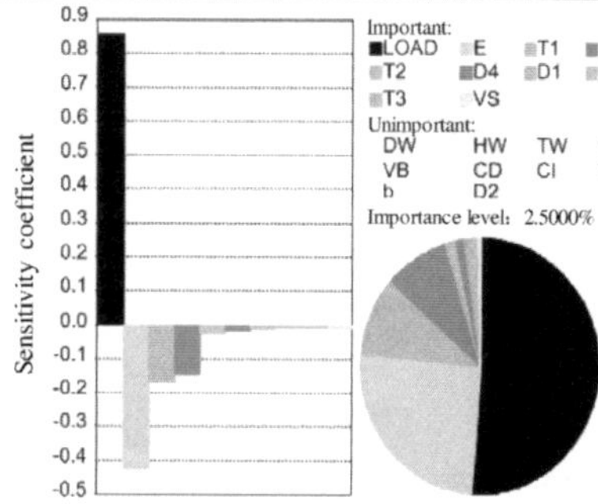

(a) Load level 1

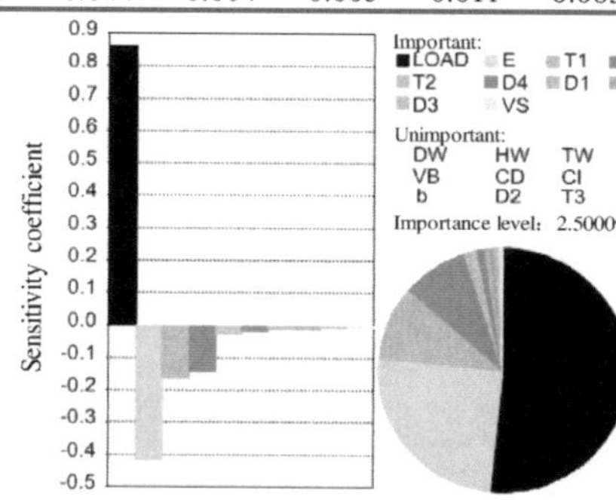

(b) Load level 2

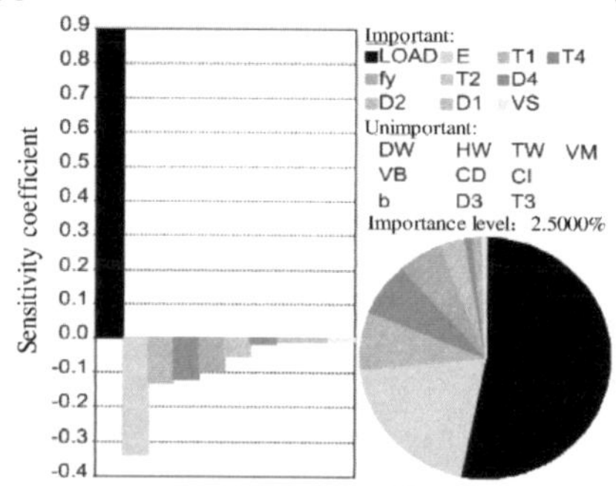

(c) Load level 3

Fig.5 Sensitivity pie diagrams of the maximum deformation to each random variable

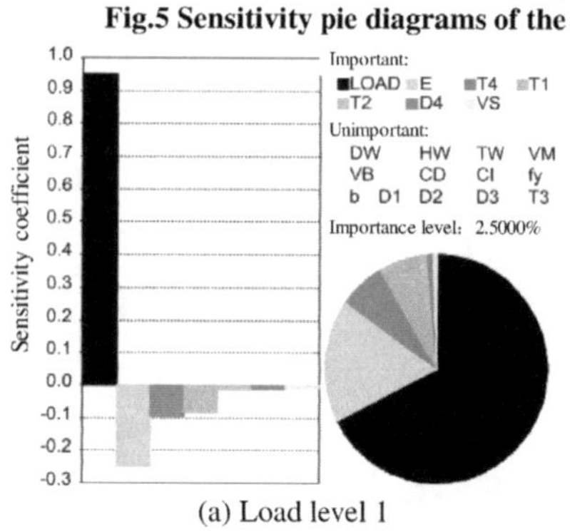

(a) Load level 1

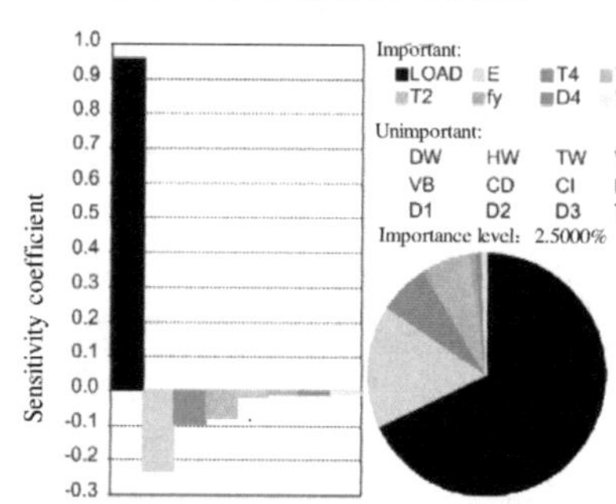

(b) Load level 2

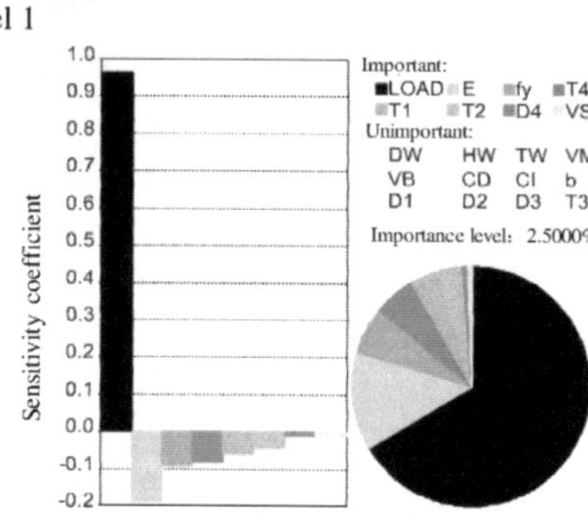

(c) Load level 3

Fig.6 Sensitivity pie diagrams of total strain energy to each random variable

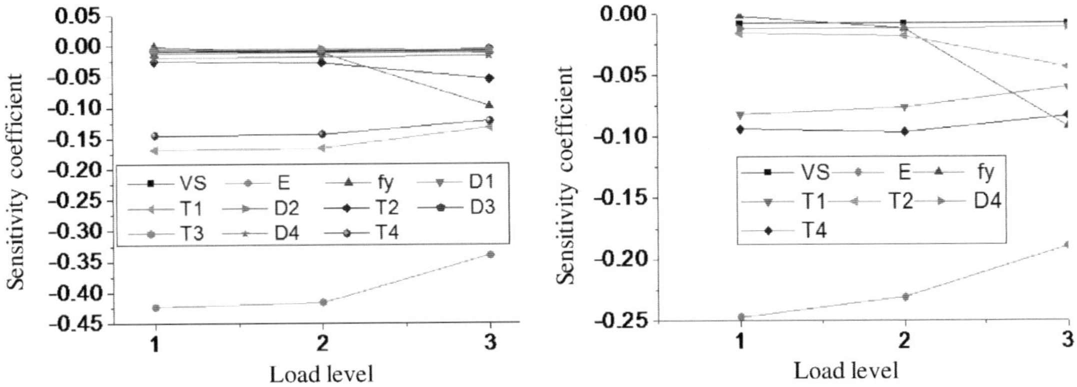

(a) Maximum deformation (b) Total strain energy

Fig.7 Trend curves of sensitivity coefficient of each random variable under different load level

Fig. 7 shows the sensitivity coefficient curve for the main random variables of the jacket assembly when subjected to the three load levels. It can be seen that as the load level increases, the maximum deformation and total strain energy gradually grow less and less sensitive to changes in the elastic modulus, and exhibit significantly increased sensitivity to changes in the yield strength and wall thickness of Class B members. The sensitivity of the deformation and strain energy to changes in the wall thickness of Class A and D members also decreases with increasing load, while their sensitivity to other random variables did not show significant change.

There is a correlation between the performance functions defined by the maximum deformation and the total strain energy of the platform. This section quantifies the extent of this correlation. The correlation coefficients under three load levels are shown in Table 5, while Fig. 8 shows the scatter plots of their correlation under Load Level 1, which is quite high. According to Table 5, the correlation coefficients are 0.971 for both Load Levels 1 and 2, and for Load Level 3, the correlation coefficient reaches 0.980. Therefore, with the increase in load level, the correlation between the maximum deformation and the total strain energy increases gradually. This is because as the load increases, more members approach the yield state, resulting in the increased impact of yield strength on the two performance functions. It can also be seen that regardless of the load level, the two performance functions considered in this study exhibit significant correlation in excess of 0.7 - 0.8.

Table 5 Correlation between performance functions

Item	Load level 1		Load level 2		Load level 3	
	D	E	D	E	D	E
D	1.000	0.971	1.000	0.971	1.000	0.980
E	0.971	1.000	0.971	1.000	0.980	1.000

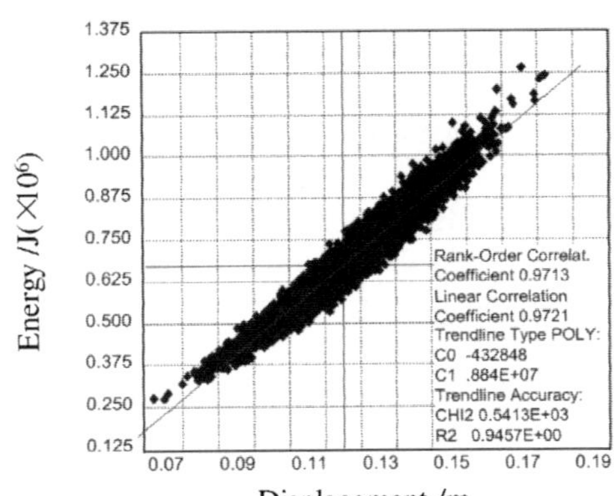

Fig.8 Scatter plots between maximum deformation and total strain energy (load level 1)

4 Conclusions

(1) In order to comprehensively evaluate the reliability of offshore platform jacket structures, a variety of performance functions can be considered for different research objectives. This study considers the two performance functions of maximum deformation and total strain energy for the jacket system, and quantitatively determines the structural reliability for each performance function.

(2) The extent of the sensitivity of the two performance functions to changes in each random variable was determined, and the random variables most affecting the performance functions were identified. It was observed that the reliability of the offshore platform jacket structure can be enhanced by rationally controlling the probability parameter of these random variables. Increasing the mean wall thickness of Class A, D, and B members, and reducing their standard error can effectively reduce the deformation and total strain energy of the overall jacket structure. With the increase in load level, the sensitivity of the structure to changes in the yield strength increases steadily. For jacket members in the plastic state, improving the mean yield strength and reducing the standard error can effectively control the maximum deformation of the jacket structure. With the increase in load level, the sensitivity of the maximum deformation and total strain energy to changes in the elastic modulus gradually decreases, and their sensitivity to changes in the yield strength significantly increases.

(3) Regardless of the load level, there is strong correlation between the maximum deformation and the total strain energy, which is an advanced correlation. With an increase in load level, the correlation between the maximum deformation and the total strain energy improves gradually, caused by more members entering into the yield state, increasing the influence of yield strength on the two performance functions

REFERENCE

[1] Ou J P, Duan Z D, Xiao Y Q. Safety evaluation of offshore platform structures: theory, method, application [M]. Beijing: Science Press, 2003,1-9.
[2] Frieze P. A., Morandi A. C., Birkinshaw M, etc. Fixed and jack-up platforms: basis for reliability assessment [J]. Marine Structures, 1997,10(2-4):263-284.
[3] Zhu Q X. Optimum design for offshore platform structural reliability [J]. China ocean Engineering, 1992, 6(3):265-278.
[4] Deng H Z, Sun Q. Reliability analysis for offshore platform structural systems [J]. China Ocean Engineering, 1997,11(1):1-10.
[5] Salau M. A., Esezobor D. E., Omotoso M. F. Reliability assessment of offshore jacket structures in Niger Delta [J]. Petroleum & Coal, 2011， 53(4):291-201.
[6] Jin W L. Reliability-based design for jacket platform under extreme loads [J]. China Ocean Engineering, 1996,10(2):145-160.
[7] Mona Hezarjaribi, Bahaari M. R., Vahid Bagheri, etc. Sensitivity analysis of jacket-type offshore platforms under extreme waves [J]. Journal of Constructional Steel Research, 2013,83(3):147-155.
[8] Bernt J. Leira, Daniel Karunakaran. Site-dependent reliability of a mobile jack-up platform [J]. Marine Structures, 1995,8(2):151-169
[9] Morison J. R., O. Brien M. P., Johnson J.W., et al. The force exerted by surface wave on piles [J]. Petroleum Transactions, AIME, 1950,2(5):149-154.
[10] Yang J, Liu S J, Xie R J. Application of ANSYS in offshore oil engineering [M]. Beijing: Petroleum Industry Press, 2010,10-15,182.
[11] Skjelbreia, L., Hendrickson, J. A., Fifth order gravity wave theory [C]//Proceedings, Seventh Conference on Coastal Engineering, Ch. 10, 1961:184-196.
[12] Jiang Y B. Reliability calculation and discussion of design method of cable-stayed double-layer cylindrical latticed shells[D]. Nanjing: Southeast University, 2006,15-18.
[13] Du C. System reliability analysis of ocean jacket platforms [D]. Dalian: Dalian University of Technology, 2006,32-37.

IOP Conf. Series: Earth and Environmental Science **86** (2017) 012010 doi:10.1088/1755-1315/86/1/012010

Opportunities and Challenges of Electric Vehicles Development in Mitigating Climate Change in China

R Liu, M H Li and H N Zhang

Beijing Institute of Science and Technology Information, Beijing, 100044, P.R China

Abstract. As a developing country, China has also undergone a noticeable climate change due to the increasing consumption of fossil fuels. The automotive market in China is estimated to be the world's second largest new automotive growth market. China is now capable of manufacturing cars totally independently, which makes car prices more attractive to middle-income families. As one of the energy solutions, Electric Vehicles Technologies cannot be considered solely in environmental aspects. One energy system should not only contribute to sustainable development, but also to the environmental, social and economic aspects.

1. Introduction

The term motorized transport has never developed without the companion of the term energy. No matter the steam engine in the 18th century or the later invented internal combustion engine (ICE) or the electric motor or hybrid cars, transport development has undergone a process of matching proper energy supplies with updated technologies. Over the past centuries and up to now, ICE has been playing a dominant role in road transport. The use of fossil fuel has within the same period been relied by the modern society. However, being a limited and contested resource, oil has rapidly been explored and consumed, which revealed a day of oil depletion. Due to that the reservation of oil is mostly in the areas that are today politically unstable, and the oil prices are also fluctuating, which is the threat to a large part of the car users. But more importantly, the atmosphere has been severely damaged by CO2 and other greenhouse gas emissions from ICE driven vehicles. The impact of the greenhouse gases will increase gradually in the temperature level of the Earth. Wherever we looked, we saw evidence of destruction: melting permafrost, higher sea levels and unpredictable climate changes. "Climate change, which has identifiable potentially catastrophic effects on the environment and human security in the broadest sense, cannot be halted, much less reversed, without the control and reduction of carbon dioxide emissions into the atmosphere."[1]According to IEA (International Energy Agency) predictions, global emission of CO2 will be 60% higher in 2030 than they are today if no counteracting political action is taken [2]. Climate change is a very important global challenge we are facing. The action that we fight pollution of emission especially come from vehicle is therefore needed nowadays and all sectors of society need to make prominent contributions.

The introduction of electric vehicles (EVs) has been a proposal to prevent climate change and attempts have been made in several countries during the 1980s and 1990s such as the USA and Denmark. However due to battery technology deficiency, the attempts were unsuccessful and given up. As the oil price increase and the atmosphere is severely damaged, the attempts of reintroducing EVs emerged in recent years in several countries such as USA, Japan, Denmark and China.

China has been undergoing a strong economic growth and at the mean time the demand of energy has also surged rapidly. As a developing country, China has also undergone a noticeable climate change due to the increasing consumption of fossil fuels. So, China is promoting electric vehicles (EVs) as another alternative CO2 reducing energy resource. The automotive market in China is estimated to be the world's second largest new automotive growth market. China is now capable of manufacturing cars totally independently, which makes car prices more attractive to middle-income families[3].

At present, the key driver of EV is the improvement of battery technology. The major factors of battery technology are the range per charge, weight, battery life and cost. The development of advanced batteries such as lithium-ion is in progress.

2. Current situation of EV development in China

In order to solve energy supply issue, an electric vehicle technological innovation research was carried out. For many years, exploration and practice of the scientists has proved that the electric vehicle system can be improved significantly. This kind of electric vehicle system has been planned as project in some countries. In the future, the environment need suggestions that emissions of vehicle exhaust will decrease in big leaps during the next 50 years. In this point of view, emission goal will be near zero emissions.

In China, the private cars stock that relies heavily on fossil fuel is an indicator of cars use. This has important impacts on global environment. Private car use is increasing most rapidly in China. The private cars stock increased almost 4.6 times between 2005 and 2014[4]. As EVs begin to penetrate the automotive market, the shares are likely to increase, and global sales of electric vehicle is on the rise. (See figure 1)

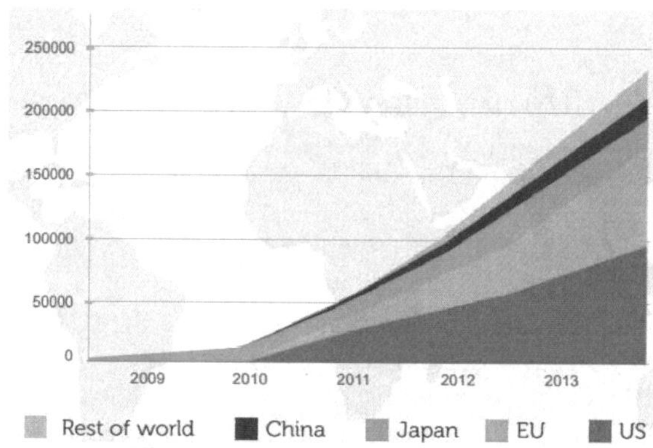

Figure1.global sales of EV [5]

From 2014 to 2020, the EV stock will increase almost 50 times from seventy-eight thousand to more than 4 million EVs by estimation (See figure 2).

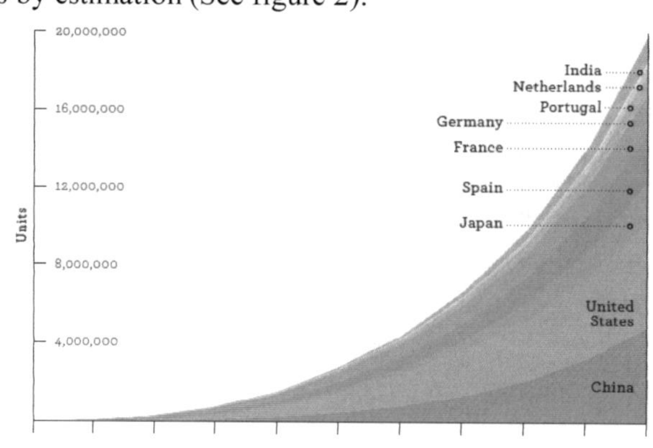

Figure 2. EV stock target(select EVI members) [6]

The developmental trend in China involves a continuous growth of automobile industry. The government considers the automobile industry as one of the seven "pillar industry" of the economy. China has set EV sales targets to signal long-term commitment to vehicle electrification. According to the report of EVI (Electric Vehicles Initiative) 《2015 global electric vehicle and charging infrastructure outlook》, from 2012 to 2014, USA(39%)、Japan(16%) and China(12%) are countries with fastest growing EV sales in the world.

In 2014, New energy vehicles sold 74763, pure electric vehicles accounts for about 60%, an increase of 2.1 times [7]. (See figure 3)

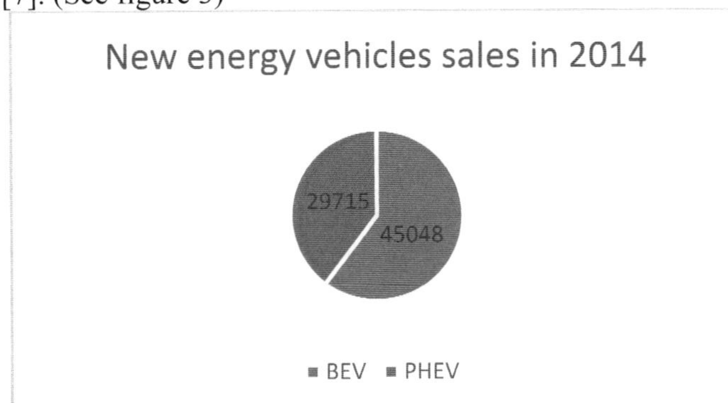

Figure 3. New energy vehicles sales in 2014

There are 230 million e-bikes、83,000 e-cars and 36,500 e-buses on the road in China by 2014[8].

3. China's Policies and Measures to Spread of Electric Vehicles

Target-setting is by no means a prerequisite for, or determinant of, successful EV deployment, but it is useful for understanding the level of ambition and support from national policymakers[9].

In order to fulfil China's Climate Change Program, the Chinese government intended to address climate change by means of sustainable development strategy. Key areas were targeted to reduce GHG emission by the strengthening of institutional, legal, economic, and technological instruments and hence to achieve energy conservation, energy structure optimization as well as ecological improvement.

The key targeted areas for potential spread of Electric Vehicles are listed below:

Table1. China's Policies and Measures to Spread of Electric Vehicles

Year	Policies and Measures	Highlights
2007	Chinese government launched '863 Initiative'	Encourage R&D
2009	《New energy automobile production enterprises and product access management rules》	Refinement phase
2010	《Notice on carrying out the pilot of the private purchase of new energy vehicles subsidy》	First subsidy policy
2012	《Energy saving and new energy automobile industry development planning(2012-2020)》	Put forward industrial planning
2012	《Electric vehicle technology development "Twelfth Five Year" special plan》	Establish three goals
2013	《First new energy vehicles pilot cities announced》	Government push to implement
2014	《Notice on further promoting the application of new energy vehicles》	Promoting the application of new energy vehicles
2014	《Announcement on new energy vehicle purchase tax	Stimulating

	exemption》	consumption
2014	《Notice of the new energy vehicle charging facilities construction Award》	Financial support
2015	《government work report》	"Internet+"
2015	《Made in China 2025》	Manufacturing power
2015	《New pure electric passenger car business management regulations》	Lower entry threshold

In order to follow up with the ambitions on the Electric Vehicles, laws and regulations are to be constituted by the Chinese government. Each entity's responsibility will be clarified; policy incentives and efforts on discipline will be intensified; legal executants will be identified. Supervision and monitoring will be enhanced on spread of Electric Vehicles. Preferential policies for Electric Vehicles products will be formulated. Tax policies will be made favourable for the sale of Electric Vehicles. The government will also guide the public to purchase energy-saving and low carbon emission vehicles and disseminate the idea of conservation-orientation. Scientific research and technological development shall be promoted; constructions of talents and financial support are to be strengthened. Finally, public awareness will increase by using the promotion function of the government. Awareness will penetrate through all levels through a top-down procedure. Besides, publicity, public participation, education and training will be reinforced by the government. International cooperation and communication shall be reinforced in order to promote public awareness on climate change issues as well as exchange experience of other countries on climate change publicity and spread of Electric Vehicles. [9]

4. The development of EV key- technologies in China

Key activities for countries seeking to help technological innovation reach full market potential. As seen in Figure 4, Based on analysing the geographical distribution of electric vehicles patent, which can reflect the technical strength of individual countries. We can find that Japan is in a leading position, up 55.3%. China ranked second, accounted for 18.4%, but there is considerable distance compared with Japan.

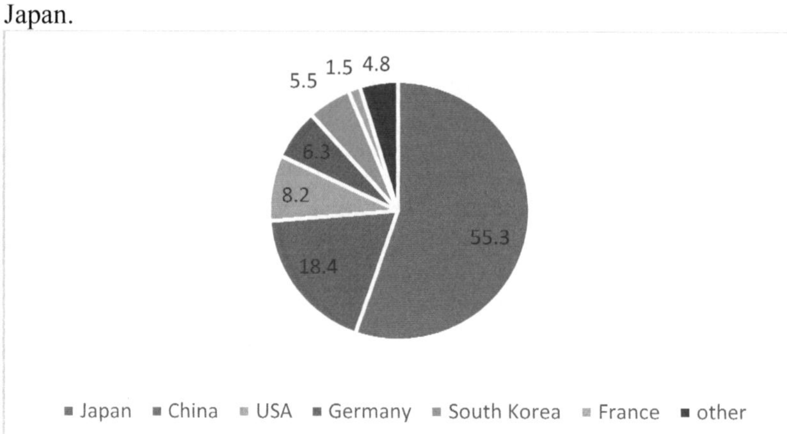

Figure 4. The geographical distribution of electric vehicles patent over the world

From 1971, Japan has started to research EV technologies. The number of patents in nearly three years accounted for 17% of his total patents. However, China started to research EV technologies since 1988. The number of patents in nearly three years accounted for 51% of China's total EV patents. Otherwise, China's enterprises still far compared with Japan, the number of Toyota's EV patents related to 3743, but China enterprise Chery only has 134 EV patents.

IOP Conf. Series: Earth and Environmental Science **86** (2017) 012010 doi:10.1088/1755-1315/86/1/012010

The earliest patent application for electric vehicles in China is in 1998. Before 2001, there is a relatively small number of patents in China. After 2001, the amount of patent applications and publications is beginning to show a steady upward trend, especially the rapid increase in 2009 and beyond. (Figure 5)

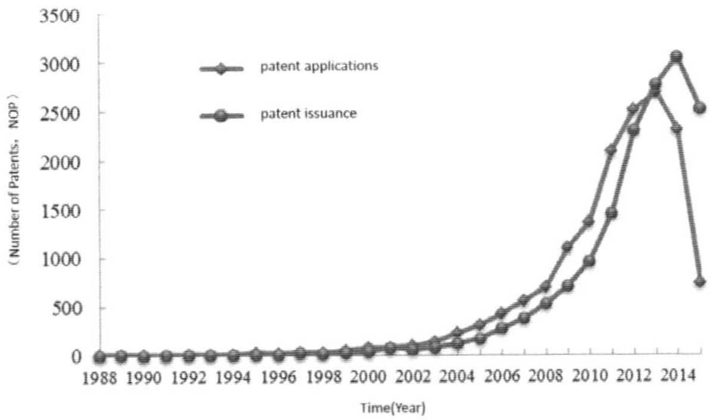

Figure 5. The development trend of EV patent in China

Along with rapidly development of EV in China, EVs have the potential to unlock innovation and create new advanced industries that spur job growth and enhance economic prosperity.

5. The Challenges of EV development in China

Despite the advances that vehicle electrification has made in the past two years, there are still significant barriers that stand in the way of widespread adoption. Technological, financial, market, and policy challenges could hinder market transformation if not addressed through further RD&D investments, public-private collaboration, and innovative policy and business solutions. The electric vehicles market in China will have the following specific characteristics:

(1)Main driver for electric vehicles market in China

Environmental protection is the main driving force for bringing electric vehicles into China's market, in particular mitigating climate change. And another important driving force is energy shortage. Application of electric vehicles in China's market can reduce the oil import dependency.

(2)Main competitor for electric vehicles in China

In the industry sector, electric vehicles will also face fierce competition from traditional cars as well as other newly developed technologies such as hybrid cars and hydrogen cars. Although electricity is cheaper than oil, electric vehicles system is so immature that higher price of electric vehicles is resulted from the higher initial investment. Cheap and abundant domestic traditional cars are the key competitors to EV.

(3)Main hurdle for electric vehicles in China

Most renewable energy in scattered areas is only a small fraction (9.3%) of Chinese energy supply, which is hampering the development of construction of nationwide grid for electric vehicles based on clean energies.

Otherwise, the most significant technological challenges currently facing EVs are the cost and performance of battery. Most EVs will remain more expensive in the near term than their petrol vehicle equivalents even when combined with government purchase subsidies offered in many countries.

(4)Huge need for investment in China

Even electric vehicles system has enough investment to develop the electric cars, recharge stations and batteries, China would need much more investment to extensively expand the national market and construct nationwide grid for EV. It would be a huge investment and long term plan.

(5)Policy framework for electric vehicles in China

The policies are also very important and affect future growth of electric vehicles market in China. Although Chinese government already has the plan to improve environment and develop the electric vehicles project, the regulatory framework is not clearly defined yet and policy has low priority assignment. For example, Chinese government can make policy to reduce the import tariff and VAT (value added tax) on electric vehicles technologies and equipment for the use of the renewable energy.

6. CONCLUSION

To sum up, we think that there is a market and economic feasibility of developing EVs in China. With all the driving force on position, what is still missing is much more investment on constructing both renewable energy plants and EVs for the purpose that EV market can be eventually formed.

As we all know, it is the common topic to modernize the society and protect the environment. Here, electric vehicles system is the double-win method relating to this topic. In the last decade, people always thought a lot of economic development and financial benefits. However, they often ignore the energy storage and environment breakage. At the same time, more and more problems came into human's concern, which violates the purpose of sustainable development. Let us think about the electric vehicles system. The development of electric vehicles can offset its dependence on imported oil. The other side, vehicle exhaust is the main greenhouse gas. If we can abase the vehicle exhaust, we can achieve environmental protection. Electric vehicles system is definitely one way of sustainable development. So far we do not have many large-scale electric vehicles demonstration projects because of the high costs. In the world, there are only a few pilot projects in some specific areas. It is not enough for electric vehicles development and progress. If we want to spread it out, we need to make more large-scale electric vehicles demonstration projects. For the government, they need to change policy to enhance the promotion of electric vehicles technology in industry in practice. For more and more demand of market, it will be indispensable in the future. Concerning the Better Place project, not only Demark but also the other countries started to test and practice electric vehicles system. For modernizing and protecting environment, it is profound in the future.

At the government part, Chinese government also plays a very important role in this whole project. Collaboration on research development and demonstration occurs among the countries. National policy and international policy collaboration also took place decades ago, which includes the efforts of technologies and equipments from the developed countries to developing countries. 'These efforts include, in particular, bilateral cooperation, and more collective efforts through regional cooperative frameworks such as the Chinese battery maker BYD, State Grid Corporation of China (SGCC).' [10] As one of the case study, there are some organizations, like Better Place who attempted to research and experiment to some local sample factories as well as the technology collaboration and international collaboration with China where it takes place. For the import of the key technologies and facility, exemption or reduction of import VAT (value added tax) should be granted. Furthermore, most renewable energy technologies are mature from abroad, and these energy technologies are open for commercial and industrial investment.

Moreover, energy system is also the main driver to promote the development of EV technologies. At present, gasoline and diesel provide the most of the transport energy. EV can use electricity to provide power for transportation. Electricity has already provided power for some rail network. In future, electricity will be used to charge batteries in fully electric vehicles. The demand for mitigating climate offered by future electricity based EV system would therefore be dependent on the clean energy, such as renewable energy, nuclear energy and coal energy with carbon capture and storage. The renewable energy is widely distributed in different parts of the China, which can form a mix grid. And EV might be recharged principally at night when electricity demand is lower. So EV system has less impact on the requirement for additional electrical capacity and EV can be used to absorb some surplus electricity form discontinuous forms of power generation. In the future, electricity based on the EV could have important impacts on Chinese electricity demand. Government shall also intervene energy system so that electricity power generations are moving to lower carbon electricity generations.

More and more countries have begun to attach importance to the reality of global climate change and attempted to modify their power framework towards a more environmentally friendly pattern. They will sooner or later compel the applications of electric vehicles in their market and society.

IOP Conf. Series: Earth and Environmental Science **86** (2017) 012010 doi:10.1088/1755-1315/86/1/012010

References

[1] HowseR and Eliason A2008*Domestic and International Strategies to Address Climate Change: An Overview of the WTO Legal Issues* (Cambridge: Cambridge University Press) p 117
[2] The Danish Ministry of Transport and Energy 2005*Energy Strategy 2025*[EB/OL]http://www.energetska-strategija.hr/doc/stdr,2005
[3] Wikipedia*Automobile industry in China*
[4] National Bureau of Statistics 2007 *China Statistical Yearbook(1996-2006)* (Beijing: China Statistics Press)p56
[5] UNEP Climate Action 2015 *Sustainable Cities for Climate Action*[EB/OL]http://www.cop21paris.org/knowledge-centre/infographics/sustainable-cities-for-climate-action-an-infographic
[6] TriggT and Telleen P 2013 *Global EV Outlook-Understanding the Electric Vehicle Landscape to 2020* pp9-10
[7] Minstry of Industry and Information Technology of the People ' s Republic of China 2015*Economic operation situation of automobile industry in 2014*[EB/OL]http://www.miit.gov.cn/n1146312/n1146904/n1648362/n1648363/c3488220/content.html
[8] IEA 2015 *Global EV Outlook 2015* p23
[9] National Development and Reform Commission 2007 *China's National Climate Change Programme*
[10] Philibert C and Podkanski J 2005 *International energy technology collaboration and climate change mitigation* P17

Spatial Distribution of Transverse Dispersion Coefficients at U-shaped Bend Confluence with 60° Junction Angle

X X Zhao[1], L Gu[1,2,3,*], B Dai[1]

[1]Key Laboratory of Integrated Regulation and Resource Development on Shallow Lake of Ministry of Education, College of Environment, Hohai University, Nanjing 210098, P.R. China.
[2]National Engineering Research Center of Water Resources Efficient Utilization and Engineering Safety, Hohai University, Nanjing 210098, P.R. China.

Abstract. The confluence of bend channel and the tributary is fairlycommon in nature. At present the study about the dispersion capacity of pollutants in the bend confluence is very limited. In this paper, a three-dimensional numerical model was established to explore the hydraulic characteristics such as velocity distribution and circulation structure of the different sections in the U-shaped bend confluence. Then the transverse dispersion coefficient was calculated using the dispersion tensor method and N-zone model on the base of flow structure data, and its spatial distribution along the bend is analyze. The results show that there is a significant difference in the flow structure of each section at the U-shaped bend confluence, especially before and after the junction of tributary.The peak value location of the transverse dispersion coefficient at the cross section appears near the inner bank. Along the bend channel, the transverse dispersion coefficient tends to increase first and then decrease, and the maximum value of the section appears before the bend apex where the tributary feeding into the main channel.

1. Introduction

An understanding of the fate of pollutants is essential for the management of river environments. Numerous investigators have contributed to the understanding of mechanisms of transverse dispersion in rivers. In 1978, Fischer et al. [1] estimated the mixing coefficients in a regular straight channel. For a laboratory straight flume, the dimensionless transverse dispersion coefficient ranges from 0.1 to 0.26 [2]. For curved rivers, the transverse dispersion coefficient was positively correlated with the bending angle and the width-depth ratio, but negatively correlated with the radius of curvature [3-6]. Rutherford [7] reported that the dimensionless value of the transverse dispersion coefficient for a curved channel is 0.3-1.0 and that the value for a strongly curved channel is 1.0-3.0. Boxall et al. [8] demonstrated that the secondary circulation was the main factor affecting the transverse dispersion of a curved channel.

However, the dispersion coefficient of pollutants in a channel confluence differs from the straight and curved rivers due to the distinctive flow structure of a confluence. Confluences

[3] Corresponding Author: L Gu, E-mail: guliqc@hhu.edu.cn
[4]This work was financially supported by the National Natural Science Foundation of China [grant numbers 51379058]

can be widely found in the natural river systems and artificial channels. When a tributary feeds into a mainstream, two flows meet and mix with each other, and the flow structure of the original main channel changes drastically. Thus the mixing process and the dispersion capacity of pollutants in the main channel will also change significantly. However, at present the major studies about the confluence focus on the flow structure of the junction, especially for the junction of two straight channels [9-12]. Best [10] proposed a generalized model of flow for the open channel junction, which was divided into six different zones, namely, regions of flow stagnation, flow deflection, flow separation, maximum velocity, flow recovery, and shear layers. Recently, some flume experiments about the flow structure for U-shaped bend confluence were carried out [13-15]. Moreover, many researchers used the numerical models to explore the complex flow structure for confluences [16-19]. Above all, the research about the dispersion capacity of pollutants in the bend confluence has not been reported.

In this study, a three-dimensional numerical model was established to investigate flow structures and dispersion characteristics of U-shaped bend confluences. The hydrodynamic model was validated using Gao et al. [14] laboratory experiments data. Then the flow structures, especially the circulation structure of the bend were investigated, and the transverse dispersion coefficient was calculated by the dispersion tensor method and the modified N-zone model. Finally, the spatial distribution characteristics of the transverse dispersion coefficient along the bend was analyzed.

2. Setting up of three-dimensional mathematical model for U-shaped bend confluence

The simulated model domain was based on Gao et al. [14] laboratory experiments. The widths of main channel and tributary are 1 m and 0.3 m, respectively. In order to facilitate the full development of flow, the upstream and downstream flumes will be extended to 6m in the numerical model. The simple layout of the model was shown in figure 1.

The bend has a turning angle of 180°. The inner wall radius is 1.5 m and the outer wall radius is 2.5 m. The bottom section of the bend is of partial V-shape. The mainstream discharge Q_1 is 0.03 m^3/s, and tributary discharge Q_2 is 0.009 m^3/s. Define the discharge ratio $Q_r=Q_2/Q_1$. In additional, the junction angle of two channels is 60°, and the downstream water depth is 0.18 m.

2.1. Control equation and solution

The mass conservation equation and the momentum equations of Reynolds-averaged Navier-Stokes in three dimensions are as follows (in tensor form):

$$\frac{\partial \rho}{\partial t}+\frac{\partial \rho u_j}{\partial x_j}=0 \tag{1}$$

$$\frac{\partial \rho u_i}{\partial t}+\frac{\partial \rho u_j u_i}{\partial x_j}=-\frac{\partial p}{\partial x_i}+\frac{\partial}{\partial x_j}\{(\mu+\mu_t)(\frac{\partial u_i}{\partial x_j}+\frac{\partial u_j}{\partial x_i})-\frac{2}{3}\rho k\delta_{ij}\}+(\rho-\rho_a)g_i \tag{2}$$

Where ρ is the fluid density, ρ_a is the density of air, p is the static pressure, g is the gravitational force, k is the turbulent kinetic energy, δ_{ij} is the Kronecker delta, μ is the molecular viscosity of fluid, and μ_t is the turbulence eddy viscosity of fluid.

In order to simulate the flow structure of U-shaped bend confluence more accurately, the three-dimensional RNG k-ε turbulence model of water-gas two-phase flow is used, and the free surface is tracked by VOF. In this study, the phases of air and water were represented by using the subscripts a and$w$ respectively. Accordingly, the volume fractions of air α_a and water α_w sum to unity. Other details can be referred to Dai [20].

2.2. Boundary conditions

(a) Inlet and outlet boundary conditions: The velocity value and direction was specified at the upstream water inlet of mainstream and tributary.The pressure was specified at the upstream air inlet and downstream outlet.

IOP Conf. Series: Earth and Environmental Science **86** (2017) 012011 doi:10.1088/1755-1315/86/1/012011

(b) Wall boundary conditions: Non-slip boundaries were applied at all sidewalls and channel floor. The channel floor roughness height was 0.009.

2.3. Mesh generation

The total number of grids is about 600,000 and the local computational grids of bend is shown in figure 2. The mesh is refinednear bend and near the interface of water andair.

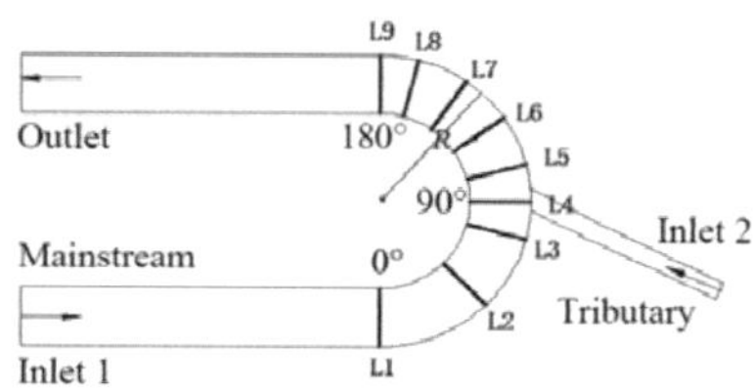

Figure 1. Sketch map of a simple U-shaped bend confluence.

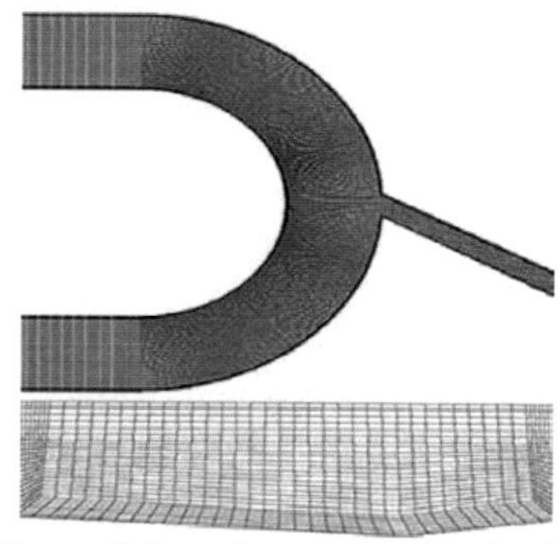

Figure 2. Local computational grids.

2.4.Model validation

Experimental results of Gao et al.[14] were used to validate the numerical model.The measured values and calculated values of the water surface elevation and longitudinal velocity were compared. The locations of some typical cross sectionswere shown in figure 1 and Table 1. All the section plots in this paper were looked downstream

Table 1. The cross section corresponds to the center angle.

Cross-section	L1	L2	L3	L4	L5	L6	L7	L8	L9
Central angle	0°	45°	75°	90°	105°	125°	145°	165°	180°

2.4.1.Surface elevation. The surface elevation of the measured and simulated values was shown in figure 3. It was apparent that water surface elevation of outer bank was significantly higher than the inner bank before the tributary into the U-shaped bend. The water surface elevation will rapidly decline after the confluence, and then slowly rising in the downstream. The water surface elevation was relatively low near the inner bank. It can be seen that the simulated surface elevation coincides with the measured value basically, and the deviation was less than 5%.

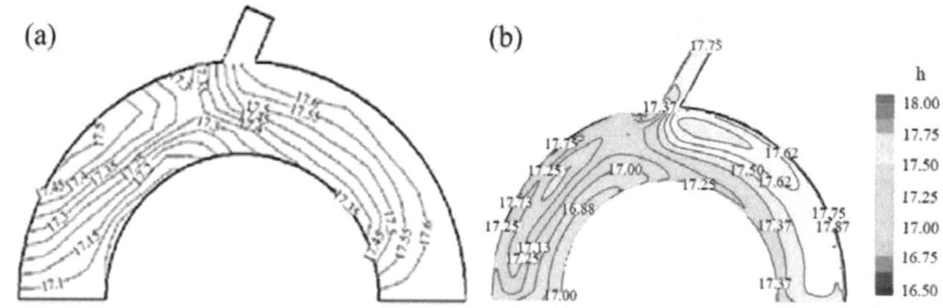

Figure 3. The surface elevation of the measured and simulated value, (a) measured value (b) simulated value(the unit is m).

*2.4.2. Longitudinal velocity distribution.*The longitudinal velocity comparison of the inner bank ($R = 1.55$), centerline ($R = 2.05$), and outer bank ($R = 2.45$) for cross-section L4 was shown in figure 4. z^*, u^* was the dimensionless depth and velocity. Obviously, the simulated value had a good agreement with measured value.

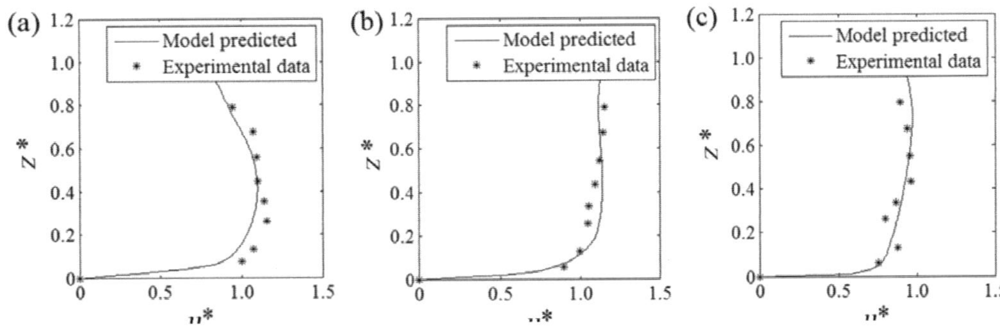

Figure 4. Comparison of measured and calculated longitudinal velocity for cross-section L4, (a),(b),(c) are the verticals at$R$=1.55m,R=2.05m and$R$=2.45m, respectively; R is the distance from the center of the bend to the measuring point.

3. Spatial distribution of transverse dispersion coefficients

The detailed three-dimensional flow structure of the whole U-shaped bend confluence was obtained by the above mathematical model for the test with the junction angle 60° and discharge ratio Q_r=0.3. Then, the transverse dispersion coefficient of the U-shaped bend confluence was calculated by the two-dimensional dispersion tensor method proposed by Fischer [2], and its spatial distribution characteristics along the bend was analyzed.

3.1. Calculation Method of Dispersion Coefficient

A dispersion tensor method by Fischer (1979) [2] was used to calculate the transverse dispersion coefficient at the typical verticals of the cross-section, where the equation for transverse dispersion coefficient was as follows,

$$D_{yy} = -\frac{1}{h}\int_0^h v'\int_0^z \frac{1}{\varepsilon}\int_0^z v'dzdzdz \tag{3}$$

whereD_{yy} is a component of the dispersion tensor in the y direction, h is the local depth of flow, v' denotes the vertical deviations of point velocities with respect to the depth-averaged velocity v, and ε is a vertical turbulent diffusion coefficient.

Then the modified N-zone model [21] was used to calculate a cross-sectional dispersion coefficient value by Fourier transform of the point values at every vertical line. The modified N-zone model [21]was based on the model of Chickwendu [22]. However the transverse velocitywas divided into N different regions in the lateral direction, which is different from the model of Chickwendu [22]. Combined with the dispersion tensor method and the modified N-zone model, the variation of transverse dispersion coefficient was studied in the lateral and longitudinal direction, respectively. The specific calculation method can be referred to Jiao [21] and Gu et al.[23].

3.2. Result analysis

The circulation structure and lateral distribution of transverse dispersion coefficient of cross section L2, L4 and L6 were shown in figure 5. At section L2, the transverse dispersion coefficient increases first and then decreases, and the maximum value appears at y=0.45 m near the centerline of the section (figure 5(b)). The flow structure of this section was shown in figure 5(a), where a big clockwise circulation exists at the whole section and the big transverse velocity gradient of the vertical is at the center of the circulation. This strong velocity shear leads to the increase of the transverse dispersion coefficient, and thus the peak value occurs near the middle of the section (figure 5(b)).

L4 were the cross-section where the tributary feeds into the mainstream. It was obvious from figure 5(d) that the transverse dispersion coefficient has two peak values in this section, and the peak near the inner bank was larger than the one near the outer bank. The reason was that a larger clockwise circulation occurs near the inner bank, and the larger lateral velocity difference of the verticals makes the transverse dispersion coefficient increase. On outer bank, due to the inflow of tributary, lateral velocity was big, however the difference between the lateral velocitywas not big. Thus there was a smaller peak on outer bank (figure 5(c)).

On the section L6, the clockwise circulation and counterclockwise circulation appear in the inner bank and the outer bank, respectively. Due to the two circulation structures, the transverse dispersion coefficient appeares two peaks at cross section L6, and the value of the outer bank was significantly smaller than the inner bank (figure 5(e) and (f)). The peaks value of transverse dispersion coefficient for cross section L2, L4 and L6 all appear near the inner bank. However the transverse dispersion coefficient of L6 with the double circulation structure was significantly smaller than that of L2 and L4 with a single clockwise circulation. (figure5(b),(d) and (f)).

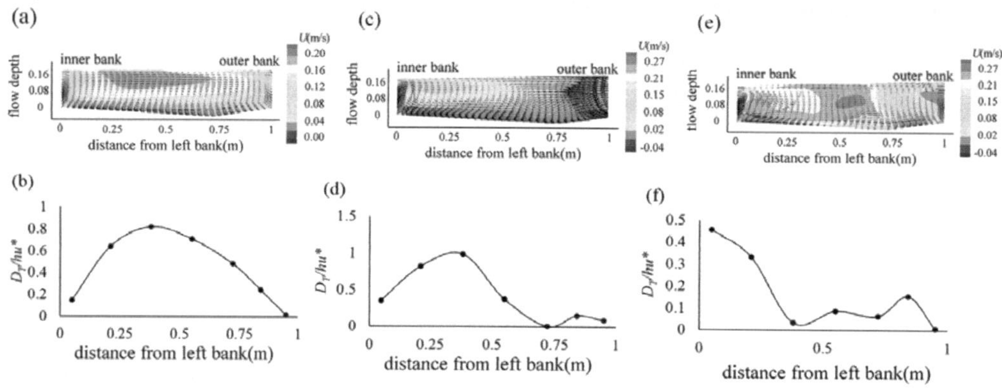

Figure 5. The circulation structure and lateral distribution of transverse dispersion coefficient of cross section L2, L4 and L6: (a) and (b),(c) and (d),(e) and (f) are corresponding to L2,L4,L6, respectively.

The longitudinal distribution of the transverse dispersion coefficients of different sections was depicted in figure 6. It can be seen that the transverse dispersion coefficient increases first and then decreases along channel. The maximum value appears at the positions of the cross section L2 and L3, whose values were a little bigger than the bend apex L4 section. The flow structures of cross sections L2 and L3 were similar to L4, where a single clockwise circulation exists at the whole cross section, and a large lateral velocity difference results in a larger transverse dispersion coefficient value.

Figure 6. The longitudinal distribution of the transverse dispersion coefficients.

4.Conclusion

The aim of this research was to investigate the spatial distribution of transverse dispersion coefficient in the U-shaped bend confluence. By analysis, it was found that there is a significant difference in the flow structure of each section at the U-shaped bend confluence. Before the tributary into bend, only a single clockwise circulation was formed in the cross section, and after the junction, the reverse double circulation structure is formed in the cross section with the clockwise cell at inner bank and counterclockwise cell at outer bank. The circulation structure makes the spatial distribution of transverse dispersion coefficients change. In the lateral direction, the peak value position of the transverse dispersion coefficient appears near the inner bank. In the longitudinal direction, the transverse dispersion coefficient tends to increase first and then decrease along channel, and the maximum value of the section appears before the bend apex with the tributary inflow.

5. References

[1] Fischer H B 1978 *J.Geophys.Res* **83** 2373-75
[2] Fischer H B, List E J, Koh R C Y, Ouml, Omberger R G and Brooks N H 1979 *Mixing in inland and coastal waters* (London: Academic Press)
[3] Beltaos S 1979 *Can. J. Civ. Eng* **6** 575-591
[4] Xing Z and Jiang N 1994 *Chongqing. Environ. Sci* 38-43 (In Chinese)
[5] Albers C and Steffler P 2007 *J. Hydraul. Eng* **133** 186-196
[6] Garutti E 2011 *J.Hydraul. Eng* **137** 1126-34
[7] Rutherford J C 1994 *River mixing* (England: Wiley)
[8] Boxall J B, Guymer I and Marion A 2003 *J. Hydraul. Res* **41** 153-165
[9] Best J L and Reid I 1984 *J. Hydraul. Eng* **110** 1588-94
[10] Best J L 1987 *Special Publications* **39** 27-35
[11] Weber LJ, Schumate E D and Mawer N 2001 *J. Hydraul. Eng***127** 340-350
[12] Szupiany R N, Amsler M L, Parsons DR and Best JL 2009 *Water Resour. Res* **45** 641-8
[13] Li W Q2012 *Water transport engineering* 62-68 (In Chinese)
[14] Gao Y, Huang S H and Li Q 2012 *Hydropower and Energy Science* **30** 90-93 (In Chinese)
[15] Fu Z M, Gu Z P, Zheng J T and Lie G P 2013 *Water transport engineering* 46-51(In Chinese)
[16] Biron P M, Ramamurthy A S and Han S 2004 *J. Hydraul. Eng* **130** 243-253
[17] Constantinescu G, Miyawaki S, Rhoads B, Sukhodolov A and Kirkil G 2011 *Water Resour. Res* **47** 5507
[18] Shakibainia A, Tabatabai M R M and Zarrati A R 2010 *Can. J. Civ. Eng* **37** 772-781
[19] Wang X and Cheng L 2000 *Computers & Fluids* **29** 415-433
[20] Dai B 2014 Master Thesis Nanjing: *Hohai University*(In Chinese)
[21] Jiao Z N 2015 Master Thesis Nanjing: *Hohai University* (In Chinese)
[22]Chikwendu S 1986 *J. Fluid Mech* **167** 19-30
[23] Gu L, Jiao Z N and Hua Z L2014 *Water Sci. Technol* **70** 256-264

Ongoing Relative Performance Evaluation for a CO_2 EOR Asset in a Worldwide Peer Group

C F Zhao[1,4,5], **X S Li**[1], **G H Wang**[2] **and L Li**[3]

1.EOR Research Institute, China University of Petroleum – Beijing, China
2.CNPC Middle East Company, Beijing, China
3.Oil and Gas Development Management Center, Sinopec Shengli Oilfield Company,Dongying, Shandong,China
4.Corresponding author: zhaochuanfeng@cup.edu.cn

Abstract. Operators of a CO_2 EOR asset need to know the relative performance level of their asset against its peers. The ongoing relative performance evaluation method is appropriate for this purpose. We first choose 52 CO_2 assets around the world as the peer group, and then define the four ranking levels in terms of CO_2consumption ratio. Only the final values of CO_2consumption ratio for the group are obtained, and therefore cannot be used for an ongoing evaluation during a CO_2 EOR asset's life circle. Consequently, numerical reservoir simulation is employed to quantify the process values corresponding to the four ranking levels. Type curve plots are generated on the basis of the process values and utilized for the ongoing relative performance evaluation of a CO_2 EOR asset in China.

1. Introduction

CO_2, known as the greenhouse gas if directly emitted to the air, can be injected into an oil reservoir for the purpose of enhanced oil recovery (hereafter abbreviated as EOR). Operators of a CO_2 EOR asset need to make decisions, such as expanding or reducing or even suspending their investment, according to their knowledge about the development process of their CO_2 EOR asset and its relative performance level against its peer assets. A method, called ongoing relative performance evaluation, stands out for this purpose. A large number of examples using this method have been found in fields of human resources, but none in oil industry. In this paper we present its major steps involved in the assessment of a CO_2 EOR asset in China.

2. Steps of ongoing relative performance in a peer group

A CO_2 EOR asset in China is producing crude oil during its life cycle. The aim is to conduct an ongoing relative performance evaluation for it in a worldwide peer group.Suppose there are N CO_2 EOR assets in a peer group not including the target asset. The life cycle of the ith asset is T_i, and the evaluation index of the ith asset is $X_t^{(i)}$, where $0<t\leqslant T_i$, $1\leqslant i\leqslant N$.

2.1. Choosing a peer group

[5]Supported by Science Foundation of China University of Petroleum, Beijing (No. ZX20150135).

CO_2 EOR is a promising oil recovery technique, but lower oil prices and insufficient CO_2 sources put limitations to its application. Only 52peer assets have been found in publically published articles, most in USA and China. Maybe there are other peer assets not publically reported. Unfortunately we have no way to gather their information.Therefore, the capacity of the peer group is *N*, that is 52.

2.2. Choosingan evaluation index

Although many indices are available for a CO_2 EOR asset, most are inappropriate for evaluation purpose for they are strongly related with the oil reserves and CO_2 injection volumes. It is unfair to assess an asset with small reserves and injection volumes against another with larger ones. However, CO_2consumption ratio is an exceptional index. With unit oft/t, this index isdefined as the mass of injected CO_2 needed toobtainan oil incremental of 1.0t.A lower value of this index indicates higher utilization efficiency of CO_2. Its ability to eliminate the differences in reserves and injection volumes between assets makes it appropriate for a relative performance evaluation.

The values of CO_2consumption ratio for the peer group distribute within the range of 0.18~7.69, with the arithmetic average of 2.8 and the median of 2.6 (figure 1)[1~12].

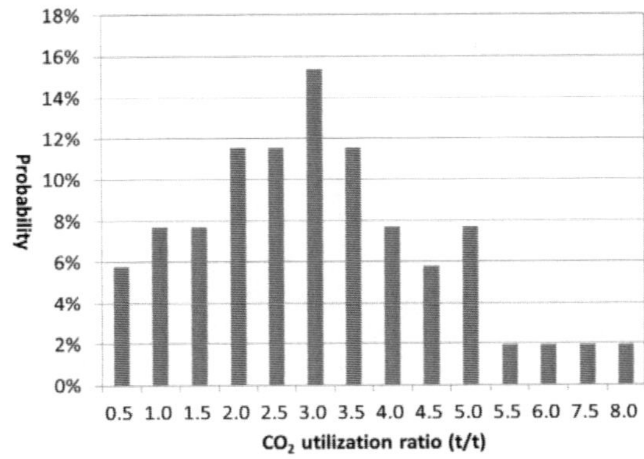

Figure 1. Probability distribution of CO_2consumption ratio in the peer group.

2.3. Defining ranking levels

According to the general classification method in Probability Theory and Mathematical Statistics, a cumulative probability distribution curve can be divided into four ranges, that is, [0.00%, 17.37%), [17.37%, 50.00%), [50.00%, 82.63%), [82.63%, 100.00%], respectively representing A, B, C and D levels (figure 2). The three boundary values, $V^{(1)}$, $V^{(2)}$ and $V^{(3)}$, are successively 1.2t/t, 2.6t/t and 4.2t/t. The lower the consumption ratio of an asset is, the higher its relative remark and its ranking level.

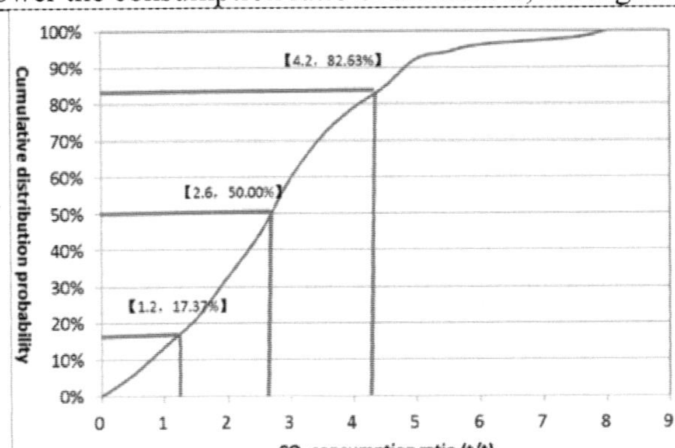

Figure 2. The cumulative probability distribution curve of CO2 consumption ratio in the peer group.

2.4. Type curve plots for ongoing relative performance evaluation

Suppose the sequence number of the target asset is 0. Now it is expected that we can conduct the ongoing relative performance evaluation for the target asset by comparing $X_t^{(0)}$ against the aggregate $\{X_t^{(1)}, X_t^{(2)}, \ldots, X_t^{(N)}\}$. However, a problem arises from the fact that all the values of CO_2 consumption ratio are final values at the end of the life cycles of these peers, that is, $X_{T_i}^{(i)}$ (i=1, 2, …, *N*). These values are justified to be used for a post relative performance evaluation rather than an ongoing evaluation.

Though the process values during life cycle are unavailable, we have luckily obtained the three boundary values, $V^{(1)}$, $V^{(2)}$ and $V^{(3)}$. Numerical reservoir simulation can be used to synthesize process values corresponding to the three boundary values. Then the ongoing evaluation is feasible.

2.4.1. Numerical reservoir simulation

Numerical reservoir models are built to represent the potential ranges of reservoir and fluid properties of the target asset. The number of simulation scenarios should be large enough to ensure that almost allits possible geological realizationscan be covered. Finally three simulation scenarios reveal themselves from the scenario pool respectively corresponding to the three boundary values.As a result, we now have three process boundary value series, $V_t^{(1)}$, $V_t^{(2)}$ and $V_t^{(3)}$ ($0<t\leqslant T_0$). $V_{T0}^{(1)}=V^{(1)}$, $V_{T0}^{(2)}=V^{(2)}$,$V_{T0}^{(3)}=V^{(3)}$. In fact, process boundary value series of three more indices, such as recovery percentage during CO_2 EOR life cycle, EOR incremental during CO_2 EOR life cycle over waterflooding and oil displacement efficiency during CO_2 EOR life cycle, can be gained from the simulation results including $R_t^{(1)}$, $R_t^{(2)}$, $R_t^{(3)}$, $I_t^{(1)}$, $I_t^{(2)}$, $I_t^{(3)}$, $D_t^{(1)}$, $D_t^{(2)}$, $D_t^{(3)}$. These indices can be employed for ongoing relative evaluation of economic benefit, technical effect and effective EOR mechanism.

2.4.2. Type curve plot for ongoing relative evaluation of economic benefit

With recovery percentage during CO_2 EOR life cycle as the horizontal ordinate and CO_2 consumption ratio as the vertical ordinate, we can obtain the type curve plot for relative evaluation of economic benefit (figure 3).

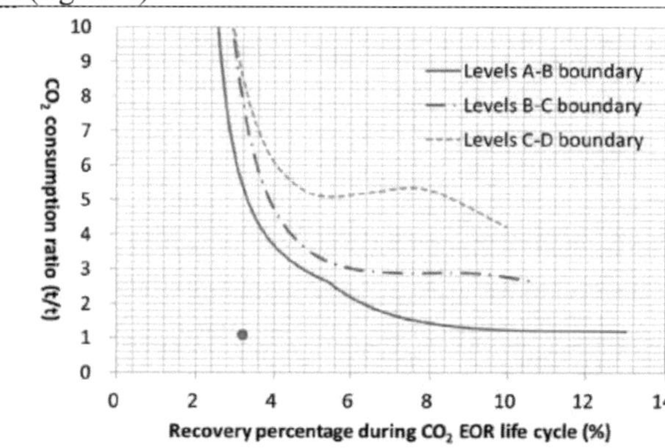

Figure 3. Type curve plot for ongoing relative evaluation of economic benefit.

2.4.3. Type curve plot for ongoing relative evaluation of technical effect

With recovery percentage during CO_2 EOR life cycle as the horizontal ordinate and EOR incremental during CO_2 EOR life cycleover water flooding as the vertical ordinate, we can obtain the type curve plot for relative evaluation of technical effect (figure 4).

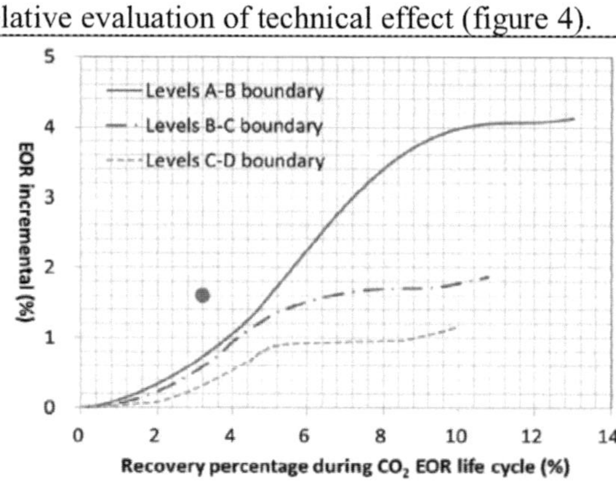

Figure 4. Type curve plot for ongoing relative evaluation of technical effect.

2.4.4. Type curve plot for ongoing relative evaluation ofeffective EOR mechanism

The primary EOR mechanism of a CO_2EOR asset lies in improving the oil displacement efficiency rather than the sweep efficiency. With recovery percentage during CO_2 EOR life cycle as the horizontal ordinate and oil displacement efficiency during CO_2 EOR life cycle as the vertical ordinate, we can obtain the type curve plot for relative evaluation of effective EOR mechanism (figure 5).

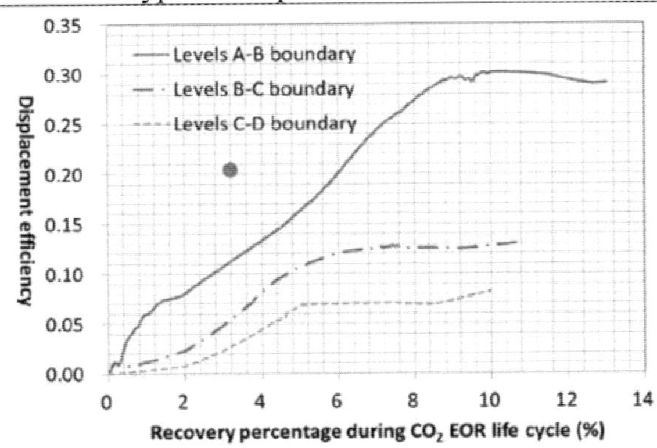

Figure 5. Type curve plot for ongoing relative evaluation of effective EOR mechanism.

3. Field case

Up to December, 2016, the recovery percentage of the original oil reserves in the target oilfield was 9.6%. The recovery percentage during CO_2 EOR life cycle was 3.2%, and the EOR incremental was 1.6%. The oil displacement efficiency was calculated as 0.202 with the methods in references 13~16. The gas consumption ratio was 1.1t/t.

In figure 3, a point is pictured on the type curve plot for ongoing relative evaluation of economic benefit. In figure 4, a point is pictured on the type curve plot for ongoing relative evaluation of technical effect. And in figure 5, a point is pictured on the type curve plot for ongoing relative evaluation of technical effect. It can be seen that this point lies at the region corresponding to the ranking level of A, indicating that the target asset was ranked top 17.37% in the peer group.

4. Conclusions

The ongoing relative performance evaluation method can be used to assess one CO_2 EOR asset against its peers. This method involvessteps including choosing a peer group and an evaluation index, defining ranking levels, and conducting evaluation.

With the help of numerical reservoir simulation, three type curve plots are established for ongoing relative performance evaluation of economic benefit, technical effect and effective EOR mechanism. The ongoing relative performance evaluationindicates that the target CO_2 EOR assetis ranked as Level A in terms of economic benefit, technical effect and effective EOR mechanism, and top 17.37% in the worldwide peer group.

References

[1] WR Brock and LA Bryan. Summary Results of CO_2 EOR Field Tests, 1972-1987. SPE 18977, SPE Joint Rocky Mountain Regional/Low Permeability Reservoirs Symposium and Exhibition, March 1989

[2] RE Hadlow. Update of Industry Experience With CO_2 Injection. SPE 24928, the 67th Annual Technical Conference and Exhibition of SPE, October 1992

[3] LW Lake and MP Walsh. Enhanced Oil Recovery (EOR) Field Data Literature Search. Technical Report for Danish Energy Agency, 2008

[4] JPMeyer. Summary of Carbon Dioxide Enhanced Oil Recovery (CO_2 EOR) Injection Well Technology. Background Report for the American Petroleum Institute, 2008

[5] RESweatman, ME Parker, and SLCrookshank. Industry Experience With CO_2 Enhanced Oil Recovery Technology. SPE 126446 for the 2009 SPE International Conference on CO_2 Capture, Storage and Utilization, November 2009

[6] PLThiez and YLGallo. Use of CO_2 for Enhanced Oil Recovery and Storage. AAPG Search and Discover Article #90099 for the 2009 AAPG European Region Annual Conference, November 2009

IOP Conf. Series: Earth and Environmental Science **86** (2017) 012012 doi:10.1088/1755-1315/86/1/012012

[7] SSalem, and TMoawad.Economic Study of Miscible CO_2 Flooding in a Mature Waterflooded Oil Reservoir. SPE 168064 for the SPE Saudi Arabia section Annual Technical Symposium and Exhibition, May 2013

[8] Weiming Qian, Fengdong Zhang, Jiaren Zhang, et al. Field test and effectiveness evaluation of CO_2 immiscible displacement in Chujialou oilfield. Special Oil & Gas Reservoirs, 2001, 8(4): 79~82

[9] Junsheng Zhao, Weiming Qian, Chunyan Lang. Field Test of CO_2 Huff-Puff in Subei Low Permeability Reservoir. Fault-block Oil and Gas fields, 2003, 10(1): 73~75

[10] Fengdong Zhang, Zhenliang Zhang. The Field Experiment and Result Analysis of CO_2 Miscible Displacement in Caoshe Oilfield of the North Jiangsu Basin. Petroleum Geology and Experiment, 2010, 32(3): 296~300

[11] Shuangming Ni, Mei Hu, FeiXie. The Performance of a CO_2 Flooding Pilot Test. Oil-Gasfield Surface Engineering, 2013, 32(2): 20~21

[12] Yining Wang, Xiaodong Wu, Shaobo Zhang, et al. A CO_2 Immiscible Flooding Pilot Test and the Performance Evaluation. Journal of Oil and Gas Technology, 2013, 35(4): 136~139

[13] Shuheng Cui, Jianfeng Fu. Carbon Dioxide Flooding Sweep and Displacement Efficiency Study. Inner Mongulia Petrochemical Industry, 2011(6): 154~156

[14] ROBaker and LLMcClernon. Estimation of Volumetric Sweep Efficiency of a Miscible Flood. Journal of Canadian Petroleum Technology, 1998, 37(2): 40~46

[15] DFerguson. Investigation of Efficiency Improvements During CO_2 Injection in Hydraulically and Naturally fractured Reservoirs. Semi-Annual Technical Progress Report for DOE Contract No.: DE-FC26-01BC15361, April 2004

[16] DSenocak. Evaluation of Sweep Efficiency of a Mature CO_2 Flood in Little Creek Field, Mississippi. Master Thesis of Louisiana State University, December 2008

Evaluation of Intensive Land Use in Shuikoushan Economic Development Zone, Hunan Province, China

W He, F M Kuang and P Wang

College of City and Tourism, Hengyang Normal University, Hengyang Hunan 421002, China

Email: wangpengnju@163.com
Abstract: Taking Shuikoushan Economic Development Zone as the research objective, the degree of intensive land use was evaluated, by means of multi-factor assessment method. The results showed that the degree of intensive land use of Shuikoushan Economic Development Zone reached 84.89%. The land use status was good and the management performance of land use was superior while the land use efficiency was inferior on the contrary. Then, suggestions and measures for the future land intensive use were provided.
Keywords: Economic Development Zone, Intensive Land Use, Shuikoushan, Hunan Province

1 Introduction

Since been set up in 1984, development zone had been playing an active role in promote region and city economic development[1]. But there still had problems in land use in development area, such as extensive utilization and idleness[2]. Alter the land use of development zone from extension development to intention development. Continuous and normalized evaluation on land intensive use was an important measure to solve the land use problem of development zone for governments at all levels [3]. Evaluating the degree of land intensive use of development zone scientifically and rationally, its significance to supervise land use and realize sustainable development of development zone, and could be the scientific basis for the extension and upgrade of development zone[4]

Taking Shuikoushan Economic Development Zone (SEDZ) as the research object, setting December 31, 2013 as evaluation point, the degree of intensive land use was evaluated. Then made a thorough analysis on the evaluation results, in order to reveal the experience and problems in the land use of Economic Development Zone, and to provide basis for land management policy making and land intensive use mechanism constructing.

2 General situations of study area and data sources

SEDZ, located in the north side the Changning city, south of the Xiang river, founded in June 1992. It was a provincial development zone approved by The People's Government of Hunan Province in March, 1994, reserved as provincial development zone by National Development and Reform Commission in September, 2006.

According to *Rules for the evaluation of land intensive use in development zones 2014 (trial implementation)* (hereinafter referred to as *Regulation*) and *Development zone direction area demarcation implementation program* promulgated by Ministry of land and Resources of People's Republic of China, one development direction area was delimited on the eastern side of SEDZ in 2014. The research area included the development direction area.

The total area of SEDZ was 647.12 hm^2 (primary area 218.06 hm^2; development direction area 429.06 hm^2). In 2014, the proportion of completed urban construction land was 53.22%, uncompleted urban construction land 46.78%, and no Non-construction land.

The basic data was provided by Administrative Committee of Shuikoushan Economic Development Zone, Changning Land and Resources Bureau, Changning Urban and Rural Planning Bureau, and field investigation.

3 Land intensive use evaluation

3.1 Evaluation Index System

According to *Regulation*, combined with actual situation of SEDZ, the evaluation type of SEDZ was industry-oriented. The evaluation index system comprised three main components: land use status, land use efficiency and management performance; contain three levels, which were objective, sub-objective, and index. There were three objectives, five sub-objectives, ten indexes (nine indexes for development direction area) (Table 1).

3.2 Evaluation index weights and ideal values

In reference to unified index weights and ideal values in Hunan Province in 2014, and combined with SEDZ's actual condition, the weights and ideal value were decided by the target method (Table 1).

3.3 Calculation of intensive land use values

Evaluation was done by multi-factor comprehensive evaluation method. For unified dimension, the evaluation indexes were standardized by ideal value ratio method. According to the index weight of evaluation results, we should calculate the comprehensive score of SEDZ land intensive utilization by weighted sum. Score ranged from 0 to 100%, if the score was high, the level of land intensive use was high, and potential of the land intensive use was small.

3.4 Result analysis

The intensive land use score of SEDZ was 84.89%(Table 1), basically achieved the requirement of land intensive use, but there was a certain gap between the intensive degree score and the ideal value, and the SEDZ's land resources potential should be further tapped.

3.4.1 Land use status

SEDZ was at a higher degree of land use. In primary area, the standardized value of land supply rate was 98.92%, and the standardized value of land complction rate achieved ideal value (100%). In development direction area, the standardized value of land development rate was 97.2%.

In status of land use structure, the industrial land rate of SEDZ was high. The standardized values of industrial land rate of both primary area and development direction area were more than 96%, even the standardized values of industrial land rate of development direction area exceeded the ideal value of the unification of Human Province.

The land use intensity scores of both primary area and development direction area were not very high, for primary area was 85.87%, for development direction area was 65.55%. Only the building density of primary area exceeded the ideal value of the unification of Human Province, other indexes still had a certain difference compared with ideal.

3.4.2 Land use efficiency should be improved

Land use efficiency of SEDZ should be improved. In primary area the land use efficiency score was only 47.31%, in development direction area the score was 79.5%. There was a big gap between the land use efficiency indexes and the ideal values.

3.4.3 Management performance was well, no idle land

Through investigating, up to the evaluation point, there was no idle land in SEDZ, Management performance attained the ideal value.

4 Conclusions

The score of intensive land use score of SEDZ was 84.89%, indicated a significant intensive land use level of SEDZ. The intensive degree of the primary area (85.54%) was higher than the development direction area (82.29%). In SEDZ, land use was in good condition, management performance was good, and there was no idle land, but the efficiency of the land should be improved.

Acknowledgements

This research was financially supported by Science Foundation of Hengyang Normal University (No.13B51) and Hengyang Social Science Foundation (No.2016D101).

References:

[1] Guo K L, 2016. A Research on Evaluation and Optimization of Land Intensive Use in Development Zones ——A Case Study of Gedian Economic and Technology Development Zones. *Journal of Taiyuan Normal University (Natural Science Edition)*, 15(1):59-65.

[2] Jia K, Chang, Y., 2016. The Evaluation of Land Intensive Use in Economic Development Zones of Shandong Province. *Journal of Anhui Agricultural Sciences*, 44(1):285-288.

[3] Zhang X P, Lu D D, 2002. Land Use in the Development Areas and Its Interactive Relationship with Regional Development. *Resources Science*, 24(5):32-38.

[4] Zhang X T, Ding J Q, Lei X, 2015. Evaluation Research on Intensive Use of Development Zone Land Under Background of "New Normal". *Journal of Northwest A&F University(Social Science Edition)*, 16(2):35-42.

Table 1. Land intensive use evaluation index system, weight, ideal value and intensive degree in SEDZ

Evaluation scope	Weights	Objective	Weights	Sub-objective	Weights	Index	Weights	Present value	Ideal value	standardized value	Sub-objective score	Objective score	Evaluation range score	Comprehensive score
Primary area	0.8	land use status	0.75	Degree of land development	0.2	Land supply rate	0.4	98.92	100	98.92	99.57	91.26	85.54	84.89
						Land completion rate	0.6	100	100	100				
				Status of land use structure	0.25	Rate of industrial land	1	57.88	60	96.47	96.47			
				Land use intensity	0.55	Comprehensive volume ratio	0.25	0.74	1	74	85.87			
						Building density	0.15	68.83	68.83	100				
						Comprehensive volume ratio of industrial land	0.28	0.76	0.9	84.44				
						Industrial land building density	0.32	49.38	55	89.78				
		land use efficiency	0.15	Land input and ouput efficiency of output	1	Intensity of investment in fixed assets of industrial land	0.5	1635.93	3000	54.53	47.31	47.31		
						Tax per industrial land	0.5	160.38	400	40.1				
		management performance	0.1	Land use regulatory performance	1	Rate of idle land	1	0	0	100	100	100		
Development direction area	0.2	Status of land use	0.75	Degree of land development	0.2	Land development rate	1	92.34	95	97.2	97.2	80.49	82.29	
				Status of land use structure	0.25	Rate of industrial land	1	62.22	62.22	100	100			
				Land use intensity	0.55	Comprehensive volume ratio	0.25	0.62	1	62	65.55			
						Building density	0.15	28.76	32	89.88				
						Comprehensive volume ratio of industrial land	0.28	0.61	0.9	67.78				
						Coefficient of building occupation of industrial land	0.32	30.23	55	54.96				
		Land use efficiency	0.15	Land input and output efficiency output	1	Intensity of investment in fixed assets of industrial land	0.5	2559.85	3000	85.33	79.5	79.5		
						Tax per industrial land	0.5	294.66	400	73.66				
		Management performance	0.1	Land use regulatory performance	1	Rate of idle land	1	0	0	100	100	100		

IOP Conf. Series: Earth and Environmental Science **86** (2017) 012014 doi:10.1088/1755-1315/86/1/012014

The Response and Repairing of Three Kinds of Crops on Xi'an's Sewage Irrigation Area Soil

H Xin, Z Zhimei* , H Lei, L Huan and Z Tian

1.Shaanxi Key Laboratory of Earth Surface System and Environmental Carrying Capacity, Northwest University, Xi'an 710127,China
2.College of Urban and Environmental Science, Nrothwest University, Xi'an 710127,China

National Natural Science Foundation of China (Grant No.30700648)
E-mail: zhzhmx@126.com
Abstract. This paper focuses on the XiChaZhai village's vegetable soil which is located in the northern suburbs of Xi'an and on its vegetables, thus analyzes the quality of sewage irrigation region soil and its influence on vegetables through the measurement of Cu, Zn, Pb, Cd's content in samples. The results show that the research area soil contains apparently excessive heavy metals, and there exists significant differences of different elements' integrated intensity in soil, the content declines in sequence from Cd, Zn, Pb to Cu. The four heavy metals' contents in sewage irrigation region soil vary greatly from that in non-sewage irrigation region soil($P<0.01$) . *Raphanus sativus* and *Ottelia acuminate* have favorable effects on Cd and Cu's accumulation. Three crops having repairing effects on Xi'an sewage irrigation region soil are *Raphanus sativus*, *Ottelia acuminate* and *Brassica chinensis*, in that order. Different crop tissues differ in the accumulation of heavy metal, the order according as roots, stem and leaves, fruits. Therefore, based on differences of various crops on heavy metals' absorption and translocation, appropriate crops should be scientifically planted in heavy metal contaminated area soil.

1. Introduction

It has been nearly one hundred years since the usage of industrial wastewater and domestic sewage on agricultural irrigation appeared around the world. In the increasingly mature process of wastewater irrigation technologydevelopment, especially in America, Japan and Israel are most representative. At present, in China's farmland Irrigation, about 7.3% of which is at sewage irrigation level, and the proportion is increasing.

Long-term wastewater irrigation not only nourishes but also changes the property of soil, and turns the common soil into sewage irrigated soil. In the spirit of saving water, people irrigate farmland with sewage, while at the same time some environmental issues begin to show especially out. Research of heavy metals in sewage irrigation area carried out by Sun Libo et al found that, when heavy metals Cr, As, Hg etc. accumulated in the soil, combined with long-term wastewater irrigation, these contaminants will damage the biotic environment through underground water and food chain, and then threaten human health. For example, wastewater irrigation turns wheat color into black and rots the vegetables which become harder to store.

In recent years, research has found that repairing efficiency can be improved through enhanced measure. While as a kind of assistance measure, enhanced measure does not bring high repairing efficiency, on the contrary, it increases the pollution risk. Several scholars including Wang Fenghua et al pointed out that, Cr's toxicity and bio-availability in soil can be effectively reduced by means of microorganism's reduction and bio-sorption etc., which points out a new way for repairing Cr-contaminated soil. Dai Yunchao et al. found that, ameliorant addition can furthest reduce Cd content in pakchoi cabbage and in soil. Although people have found more than 400 kinds of hyper-accumulator, the majority of them focus on repairing Ni-pollution in the soil. In addition, ramie can well repair heavy metal contaminated soil. On the repairing of Zn and Cd, Baker and Brown et al found that Thlasi caerulescens has a higher efficiency. In phytoremediation technology studies, Yang Xiao-e and Wei Shuhe et al. respectively studied Zn-hyperaccumulator Sedum alfredii and Solanum nigrum which can well repair Cd-contaminated soil. Liu Jiali et al. found that heavy metals' content of vegetables in sewage irrigation area is related to various factors, including vegetable kinds, air mass and water quality. It refers not only to kinds of vegetable planted, but also to air pollution status, wastewater quality and property of sewage irrigation area. This paper focuses on the XiChaZhai village's farmland which is located in the northern suburbs of Xi'an, studies heavy metals content respectively in roots, stems, leaves and fruits of various crops planted here, and contrastively analyzes the accumulation and repairing effects of different crops and the potential ecological hazard evaluation. In order to understand the crops' response regularity on heavy metals, and provide theoretical foundation for evaluating soil environmental quality of Xi'an's sewage irrigation area and choosing crops for planting in sewage irrigation area.

2. The research area and method

2.1. General situation of the research are

Sewage irrigation has a long history and a relatively larger scale in Xi'an. Xi'an's sewage irrigation area are mainly including Fenghui sewage irrigation area, which is located in Weiyang District ——a suburb northeast of the city of Xi' an, and is located at 108 degrees, 47 minutes and 8 seconds to 109 degrees, 2 minutes and 21 seconds east longitude, and 34 degrees, 14 minutes and 50 seconds to 34 degrees, 26 minutes and 22 seconds north latitude, covers an area of 251 km2. And its soil belongs to loess parent material.

2.2. Sample collection and treatment Sample collection and treatment

We choose Xichazhai Village vegetable garden and clean soil apple garden as the research area, which is located in Shihua Road, Weiyang District, northern suburbs, Xi'an. For its long-term sewage irrigation history and a certain scale. In August 2015, the soil and crops of research area were gathered respectively, they are in the same area and they were irrigated by commensurate wastewater at the same time, without going through fertilizer and pesticide treatment, (Therein, the gathered background apple garden soil is from the same area which was irrigated by underground water, and without going through pesticide spraying.) the surface layer（0 - 20 cm) of soil was taken and mixed to constitute the mixed soil sample. In the same sampling plot, roots, stems, leaves and fruits of three kinds of crops ——including *Ottelia acuminata, Raphanus sativus* and *Brassica chinensis* (from 5-10cm in diameter) were respectively gathered to constitute the composite sample, which separates the aboveground part from underground part. Then let the samples dry and grind them, sifting through a 100mm nylon sieve and keeping them under seal in a polyethylene bag.

2.3. Experimental method

The soil heavy metals adopts the resolution method of HCl-HNO3-HF-HClO4 full-decomposition, and the roots, stems, leaves and fruits of crops adopt the resolution method with the ratio of nitric acid/ perchloric acid being 9:1, then make them into solutions and measure four heavy metals——including Cu, Zn, Pb and Cd's contents through atomic absorption spectrometry method.

2.4.Data Processing

Heavy metals' contents in soil and their differences among sewage irrigation area, non-sewage irrigation area and background soil were analyzed through SPSS 11.0 software calculation. Takes soil background value of Shaanxi province as the background soil standard. (See Table 1)

IOP Conf. Series: Earth and Environmental Science **86** (2017) 012014 doi:10.1088/1755-1315/86/1/012014

Assessment of potential ecological risk method: Nemerow pollution index method. The Nemerow pollution index evalution criteria of soil can be seen in Table 2.

Table 1 Soil background value of Shanxi province (mg/kg)

Project	Cu	Zn	Pb	Cd
Background value of soil element	32	74.2	29.5	0.15

Table 2 The Nemerow pollution index evalution criteria of soil

GradNemerow	Pollution index (P_N)	Class pollution
I	$P_N \le 0.7$	Clean (Safe)
II	$0.7 < P_N \le 1.0$	Slight clean level
III	$1.0 < P_N \le 2.0$	Mild pollution
V	$2.0 < P_N \le 3.0$	Medium pollution
IV	$P_N > 3.0$	Heavy pollution

3. Results and analys

3.1.The heavy metals content of experimented

The heavy metals content of experimented soil is shown in table 3. Taking the heavy metal background value in Shanxi as the evaluation standard, the result reveals that the average contents of Cu, Zn, Pb, Cd in soils planted with *Ottelia acuminata*, *Raphanus sativus*and*Brassica chinensis* are all higher than the background value with the maximum value of Cu, Zn, Pb, Cd being respectively 1.25, 1.11, 1.15 and 3.82 times of the background value; In the soil of the apple orchard, only the content of Cu slightly exceeds the standard and the contents of other 3 heavy metals are all within the background value.

Table 3 the heavy metals content of experimental soil(mg /kg)

Index	Heavy metal content			
	Cu	Zn	Pb	Cd
Soil A	36.160±3.110	78.364±5.001	22.621±4.081	0.452±0.089
Soil B	26.721±7.333	77.382±8.191	24.624±4.060	0.492±0.081
Soil C	22.324±3.202	75.437±6.533	20.235±1.623	0.457±0.083
Soil D	34.240±2.500	59.334±9.912	28.723±6.942	0.145±0.019

The soil A represents *Ottelia acuminata* soil, soil B for *Raphanus sativus* soil, soil C for *Brassica chinensis* in soil, soil D for in apple orchard soil , soil E for background soil, the data is presented as mean plus or minus standard deviation, The same below.

From the table above, it can be learned that the content of Cd in the experimented soil severely exceeds the standard, and that of Cu, Zn, Pb also exceeds slightly.

3.2 Studies on the dfference of the heavy metal content in experimented Soils

By taking F test, this paper studies on the difference of heavy metal content in the area of the sewage irrigation and the non-sewage irrigation area and the background soil. The result shows that the contents of 3 heavy metals (Zn, Pb, Cd) in the sewage irrigation area planted with Ottelia acuminata differ a lot with that in the non-sewage irrigation area and the soil background value ($P<0.05$). The contents of 4 heavy metals in the sewage irrigation area planted with Raphanus sativus are all obviously higher than that in the non-sewage irrigation area and the background value; the contents of Zn and Cd in soils planted with Brassica chinensis are significantly higher than that in the non-sewage irrigation area, but the contents of other 2 heavy metals have no remarkable difference between these two areas.

Table 4 Study on the difference of heavy metal content in the area of the sewage irrigation and the non-sewage irrigation area and the background soil (mg /kg)

Index	Heavy metal	Soil D Cu	Zn	Pb	Cd	Soil E Cu	Zn	Pb	Cd
Soil A	Cu	0.079	-	-	-	0.228	-	-	-
	Zn	-	0.000**	-	-	-	0.027*	-	-
	Pb	-	-	0.050*	-	-	-	0.000**	-
	Cd	-	-	-	0.000**	-	-	-	0.000**
Soil B	Cu	0.008**	-	-	-	0.049*	-	-	-
	Zn	-	0.000**	-	-	-	0.000**	-	-
	Pb	-	-	0.002**	-	-	-	0.004**	-
	Cd	-	-	-	0.000**	-	-	-	0.000**
Soil C	Cu	0.057	-	-	-	0.086	-	-	-
	Zn	-	0.001**	-	-	-	0.000**	-	-
	Pb	-	-	0.232	-	-	-	0.000**	-
	Cd	-	-	-	0.000**	-	-	-	0.000**

$*P<0.05$; $**P<0.01$.

From above data, it can be concluded that Raphanus sativus can well accumulate the 4 heavy metals and Ottelia acuminate can accumulate 3 of these heavy metals (Z, Pb, Cd).

3.3 The effect of different parts of plants on the accumulation of heavy metals

From Figure. a, we can see that the roots and stems of these three plants absorb nearly the same content of Cu, the maximum reaching 5.234 mg/kg by the roots of *Ottelia acuminate*; the leaves of *Brassica chinensis* comparatively absorb less Cu than that of the other two plants; the fruits of *Ottelia acuminate* and *Brassica chinensis* absorb almost the same amount of Cu; the fruits of *Raphanus sativus* absorb relatively less Cu than that of the other 2 plants, reaching the minimum of 0.926 mg/kg; therefore, the accumulation of Cu in soils planted with different plants can be ranked as as *Ottelia acuminate, Raphanus sativus, Brassica chinensis* (from high to low); the distribution in tissues of different plants can be ranked from high to low as root, stem,leaf and fruit.

From Figure. b, we can know that the accumulation of Zn in soils planted with different plants is *Ottelia acuminate, Raphanus sativus, Brassica chinensis* (from high to low); the accumulation amount of Zn by the roots of the three plants is *Ottelia acuminate, Brassica chinensis, Raphanus sativus* (from high to low), and the maximum reaches 30.568 mg/kg by the roots of *Ottelia acuminate*; the stems and fruits of the three plants absorb nearly the same amount of Zn; the leaves of the three plants absorb relatively less Zn than the stems and roots do; therefore, the accumulation of Zn in different parts of different plants can be ranked as root, fruit, stem, leaf (from high to low); the repair effect on Zn is *Raphanus sativus, Ottelia acuminate, Brassica chinensis* (from high to low).

From Figure. c and Figure. d, we can see that the accumulation of Pb and Cd by the roots, stems, leaves and fruits of these three plants is relatively low. The amount of Cd absorbed by these three plants is slightly greater than that of Pb.

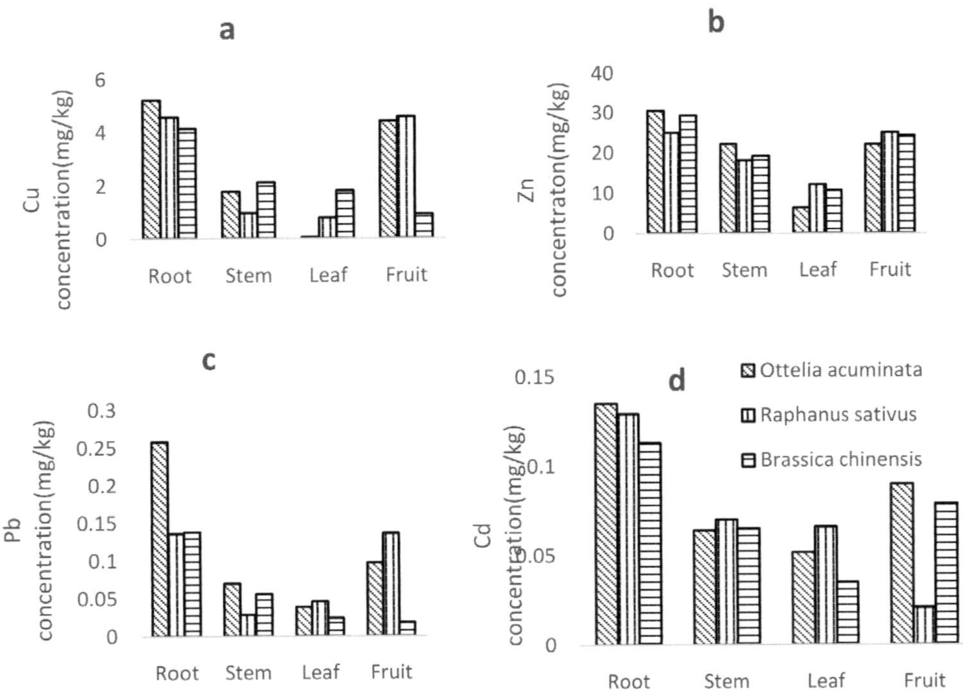

a represents the content of Cu, b is Zn, c is Pb and d is Cd

Figure.1 Distribution of heavy metals in different tissues of plants

3.4.The potential ecological hazard

From table 5, it can be learned that the Nemerow Pollution Index of Cd in soils planted with *Ottelia acuminate* is higher than 3, which is heavy pollution, and the indexes of Cu, Zn and Pb are between 0.98 and 1.24, being at the level of clean but already at the edge of the warning line. In soils planted with *Raphanus sativus*, the pollution indexes of Cu and Zn are both higher than 1, reaching the level of mild pollution; the pollution index of Pb is between 0.82 and 1, which is at the level of clean but is still at the edge of the warning line; the index of Cd is, however, higher than 4, reaching the level of heavy pollution. In soils planted with *Brassica chinensis*, the index of Cd is higher than 3, which is heavy pollution; the other 3 are at the level of mild pollution. Besides, the heavy metal pollution indexes in the apple orchard are all lower than 1, which means they are stable and will not threaten the environment.

In this soil sampling, the soils of *Ottelia acuminate, Raphanus sativus* and *Brassica chinensis* are collected from the sewage irrigation area and the soils of apple orchard from tap water irrigation area with a cleaner environment. According to Nemerow comprehensive pollution index, the soils of *Ottelia acuminate, Raphanus sativus* and *Brassica chinensis* are all suffered from heavy pollution; the heavy metal pollutants are mainly resulted from the irrigation using industrial and sanitary wastewater, which intensifies non-point sources pollution in soils. The soils in the apple orchard, however, are safe. In this soil sampling, the soils of *Ottelia acuminate, Raphanus sativus* and *Brassica chinensis* are collected from the sewage irrigation area and the soils of apple orchard from tap water irrigation area with a cleaner environment. According to Nemerow comprehensive pollution index, the soils of *Ottelia acuminate, Raphanus sativus* and *Brassica chinensis* are all suffered from heavy pollution; the heavy metal pollutants are mainly resulted from the irrigation using industrial and sanitary wastewater, which intensifies non-point sources pollution in soils. The soils in the apple orchard, however, are safe.

Table 5 Pollution index of heavy metals in the study area

Index	Heavy metals	Pollution index	Pollution situation
Soil A	Cu	1.172±0.0230	Mild pollution
	Zn	1.061±0.098	Mild pollution
	Pb	1.005±0.045	level of clean
	Cd	3.423±0.076	heavy pollution
Soil B	Cu	1.070±0.045	Mild pollution
	Zn	1.278±0.020	Mild pollution
	Pb	0.912±0.0431	level of clean
	Cd	4.230±0.0234	heavy pollution
Soil C	Cu	1.211±0.0789	Mild pollution
	Zn	1.472±0.067	Mild pollution
	Pb	1.162±0.056	Mild pollution
	Cd	3.53±0.043	heavy pollution
Soil D	Cu	0.890±0.076	level of clean
	Zn	0.685±0.024	Clean (Safe)
	Pb	0.32±0.01	Clean (Safe)
	Cd	0.19±0.02	Clean (Safe)

4.Conclusion and discussion

(1) The content of heavy metals in soils of the experimented area remarkably exceeds the standard. The content of Cd exceeds severely and that of Zn and Cu also exceeds slightly. The accumulation of different elements in soils differs a lot and can be ranked as Cd, Zn, Pb, Cu (from high to low).

(2) The content of the four heavy metals in the sewage irrigation area is significantly different from that in the non-sewage irrigation area. *Raphanus sativus* can well accumulate Cd and *Ottelia acuminate* can enrich and accumulate Cu.

(3) The soils of *Ottelia acuminate, Raphanus sativus* and *Brassica chinensis* are all heavily polluted areas, but the soil of apple orchard is currently within a safe range.

(4) The accumulation amount of heavy metals in different parts of plants is root, stem, leaf, and fruit (from high to low). The repair effect on the soil of the studied area is *Raphanus sativus*, *Ottelia acuminate,Brassica chinensis* (from high to low), therefore, more *Raphanus sativus* and *Ottelia acuminate* should be planted in Xi'an's sewage irrigation areas since they are able to repair the heavy metal-contaminated soil and their fruits, with a relatively lower accumulation of heavy metals, are safe to eat.

(5) The heavy metals in Xi'an's sewage irrigation area have obviously accumulated in vegetables but have not yet influence the appearance of vegetables. It is similar to the study result of this paper. The impact of the accumulation of heavy metal on vegetables and its physicochemical property still remain to be further studied.

References

[1] Liu Runtang,Xu Jianzhong.Sewage water irrigation problems and solutions in China[J]. China Water Resources,2002,10(1):123-125.

[2] Pang Jiangli,Huang Chunchang,Sun Gennian.Heavy metal elements content of sewage irrigated soil in Xi'an and it's influence to quality of vegetable[J].Journal of Shaanxi Normal University,2000,22(3):55-57.

[3] Sun Libo, Guo Guanlin,Zhou Qixing,.Assessment of heavy metals and POPs contamination in a sewage irrigation area[A].Chinese Journal of Ecology,2006,25(1):29-33.

[4] POCIECHA M, LESTAN D. Novel EDTA and process waterrecycling method after soil washing of multi-metal contaminated soil [J]. Journal of Hazardous Materials, 2012,20(2):273-279.

[5] Wang Fenghua,Luo Xiaosan,Lin Aijun.Remediation of Chromium-Contaminated Soil[A].Asian Journal of Ecotoxicology,2010,5(2):153-161.

[6] Dai Yunchao,Lv Jialong,Diao Zhan.Effects on Soil Amendments on Cd Bioavailability to and Uptake by *Brassica chinensis*in Different Dd-contaminated Soil[A].Journal of Agro-Environment Science,2015,34(1):80-86.

[7] Liu Jianv,Zhou Qixing,Sun Ting,.Feasibility of applying ornamental plants in contaminated soil remediation[A].Chinese Journal of Applied Ecology,2007,18(7) :1617-1613.

[8] Wang Kairong,Gong Huiqun.Absorption and the effect of cleaning up of ramie on soil cadmium[J].Acta Scientiae Circumstantiae, 1998,18(5):510-516.

[9] She Wei,Jie Yucheng,Xing Hucheng.Uptake and Accumulation of Heavy Metal by (Boehmeria nivea) Growing on Antimony Mining Area in Lengshuijian City of Hunan Province[J].Journal of Agro-Environment Science, 2010, 29(1):91-96.

IOP Conf. Series: Earth and Environmental Science **86** (2017) 012014 doi:10.1088/1755-1315/86/1/012014

[10] BakerAJM, McGrath SP, Sidoli CMD,et al. The possibility of in situ heavymetal decontamination of polluted soils using crops of metal-accumulating plants [J]. Resources Conservation and Recycling,1994,11(1): 41-49.

[11] Brown SL, Chaney RL, Angle JS, et al.Zincand cadmium uptake by hyper accumulator Thlaspi caer-ulescensand metaltolerant Silene vulgarisgrown onsludge-amended soils[J]. EnvironmentalScience and Tech-nology,1994,29(6): 1581-1585.

[12] Yang Xiaoe,Long Xinxian,Ni Wuzhong.*Sedum alfredii H*-A new zinc hyperaccumulator[J]. Chinese Science Bulletin,2002,47(13): 1003-1006.

[13] Wei Shuhe, Zhou Qixing,Wang Xin.A newly discovered cadmium hyperaccumulator-*Solanum nigrumL*[J].Chinese Science Bulletin,2005,50(1): 33-38.

[14] Lv Jiali,Liu Xiaojuan,Wang Zhentao.Study on Available Contents of Cd and Pb in Sewage-irrigated Soils in Taiyuan[A].Journal of Shanxi Agricultural Sciences,2015,43(12):1632-1637.

[15] Wang Mei,Jiang Lihua,Liu Zhaohui,.Impacts of petroleum pollutants on microbial population and enzyme activity in three different types of soils in Shandong Province[A]. Acta Pedologica Sinica ,2010,47(2):200-208.

[16] Li Peijun,Gong Fanqiang,Zhang Hairong.Bioremediation of petroleum-contaminated soil and its relationship with soil enzyme activities[A].Chinese Journal of Ecology,2005,24(10):100-104.

[17] Pang Jiangli,Huang Chunchang,Sun Gennian. Heavy metal content of sewage irrigated soil in Xi'an and it's influence to tomato[J]. Soil and Environmental Science,2001,(02):94-97.

The Operation Method of Smarter City Based on Ecological Theory

C Fan

School of Economics and Management ,BeijingJiaotong University, China

Abstract: As the city and urbanization's accelerated pace has caused galloping population, the urban framework is extending with increasingly complex social problems. The urban management tends to become complicated and the governance seems more difficult to pursue. exploring the urban management's new model has attracted local governments' urgent attention. tcombines the guiding ideology and that management's practices based on ecological theory, explains the Smarter city Ecology Managementmodel's formation , makes modern urban management's comparative analysis and further defines the aforesaid management mode's conceptual model.based on the smarter city system theory's ecological carrying capacity, the author uses mathematical model to prove the coordination relationship between the smarter city Ecology Managementmode's subsystems, demonstrates that it can improve the urban management's overall level, emphasizes smarter city management integrity, believing that urban system's optimization is based on each subsystem being optimized, attaching the importance to elements, structure, and balance between each subsystem and between internal elements. Through the establishment of the smarter city Ecology Managementmodel's conceptual model and theoretical argumentation, it provides a theoretical basis and technical guidance to that model's innovation.

1.Introduction:

As the city and urbanization'saccelerated pace has caused galloping population, the urban framework is extendingwith increasingly complex social problems. The urban management tends to become complicatedand the governance seems more difficult to pursue. From this issue, exploringthe urban management's new model has attracted the local governments' urgent attention. Cities like a machine, only normal operation can make it play a good effect. The city, through the management, fully mobilizes the urban subject - people's enthusiasm and self-consciousness. The human factor isorganically combined with a variety of other things'factors, thereby promoting each system's urban normal effective operation. But the city will encounter a series of problems during the operation such as production, employment, safety, ecological environment and other issues. Itwill be solved by strengthening the city's integrated management.The traditional urban management is centered on things like technology, production and other things' management, with executive order institutional constraints as leading management, but blindly seeking for the normative systems' establishment will not alleviate urban problems.

The current state smarter city management-related standards are not clear. The social management's relevant departments has the common "fragmentation", "separation" and "repeated construction"phenomenon. Faced with these problems, it urgently needs the application of a scientific theory combining withsmarter city information means in the society's business management in order to effectively solve the social management problem, improve economic management efficiency, enhance public services, facilitate people's livelihood communication, increase system integration and resource sharing strength, bring into play of smart ecological construction overall effectiveness and improve work efficiency and management level. Based on ecological theory as the guiding ideology, it is proposed the urban ecology management theory and methodscombination, the smarter city ecology management's conceptual model, theoretical basis and technical guidance to its innovation.

2. Reconstruct The Modern Urban Management Model

2.1.Modern Urban Management's Problems Analysis

People continue to pursue the urban development concept such as the smart, green, digital and eco-city, and its essence is to bring more peace and well-beinginto the cities, but things are not always as people imagine. With the modern cities' rapid development, the city is full of contradictions and ill-posed problem.The problem not only reduces the urban residents' life quality, but also directly affects the city's economic development, social stability and so on. Urban management model change is the only way for people to solve their problems. It is a series of activities aimed at improving the function and efficiency of urban systems. Due to various disciplines emphasizing different angles and directions, it has caused some problems. Through the traditional urban management model studies, we can see that it faces the following obvious questions.

(1) Various urban systems'uneven ecology development. Urban development has made great progress; not only enormous changes in the original city has taken place, but also a large number of new cities have undergone the rapid development just like the bamboo shoots after a spring rain. Urban planning attaches importance tophysical form but neglects human nature demandswhich is observed inthe natural environment and humanization's fusion defects (lack)which results in laying more emphasis on economic development, neglecting society and the people's livelihood, urban public facilities and infrastructure's lack, green space reduction, road congestion, environmental degradation and other discordant urban problems. Formerlyin the urban planning and construction, those people more respected those breakthrough achievements at the "visible", "tangible" material and hardwarelevel which is regarded as the smart city's fundamental signs. It has resulted in the "big engineering", "big project", "large scale" and "generous" urban planning'sblind pursuit with the urban population reaching hundreds of millions and the urban nature apt to engage in "an international metropolis". People, in the high material civilization's pursuit and enjoyment , ignore the social, infrastructure and natural ecological development. Some citieseven exchange for a "temporary prosperity" of the city's economy at the expense of the natural and ecological environment.

(2) Lack in urban resources' economic development

1. Various cities are similar in industrial structure, with repeated construction prevalence and a

large number of overcapacity enterprises. Whether the cities are in north or south, and regardless of their strengths, their industry structure generally has equipments manufacturing, chemicals, electronic communications and other industries, which is very similar and the similarity coefficient between some cities reaches 0.8-0.9. Great similarity between the industrial structure restricts the economic links between cities, hinders the city's economic development and also causes a vicious competition between cities.Unobvious leading industry advantage: Urban economy leading industry did not achieve the desired results andthe growth brought about by the institutional changes is far greater than the leading industry growth strategy'simpact. Many cities made a mistake in the leading industries choice.Tertiary industry lags behind with prominent internal structural contradiction. This not only affects the urban residents' life and improves the overall economic benefits, but also exacerbates the already severe employment situation. Urban economic innovation capability is not strong, with low economy knowledge levels. Due to the science and technology's imperfect and uneven development, not scientific and enough rational government functions and many other factors, technological development and transformation, science and technology input, technology's economical contribution, and introduction, digestion and absorption to technology, most cities are far fromthe smart cities' economic goal objectives requirements.

(3) Urban society's Lower overall leve

The construction of human resources is backward, due to the fact that the industrial structure of some cities is single, the urban economic foundation is weak, the labor market development is not perfect, and the social security system of workers has not been fully established. Management level and the legal system construction is not perfect, in recent years, according to the unified requirements of the central government of the city government has carried out many reforms, but the overall effect is still not obvious. Mainly in the management of large institutions, and low efficiency, more personnel than work available management methods and procedures are not scientific. At the same time in the construction of city laws and regulations on the one hand is not perfect, the law, knowingly violate the law, breach of privilege etc. on the other hand is quite prominent, provide a hotbed for corruption, and seriously affected the overall image of the city [2]. Social security system is not perfect. In recent years, the city has carried out the construction of labor insurance, unemployment insurance, medical insurance, city social security system, but in general, the coverage is not complete, the level of assistance is low, has become a key problem in city social stability and restricting the healthy development of the city society

(4) Inadequate urban infrastructure, unreasonable internal structure

The level of urban infrastructure can not meet the needs of urban development. The city government made great efforts in city infrastructure, to some extent alleviate the serious shortage of city residents in the electricity, gas, water supply and drainage, transportation and other infrastructure of the city, but the wisdom of the city and the demand is still a big gap. Urban infrastructure investment and construction of the structure is not reasonable. The inherent law of urban infrastructure requires that all the internal systems and departments should be reasonably considered in accordance with a certain proportion of science.

(5) The urban natural environmental protection is not seriously taken

Urban natural ecological situation is still very grim. Many cities, especially large ones, in terms of air and water quality, green space, biodiversity and urban environmental situation,are still in a worsening trend.The city's artificial environment and natural coordination are not scientific enough. Because of the urban landscape design and important buildings and structures' design and construction, the leaders make more decisionswithout enough emphasis on the experts and

IOP Conf. Series: Earth and Environmental Science **86** (2017) 012015 doi:10.1088/1755-1315/86/1/012015

scholars' opinions, the residents' views ignored or omitted and this results in uncoordination between urban construction's artificial and local natural environment which has led to waste. Shortage in urban resources. Due to the natural resources'uneven distribution in the city, the most prominent is lack in the drinking water resources, mineral resources, natural gas and other natural resources which is an important factor that constrains the rapid urban development.

3. Ecological model of smart city operations

As for the issues existing in the traditional urban system construction and development aspects, the smart city construction and management should change the traditional urban management model and develop towards the direction more conducive to the sustainable development. Smart cities are essentially different from the traditional ones in terms of pattern, structure, function or other aspects, and the traditional urban system construction city is built on the basis of the industrial civilization and technological progress era's value concept , which is passive reaction to the various urban people's counter-industrial civilization problems, which is not enough for the smart city management development, so we must change our urban management mode, under the guidance based on ecological values and create its smart ecology's theoretical framework.

3.1.Autogenesis to Symbiosis

Facing the problems like economic, social, infrastructural, natural ecosystems that are not development coordinated, the environment deterioration, the economic growth's extensive mode, serious waste issues, the widening gap between rich and poor in the urban construction and development process, in order to achieve the overall healthy urban system development, change the traditional autogenic relationship in urban construction process, involving the city's material and spiritual aspects, both conducting organic matter updates for economic and ecological systems and infrastructure at the material level, but also build social and natural ecosystems in line with the ecological theory construction, establish a social and natural ecosystem in line with the ecological theory, build a mutual equality, harmonious symbiotic relationship between man and nature, person and person and various subsystems in the city. Various subsystems in smart city run along the co-evolution path, these subsystems are in common activation, joint adaptation and joint development cooperation and coordination in the process.

3.2 Artificial Plunder into a Harmonious Balance

Now that there are all sorts of urban problems, it is serious damage to the inner nature's harmony and urbanization's artificial predatory natural results. Although artificialization can improve the human beings viability and the urban environment's regulatory capacity but also causes destruction of the balanced relationship's various elements in the city system. Smart city-building should extend from the past human studies to the study on the human and nature balanced harmony. Therefore, the smart city construction should be based on dual existence of ecological and economic theory, improving the nature's inner harmony.

3.3 Part to Overall Shift

Holistic thinking and decision-making is the only way to realize the smarter city management. Integrity includes comprehensive grasp and balance in social, economic, natural environment and infrastructure facilities construction in urban and rural regions. Throughout the urban construction and development process, the overall progress of the society is the development's fundamental objective, economic growth and efficiency improvement is the way and means of development. Infrastructure conditions are to ensure development while protecting the natural environment condition is the foundation for reaching the overall development. In the construction process, a balance must be achieved in four areas, and no aspect is to be neglected.

3.4 Transition Single to Multi-objective Goals

Due to the fact that those different targets are often conflicting, economic growth ones' one-sided pursuit or environmental quality objectives are inevitably at the expense of other benefits for the price. Therefore, the smart city system construction should not be a single, separate from each other, but should be weighed in a multidimensional space. The target selection is not just one dot or aspect, but is set as a whole of dots or aspects. Table (2-1) shows the comparison results of the conventional urban construction and smarter city Ecology Managementmode.

Project	Traditional urban management model	Smarter city Ecology Managementmodel
Philosophy	Autogenic	Symbiotic collaboration
Values	Labor exploitation	Balanced and harmonious
Content goal	Single goal, the economy	More objective, complex ecosystem
Subject scope	Independent discipline	Multidisciplinary
Decision-making way	Closed, administrative intervention	Open, pluralistic participation
Management procedure	One-way,static	Cycle, dynamic
Management support means	Human manual work	IT intelligence

Table (2-1) Comparison between the conventional urban management model and smarter city Ecology Managementmodel

4. The concept of Intelligent City Management

4.1 The management mode of intelligent city from the perspective of ecology

Ecological management model is a new concept of management, it uses the basic theory of ecology and the basic point of view to observe the relationship between urban system environment. To manage the entire city ecology management system as the research object, to reveal the intrinsic relationships between the organization and the management of the ecological environment, to seek coordinated approach between the organization and management of modern ecological environment in general, is the management system integration, system and organization grasp. Smart city operation method based on the perspective of ecological theory is a new perspective to study the management of the city, it is produced in the city management mode of innovation development and further awareness of ecology, is an inevitable trend of scientific development of the city. The concept of sustainable development of 4] city and ecological city, green city, ecological city shows that the proposed management trend, showing construction ecology management mode is the inevitable further development of city management.

4.2 Role of Smarter city Ecology ManagementMode

Discussion on ecological management model is based on the ecology development; and its construction provides the following functions for reference:

1 provides a new concept of ecological management of organic wisdom City, the realization of the reduction of the concept of mechanical management to the organic holism of the management concept of change, laying the foundation for the ecological management of the world.

In 2 the relationship between understanding and dealing with the organization and management of city environment, the use of ecological thinking, to consider from the perspective of ecology, as the relationship between the organization and management of the city environment is an ecological management system, is an organic whole.

3 essentially, ecology and management have internal unity, which constitutes the ideological basis of ecological management. The urban management is the management of the urban organization in essence, and the city organization is in the specific management environment. The operation of the smart city contains the ecological, organic and unified pursuit of the relationship between the urban organization and the management environment.

5. Ecological Management Mode's Conceptual Model

Smarter city Ecology Managementmodel is based on the fundamental information related to technology, through corresponding manners in monitoring, analysis, integration and wisdom; it synthesizes functional departments, integrates and optimizes existing resources, seeks to provide better service, green environment, harmony society and security for sustainable urban development.

5.1. Smarter city Ecology ManagementMode's Conceptual Model Analysis

According to the theoretical research, to provide better service, green environment, harmonious society to maintain the wisdom of the city system, to ensure sustainable development of the city, the system must realize the optimization of the structure, dynamic and open, can strengthen the control and environment adaptability, input and output to improve the conversion efficiency, through the description of the system of society with the economy, population, resources and environment, mutual adaptation mechanism, appropriate population status, adaptive capacity, environmental resources efficiency, changes in technology applicability, can evaluate the status of city management. Change of city management in the process of the interaction by external forces and internal forces, the changes may be the population migration and cultural concept, the change of industry structure, traffic form and route changes, the development of new technologies and the requirements of social life and other aspects. The goal of urban management in ecological management mode is not to optimize the development of each system in the city, but to have a certain matching relationship under certain constraints. The balance and coordination between the various elements of the city, the basic requirement is not the strength development and production activities in the city more than the ecological support system of resource regeneration and purification ability, and these goals may be affected by the system capacity constraints, in order to overcome capacity constraints, requires the implementation of a series of measures, the city system constantly the development from lower stage to higher stage.

5.2. Conceptual Model Representation

In accordance with the urban system development utility maximization theory and ecological management concept and according to its ecological carrying capacity theory [12], the city complex system's development characteristics are chosen to construct the smart city system's ecological management conceptual model as follows:

$$\max U = \pi_{pt} U_{pt}(t,s) + \pi_{et} U_{et}(t,s) + \pi_{rt} U_{rt}(t,s) + \pi_{ent} U_{ent}(t,s) + \pi_{it} U_{it}(t,s) \quad (4\text{-}1)$$

Constraints:

Population Capacity Constraints $X_{pt} \le X_{p}^{max}$ (4-2)

Environmental Carrying Capacity Constraint $X_{ent} \le X_{en}^{max}$ (4-3)

Resources Carrying Capacity Constraints $X_{rt} \le X_{r}^{max}$ (4-4)

Intergenerational Development Constraints $U(t,s) \le U(t+1,s) \le U(t+2,s) \le \cdots U(t+n,s)$

(4-5)

Inter-city Balanced Development Constraint Utility Equations $\frac{\partial U^2}{\partial t \partial s} \geq 0$ (4-6)

Utility Equation $U_p(t,s) = f_1(x_{pts}, M_{ts})$ (4-7)

$$U_e(t,s) = f_2(x_{ets}, M_{ts}) \quad U_r(t,s) = f_2(x_{ets}, M_{ts})$$

$$U_{en}(t,s) = f_4(x_{ents}, M_{ts}) \; U_f(t,s) = f_2(x_{fts}, M_{ts})$$

BertalanffyEquation

$$\frac{dX_{pt}}{dt} = g_1(X_{PT}, X_{et}, X_{rt}, X_{ent}, X_{ft}, a_1); \qquad (4\text{-}8)$$

$$\frac{dX_{et}}{dt} = g_1(X_{PT}, X_{et}, X_{rt}, X_{ent}, X_{ft}, a_2); \frac{dX_{rt}}{dt} = g_1(X_{PT}, X_{et}, X_{rt}, X_{ent}, X_{ft}, a_3)$$

$$\frac{dX_{ent}}{dt} = g_1(X_{PT}, X_{et}, X_{rt}, X_{ent}, X_{ft}, a_4); \; \frac{dX_{ft}}{dt} = g_1(X_{PT}, X_{et}, X_{rt}, X_{ent}, X_{ft}, a_5)$$

The model variables have the following meanings:t is the time variable, refers to the development stage; s is the space variables; it means different cities in space; U (t, s) is the t-thgeneration of the city's total utility value; U(t,s)is a certain city's t-th generation system elements (p, e, r, en, f respectively refer to population, economy, resources, environment and infrastructure) development utility value; $X_{it}$was the i-th element of the t-th generation for the developmental vector; $M_t$is the weight of the i-th system factor for the effectiveness of the t –thgeneration; $M_t$is the demographic variables which is a function of the population's quantity and quality, namely:

$$M_t = \beta(m_N^t, \; m_Q^t) \quad (4\text{-}9)$$

The objective function indicates that the SmartCity's development pursuit is the largest city overall system total utility value, the total utility value is the weighted average of the each system's utility value. As different generations' different preferences, so the number of generations is not exactly right. For example, in cities of developing countries, the economy is at the less developed stage, the weights given to the utility values' economic growth may be higher than that of other systems' effectiveness. In the city at the highly developed stage, the weights given to the environmental effect may be greater. Thus, the weight is actually a function of the levels of the system development, namely

$\vec{\pi}$— Weights vector; $\vec{\pi} = (\pi_{1t}, \; \pi_{2t}, \; \pi_{3t}, \; \pi_{4t}, \; \pi_{5t})$; (4-10)

X_t- Each system's development level vector. $X_t = (X_{Pt}, \; X_{et}, \; X_{rt}, \; X_{ent}, \; X_{ft})$

In general, demographic, social, environmental, infrastructural and cultural subsystem is an increasing function of system $\pi_t$at time t while the economic system π_t is a decreasing function of time, i.e.

$$\frac{d\pi_{ft}}{dt} > 0, \; \frac{d\pi_{et}}{dt} < 0, \; \frac{d\pi_{rt}}{dt} > 0, \; \frac{d\pi_{ent}}{dt} > 0, \; \frac{dX_{ft}}{dt} > 0$$

In formula (4-2), (4-3), $X_p^{max} X_r^{max} X_{en}^{max}$ respectively refer to smart city population capacity, resources carrying capacity and environment carrying capacity thresholds (4-4).

Intergenerational development constraints means that the future generations' total utility is at least not less than the contemporary generation's one. Intercity development constraints take a meaningthat each city's subsystem development does not undermine the other cities' overall effectiveness.

Utility equation represents the various subsystems' utility, which is a function of each subsystem's development level and the demographic variables.Bertalanffy relationship equation indicates that in the smart city systems under ecology management mode, any change in one element is relevant with the levels of other relevant factors' development and leading to changes in them and changes in the whole system, thereby reflecting holistic thinking of the ecological theory.

6. Conclusion:

Smart safe ecological mode concept acts as an example of innovative approaches to urban management, which is not perfect, but realizes the reflection of the application value of smarter city management methods, including the overall building ideas. Based on the conceptual model, smarter city Ecology Managementmode's definition, analysis, research and demonstration, it suggests that smart city ecology management model is a multi-dimensional aspect with a stage characteristics and regional differences, and it requires the coordinated development and overall improvement of the various parts in the smarter city management.Here we can give some examples of this aspect, namely the improvement of the overall urban management's level.While managing potential, it also gets further improvement to the city's management coordination and the smarter city management process is demonstrated to be volatile.It believes that each system, in case that it does not exceed the threshold, has the substitutability between various management indicators. It emphasizes on the smart city managementintegrity, believing that the urban system optimization is based on that of the various subsystems, thus calling attention to the balance of the internal components and the structure within those subsystems. SmartCity Ecology Management Model integrates the urban resources to achieve the city administration changing from "extensive management" to "fine-based management" in order to better safeguard the urban social order, promote economic development and improve the human life quality.

References:

[1] Liu Yungang.Development Mechanism and Regulation Countermeasures Study of Chinese Resource-Based Cities. Doctoral Dissertation of Northeast Normal University 2002,6: 95-99

[2] Jing Xingrong, Zhang Jian, Fan Yanni. Eco-City and Urban Ecology System Theory, Urban Issues.2004,6:20-24

[3] Liu Xihan, Xie Yi, Cycle Economic Analysis Based On the Perspective of Ecology, Productivity Research, 2010 (11) :18-20

[4] Chen Kankan, Strategic Planning Research on Medium and SmallCities in Southwest China from Ecological Perspective [D]. Doctoral Dissertation of Faculty of Architecture and Urban Planning, ChongqingUniversity, 2012, 5

[5] He Xiaodong, ZhengYuantong. Thoughts on Constructing the Low-Carbon City in Medium and Small Cities in the Linkage Context in Western Cities [C]. Symposium on Urban Development in the Thirteenth Annual Conference of China Association for Urban Development in An Ecological Civilization Horizon, 2011

[6] Liu Yufeng, Research on Evaluation Theory and Method Research of Regional Sustainable Development, Journal of DalianUniversity of Technology, 2003,24 (2) :52-57

[7] Li Xiaoyan, Chen Hong, Theoretical Framework of Urban Ecological Transportation Planning. Journal of Chang'an University .2006 (1):79-82

[8] Yu Taofang, City Competition and Competitiveness [M] Nanjing: Southeast University Press, 2004, 3: 17.

[9] GuChaolin. Thoughts on Transformation Development and FutureCities, Urban Planning 2011, (11) 8-10

[10] Lv Bin, Capacity Analysis and Ecological Urban Planning, China Low-Carbon Eco-City Development Report [M]. China Society fo Urban Studies, 2010

[11] Sun Yanqing, Green Urban Design and Regionalism Dimension [D]. Doctoral Dissertation of TongjiUniversity, 2007

IOP Conf. Series: Earth and Environmental Science **86** (2017) 012016 doi:10.1088/1755-1315/86/1/012016

The Simulation Realization of Pavement Roughness in the Time Domain

H L XU, L He and D An

School of Civil Engineering, North China University of Technology, Beijing 100144, China

E-mail: xuhailiang@ncut.edu.cn

Abstract. As the needs for the dynamic study on the vehicle-pavement system and the simulated vibration table test, how to simulate the pavement roughness actually is important guarantee for whether calculation and test can reflect the actual situation or not. Using the power spectral density function, the simulation of pavement roughness can be realized by Fourier inverse transform. The main idea of this method was that the spectrum amplitude and random phase were obtained separately according to the power spectrum, and then the simulation of pavement roughness was obtained in the time domain through the Fourier inverse transform (IFFT). In the process, the sampling interval (Δl) was 0.1m, and the sampling points(N) was 4096, which satisfied the accuracy requirements. Using this method, the simulate results of pavement roughness (A～H grades) were obtain in the time domain.

1. Introduction

Pavement roughness was an important controlling indicator of pavement engineering design, pavement construction and operational maintenance. Pavement roughness was the excitation source for the vehicle-pavement system. In previous research, the pavement roughness was usually used as a simple deterministic function, in fact, the pavement roughness was a stochastic process [1].As the needs for the dynamic study on the vehicle-pavement system and the simulation of vibration table, it was an important guarantee that how to simulate the pavement roughness actually.

At present, the research on power spectral density function of the vehicle-pavement system usually used linear elastic model in the frequency domain. When analysis nonlinear random vibration, the most effective way is to obtain the sample of random excitation as input, then calculate the random response of system by using numerical integration. So the stochastic process of pavement roughness in the simulation of time domain becomes important problem. The simulation of time domain is required to express an indeterminist function of time. More concretely, it means obtaining a height-time model which excited by roughness to get a three-dimensional model about vehicle speed-space pavement. In the former case, the testing simulation of ride comfort about vibration on bench. In the latter case, the testing simulation of ride comfort about vehicle-pavement.

2. Power spectral density function of pavement roughness

The evaluation method of pavement roughness can be divided into two categories, which is the method of dynamic reaction; the other is cross section method. The method of dynamic reaction is based IRI (International Roughness Index) as the representative and the method of cross section is based PSI (Present Serviceability Index) or power spectrum method as the representative. At present, in the dynamic analysis of road structure, the method of power spectrum is more applied. In this paper, the power spectrum density function of pavement roughness was used as the basis of time domain simulation.

In 1984, International Organization for Standardization proposes "the draft of pavement roughness representation method" (IS0/TC108/SC2N67). The method adopts the formula (1) to express [2]as follows:

$$G_q(\Omega) = G_q(\Omega_0)(\frac{\Omega}{\Omega_0})^{-w} \tag{1}$$

In the formula (1), Ω_0 is the frequency of reference space, Ω_0 =0.1c/m。 $G_q(\Omega_0)$ is density of surface power spectrum under reference space frequency, which is called road roughness coefficient. W is frequency index, power spectrum curve is a slash in double logarithmic coordinates. General it take W=2. In fitting the measured pavement power spectrum in order to reduce the error, it can be used different fitting coefficients to segment fitting in different spatial frequency range, which should not over four. In ISO/TCI08/SC2N67, a new pavement power spectrum density is proposed, which is divided into eight levels.

3. Power spectral density function of pavement roughness

The power spectrum density function (PSD) of pavement roughness was a statistics of pavement roughness, whether it was the standard pavement spectrum or the measured pavement spectrum. So the pavement elevation was determined in the measurement range, the PSD was the only one. But when PSD was given.， the analog-design elevation of the pavement was not unique. In other words, the statistical model and time-domain model has not one-on-one mapping. Therefore, the time-history functions derived from the statistical model of pavement roughness should be seen as a sample function that meeting the time-function of roughness in the given pavement spectrum. At present, the most commonly used main methods are: secondary filtering, triangle series method, white noise filtering and so on. But they also have some problems. Secondary Filtering was required to design the filter. We need to design a reasonable filter when pavement roughness has different power spectral density function, which is short of generality. The triangle series method and White noise filtering were considered to be a stationary Gaussian Process. That obviously did not match the actual situation. What's more, the calculation speed of those methods was slow, and has some error in the obtained irregularity samples.

In this paper, due to the shortage of the above methods, the frequency spectrum was constructed by the discrete sampling of the power spectrum density. Time domain simulation of track roughness excitation function is applied in the inverse Fourier transform of frequency spectrum. The main idea of this method was the spectrum amplitude and random phase were obtained according to the power spectrum, and then the simulation of pavement roughness was obtained in the time domain through the Fourier inverse transform (IFFT).

First, introduce the period map method of power spectral density, it was estimated by time series. We assume that time series is $\{x_s\}$, $s = 0,1,\cdots\cdots,(N-1)$, record length is $T = N\Delta$, Δ means time interval. The time delay of related functions $\tau = r\Delta$ is also discrete values. So we can get equations as follows:

$$R_{xx}(t) = \frac{1}{T}\int_0^T x(t)x(t+\tau)dt \tag{2}$$

$$R_r = R_{xx}(\tau = r\Delta) = \frac{1}{N}\sum_{s=0}^{N-1} x_s x_{s+r} \tag{3}$$

In the formula: $r = 0,1,\cdots\cdots,(N-1)$， $x_s \equiv x(s\Delta)$。

IOP Conf. Series: Earth and Environmental Science **86** (2017) 012016 doi:10.1088/1755-1315/86/1/012016

We can be obtained by $S_{xx}(f) <=> R_{xx}(\tau)$

$$S_{xx}(k) = \frac{1}{T} S_{xx}(f = k\frac{1}{T}) = \frac{1}{N}\sum_{r=0}^{N-1} R_r \exp[-i(k\frac{2\pi}{N})r] \tag{4}$$

so

$$\begin{aligned} S_{xx}(k) &= \frac{1}{N}\sum_{r=0}^{N-1}\left\{\frac{1}{N}\sum_{s=0}^{N-1} x_s x_{s+r}\right\}\exp[-i(k\frac{2\pi}{N})r] \\ &= \left\{\frac{1}{N}\sum_{s=0}^{N-1} x_s \exp[i(k\frac{2\pi}{N})s]\right\}\left\{\frac{1}{N}\sum_{r=0}^{N-1} x_{s+r}\exp[-i(k\frac{2\pi}{N})(s+r)]\right\} \end{aligned} \tag{5}$$

We can make $j = r + s$, so we can get as follows:

$$\sum_{r=0}^{N-1} x_{s+r}\exp[-i(k\frac{2\pi}{N})(s+r)] = \sum_{j-s}^{(N-j)+s} x_j \exp[-i(k\frac{2\pi}{N})j] \tag{6}$$

Time series $\{x_s\}$ has been discrete cycled because of discrete Fourier transform, N as period. So we can get as follows:

$$\sum_{j=s}^{(N-1)+s} x_j \exp[-i(k\frac{2\pi}{N})j] = \sum_{j=0}^{N-1} x_j \exp[-i(k\frac{2\pi}{N})j] \tag{7}$$

Put formula (4) and formula (5) into formula (3)

$$\begin{aligned} S_{xx}(k) &= \{\frac{1}{N}\sum_{s=0}^{N-1} x_s \exp[i(k\frac{2\pi}{N})s]\}\{\frac{1}{N}\sum_{j=0}^{N-1} x_j \exp[-i(k\frac{2\pi}{N})j]\} \\ &= \frac{1}{N^2}\left|DFT[x_s]\right|^2 = \frac{1}{N^2}[X^*(k)X(k)] \end{aligned} \tag{8}$$

In the formula, $X(k)$ is spectrum of time series $\{x_s\}$, $k = 0,1,\cdots\cdots,(N-1)$

Modulus of spectrum in time series by formula (8) as follows:

$$\begin{aligned} |X(k)| &= |DFT[x(n)]| = \sqrt{N_r^2 \times S_k(k)} = N_r\sqrt{S_k(k)} \\ &= N_r\sqrt{S_x(f = k\Delta f)\Delta f} \end{aligned} \qquad k = 0,1,\ldots,N_r - 1 \tag{9}$$

Modulus of Spectrum $X(k)$ in series $x(n)$ is concluded by formula (9). The spectrum phase must be random because the time series $x(n)$ was a stochastic process.

We assume that ξ_n as independent phase series, so the mean value of each component is zero. Because the real sequence FFT is a complex sequence (real part is even symmetric, imaginary part is odd symmetry). So ξ_n as plural and $|\xi_n| = 1$. We assume that as follows:

$$\xi_n = \cos\varphi_n + i\sin\varphi_n = \exp(i\varphi_n) \tag{10}$$

In the formula, φ_n obey the uniform distribution of $0 \sim 2\pi$

Because the real part of $X(k)$ about $N_r/2$ even symmetry, the imaginary part about $N_r/2$ odd symmetry. So we only need to calculate frequency spectrum $X(k)$, $k = 0,1,\ldots,N_r/2$. We can be obtained by the formula (7) as follows:

$$X(k) = \xi_k|X(k)| = N_r\xi_k\sqrt{S_n(f = k\Delta f)\Delta f} \qquad k = 0,1,\ldots,N_r/2 \tag{11}$$

Under the symmetric condition, it obviously that we easy to get $X(k)$, $k = 0,1,\ldots,N_r - 1$.

The simulation of pavement roughness would be obtained in the time domain through the Fourier inverse transform (IFFT) of complex sequence $X(k)$.

$$x(n) = \frac{1}{N_r}\sum_{k=0}^{N_r-1} X(k)\exp[\frac{i2\pi kn}{N_r}] \qquad n = 0,1,\ldots,N_r - 1 \tag{12}$$

4. The simulation results of each pavement roughness were obtain in time-domain

According to the above method, the corresponding mathematical calculation program was written for analyzing and calculating the given power spectral density function, the pavement roughness values list can be given. In terms of power spectral density grade, the simulate results of pavement roughness (A～H grades) were obtained in the time domain. The results were intuitive to express with chart form. The simulation of A grade pavement, resulting time domain curve of roughness as follows:

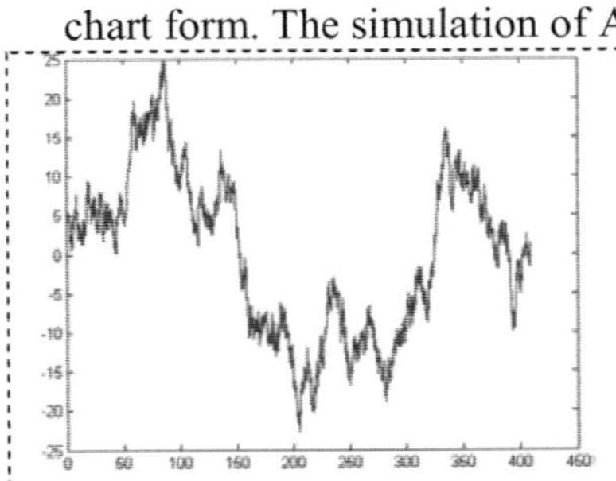

Figure 1. A grade pavement.

Figure 2. B grade pavement.

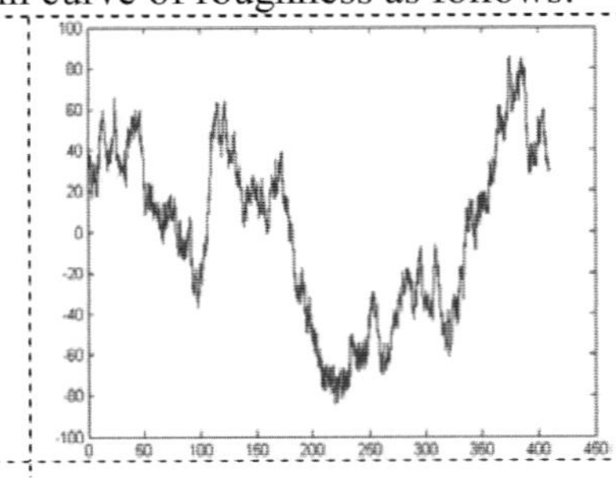

Figure 3. C grade pavement.

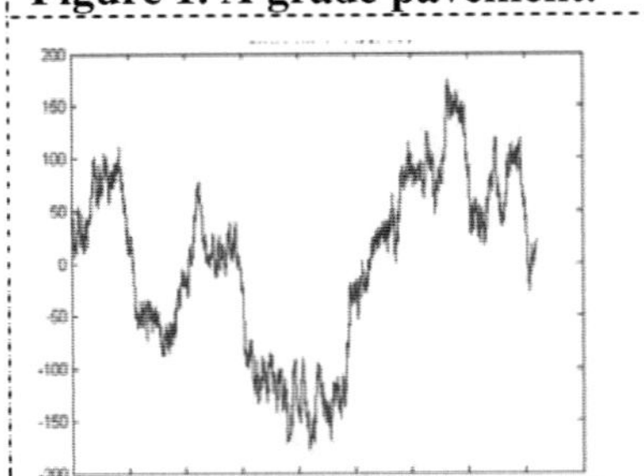

Figure 4. D grade pavement.

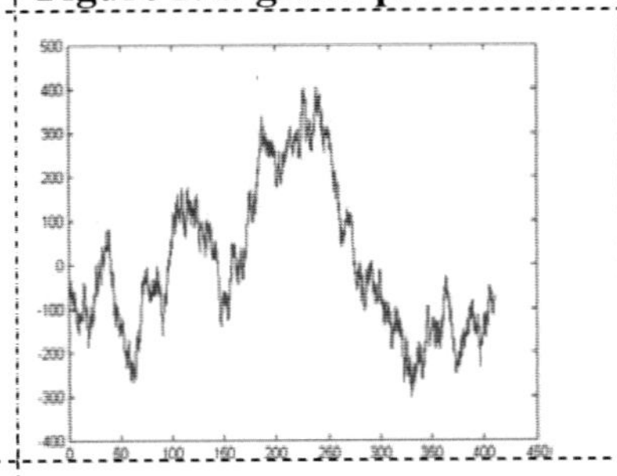

Figure 5. E grade pavement.

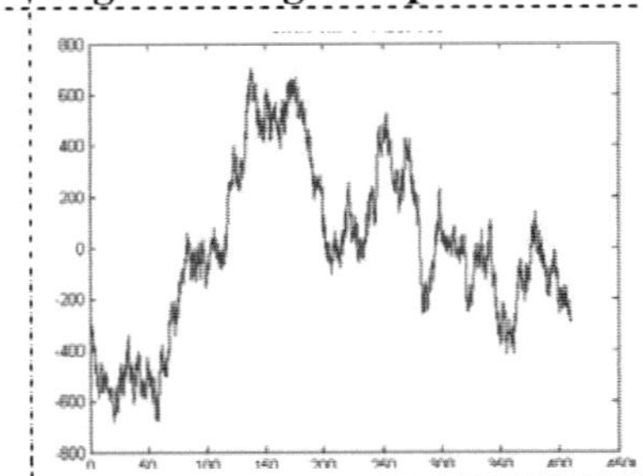

Figure 6. F grade pavement.

References

[1] Deng X J , Sun L, 1998, *Study on dynamics of vehicle-ground pavement structure system.* Beijing: People's Communications Press.

[2] Zhao J h, Wang Zh R, 2000, Guan Zhaoli. *Measurement analysis and application of pavement roughness*. Beijing: Beijing Institute of Technology press.

[3] Ontes R K, Enochson L, 1972, *Digital time series analysis.* John wiley & Sons, Inc., New York.

[4] Zhang Y L, Zhong Y F, 2004, *Time Domain Model of Road Undulation Excitation to Vehicles.* Journal of agricultural machinery, Vol.35 (2).

[5] Jin R Ch, Son J, 1999,*Simulation of road roughness and Research on nonlinear random vibration of vehicle*.Tsinghua University (Natural Science), 39 (8): 76-79.

[6] Chen G, Zhai W M, 1999 ,*Numerical Simulation of the Stochastic Process of Railway Track Irregularities*. Southwest Jiaotong University.

[7] Xu Zh X, Sun L, 1990, *Random vibration*. Beijing: Higher Education Press.

Impact of Growing-Season Meteorology on Japonica Rice Productivity in Northeastern China

W Yu, Y E Zhang, Y Wang, J J Liu, W Chen, S S Seng, J Y Zhuang, L W Gao, A-G Ahmed, Z M Li, G Q Li, S W Xu*

Agricultural Information Institute, Chinese Academy of Agricultural Sciences/Key Laboratory of Agricultural Information Service Technology, Ministry of Agriculture/Key Laboratory of Intelligent Agricultural Early Warning Technology and System, Chinese Academy of Agricultural Sciences, Beijing 100081, China

Email:yuwen@caas.cn; zhangyongen@caas.cn; xushiwei@caas.cn

Abstract. Meteorological factors have major impact on crop productivity in the world. Many researchers have evaluated the possible impact of climate change on crop yields using models. Here we use a 1980–2015panel dataset from China Agricultural Cost and Return Yearbook to investigate the meteorological impact on Chinese Japonica rice yield growth. We find that average temperature has significantly positive effect on rice yields; while maximum temperature and minimum temperature have negative effect on rice yield. Physicalinputs and time trend (technological progress) into the crop yield have an accurate estimation on crop yields. Means including irrigation or drainage could use during growing season.

Keyword: Meteorology, Japonica rice yield, Model, Northeastern China

Introduction

Climate changing have an important impact on crop yield in the world. The linkage between crop yield and climate variability has been mentioned in numerous researches using statistical models (YouL., etc., 2009; Xu S., etc., 2013; G Leng&M Huang., 2017; Wan E., 2017).With the increase of the income level of Chinese residents, more and more people choose japonica rice as the staple food. In order to discuss the effects of material input and climate on the yield of Japonica rice, this paper studies the japonica rice.

Data and method

We use time series and provincial data from 1980 to 2015 for 3 major japonica rice producing provinces in Northeastern China and the corresponding climate data such as temperature, rainfall, and solar radiation during the growing season of this period. Japonica rice input and output data are from China Agricultural Cost and Return Yearbook (1980–2015) published by China's Price Bureau.

Both Japonica rice yield and physical inputs demonstrate huge variations among the 108 observations. For example, the mean yield is about 454.3kg/mu while the rice could be as low as 165.7kg/mu (minimum), or as high as 632.0kg/mu (maximum).

Table 1 the basic statistics of the dataset.

Variable	unit	Mean	Std. Dev.	Min	Max
yield	kg/mu	454.3	83.7	165.7	632.0
seed	yuan/mu	17.0	10.0	4.1	45.4
fertilizer	yuan/mu	63.7	45.4	3.7	176.3
pesticide	yuan/mu	14.6	12.4	1.1	100.2
machinery	yuan/mu	57.1	69.6	1.2	247.3
others	yuan/mu	92.6	87.6	7.8	596.8
labor	yuan/mu	135.3	127.4	14.4	537.2
rainfall	mm	21871.5	12802.0	2260.5	49620.2
sunshine	hour	7.4	0.5	6.2	8.6
tempave	℃	15.9	3.7	8.4	20.8
tempmax	℃	21.9	3.4	14.2	26.5
tempmin	℃	10.5	4.1	3.1	15.5

The provincial climate parameters are averages of all the metrological stations within the provinces. Some basic statistics for rainfall, sunshine, average temperature, maximum temperature and minimum temperature are shown in Table 1.

We hypothesize the crop yield as a function of crop inputs, technology, land quality, and climate factors. The initial explanatory variables for the yield equation include inputs such as land, chemical fertilizer, seeds, pesticide, machinery, irrigation and other physical inputs; regional production specialization; climate variables such asprecipitation and solar radiation, average temperature, maximum temperature and minimum temperature; regional dummyvariables. In this study, the labor cost does not include in the model, for we hypothesize the labor manage the rice well. The physical inputs are measured in expenses per unit-harvested area. We includedseed, chemical fertilizer, pesticide, machinery individually and combined all the rest of inputs into an aggregated category of‘‘others’’. In the model, we hypothesize the regional production specialization variable is based on Heilongjiang Province, which is used to reflect the environment factors such assoil quality and others. Finally, a time trend is used to represent the factor due to technological change during this period (1980-2015).Finally, a Cobb–Douglas form of rice yield function is specified as follows:

$$\text{Ln Yield}= \alpha_0 + \alpha_t + \sum_{j=1}^{3} \beta_j\, lnX_{jit} + lnClimate_{it} + \sum_{l=1}^{2} \gamma_l D_l + \varepsilon_{it} \quad (1)$$

WhereLn is natural log, t = 1, 2. . .36 denotes observations from the years 1980 to 2015. Yield_itrefers to rice yield for Chinese provincei at time t (the time trend from 1980 to 2015); Xrepresents the conventional inputs per hectare of sown ricearea including seeds, fertilizer, pesticide, machinery, and other inputs such as irrigation, manure, and animal power. All other inputs are represented by real expenses per hectare of sownrice area (see Table 1 for more); Climate is the climate variables including temperature, rainfall and solar radiation duration.

Results

We find that meteorological impact is significant on rice yields during its growth seasons (Table 2):

Material input In the material inputs, the seed perimeter is 0.126, which means that the seed is positively significant for rice yield, while other input including irrigation is negatively significant, which means other material inputs more.

Meteorological factors We find the significantly negative relationships between rice yield and rainfall while significantly positive relationship between rice yield and sunshine (solar radiation). However, the average temperature has a significantly positive effect while maximum and minimum temperature has a significantly negative effect on rice yield. The significance for rainfall is at 5%level, reflecting the negative influence of rainfall on rice productivity, while sunshine has positive influence on it. Average temperature has 1% significance level. As reflects the significantly positive impact of rice growing season temperatures on rice yield in China, while maximum temperature and minimum temperature have significantly negative impact on rice yield in Northeastern China.

Time trendwe find that there is significant effect of rice yield from time series, which shows technological progress.

Region difference The based region is Heilongjiang Province, Jilin has no significance while Liaoning significantly has lower rice yield.

Table 2 Estimated Japonica rice yield function in Northeastern China, 1980-2015.

Explanatory variables	definition	Coef.	t-value	P>t	
lnseed	Logarithm of seed cost	0.126	1.86	0.067	*
lnfert	Logarithm of fertilizer cost	0.076	1.08	0.282	
lnpestcide	Logarithm of pesticide cost	-0.017	-0.22	0.828	
lnmachinery	Logarithm of machinery cost	-0.152	-1.63	0.105	
lnothers	Logarithm of irrigation, etc.	-0.112	-1.86	0.066	*
lnrf	Logarithm of rainfall	-0.848	-2.49	0.015	**
lnss	Logarithm of sunshine	2.890	1.99	0.049	**
lnta	Logarithm of average temperature	49.240	4.24	0.000	***
lnth	Logarithm of maximum temperature	-50.934	-5.08	0.000	***
lntl	Logarithm of minimum temperature	-5.237	-1.94	0.055	*
t	Trend based on 1980	0.050	7.06	0.000	***
Based region	Heilongjiang	-	-	-	
_Iregion_2	Jilin	-0.955	-1.34	0.183	
_Iregion_3	Liaoning	-2.641	-2.87	0.005	***
_cons	constant	42.947	5.14	0.000	***
	Number of obs=108				
	F(13,94)=13.07				
	R-squared=0.5946				

Note:Dependent variable = Ln(rice yield).

* 0.10 level of statistical significance.

** 0.05 level of statistical significance.

*** 0.01 level of statistical significance.

Conclusions

Generally, during the total rice growth season, the rainfall is not significant, and sunshine is negative and but significant, Average temperature is positive, but maximum and minimum temperature is negative for rice growth.

Considering the rice growing,

- The rice in jointing stage need less rainfall, but in the stage transferring and filling stage, the rice need more rainfall. Generally, the Japonica rice in Northeastern China has more rainfall during the growing season; The sunshine has the positive effect as rainfall in the rice growth seasons; For the temperature, averagely, the higher average temperature is better for rice growing, while the maximum and minimum temperature is not better for the rice yield;
- During the rice growth season, some means are need to use, for example, water irrigation or draining.
- Technology is an import factor for rice yield increasing, though the coefficient of time trend variable is low.
- Region difference is from soil quality and other environments.

Acknowledgements

This paper was supported by the CAAS Science and Technology Innovation Project (number: CAAS-ASTIP-2017-AII), funded by Chinese Academy of Agricultural Sciences; by NSFC (Number: 71233007) and by MOST (Number: 2016YFE0113100).

References

[1] G Leng,M Huang. Crop yield response to climate change varies with crop spatial distribution pattern. Science Report, 2017, 7(1):1-10

[2] Wan E., etc. The uncertainty of crop yield projections is reduced by improved temperature response functions. NaturePlants 3, 17102 (2017):1-11

[3] Xu S., Yu W., Liu S., Zhang S., Yu W. The Impact of Meteorological Factors on the Winter Rice Yield —the Case Study from Weishan County, Shandong Province. Journal of Agriculture, 2013, 3(10):45-51.

[4] You L., Rosegrant M. W., Wood S., and Sun D. Impact of growing season temperature on rice productivity in China. Agricultural and Forest Meteorology 149 (2009) 1009–1014.

The Comprehensive Benefit Evaluation of Take Shared Bicycles as Connecting to Public Transit

J Y Zhang, H Sun, P F Li and C C Li

Collage of transportation, ShanDong University Of Science And Technology

E-mail:942895320@qq.com

Abstract: Shared bicycles as an important way of connecting public transport, have few literature to evaluate its effectiveness.This paper taking Beijing city as an example, make anevaluationfor the six types of travel combinations which are commonly used by the citizens. The author selects four quantitative indicators: fuel consumption, economic cost, total time spent, and CO2 emission. And two qualitative indicators: degree of comfort and convenience. The relative efficiency of quantitative indicators is obtained by data envelopment analysis (DEA) and fuzzification and then take fuzzy synthetic evaluation with qualitative indicators.It was found that the choice of shared bicycles +subway+ shared bicycles and shared bicycles has good comprehensive benefits in medium distance travel. The findings also suggest that shared bicycles +subway+ shared bicycles is the best choice in the utilitarian trips. The conclusions not only provide suggestions for the travellers to select travel modes, but also can adjust the relevant factors to increase the proportion of green travel.

1. Introduction

As a new mode of transport in recent years, shared bicycles has been welcomed by people.Research shows that riding could reduce the use of fossil fuels, ease the land tension and completedseamlessly switch between modes. After the sharing of bicycles in China, the proportion of bicycle travel increased from 5.5% to 11.6% [1]. In recent years, domestic and foreignexpertsdid a lot of researcharound the bike's service quality, safety, environmental impact, travel efficiency and the overall contribution to urban traffic.KjartanScensminde analysed the cost-effectiveness of the bicycle network[2]; Börjesson M and Eliasson J from the time value and the environment and other external interests of the perspective of the bike cost-effectiveness[3].

There is also a lot of research focus on the bikes in the country. Liu Chen used the DEA to evaluate the urban public bicycle system synthetically[4]; Based on the system operation efficiency and user satisfaction index, Wang Wenjing analysed the cost and benefit of the bicycle life cycle[5]; Tang Wenqian calculates the economic, environmental and transportation benefits generated by the public bicycle system through the establishment of a benefit analysis model[6]

In general, due to the sharing of no-pile-sharing cycle form the scale of the time is not long, existing papers do not takeinto account the impact of the characteristics of the pile on the pedestrian, the lack of a shared bicycle on the first and last mile solution" the specific impact of the study. This article aims at people commonly used six kinds of travel modes: shared bicycles + bus + shared bicycles(B+B+B)、private car(P)、shared bicycles(S)、walk + subway + walk(W+S+W)、shared bicycles + subway + shared bicycles(B+S+B)、walk + bus + walk(W+B+W)[7].First of all, the author selects six kinds of evaluation indexes.Secondly using the DEA analysed quantitative indexes, And the take the result as the input evaluate again. Finally, take fuzzy synthetic evaluation with qualitative indicators.

2. Background

As early as 2005, the first public transportation system was established in Beijing. But in 2010, because the system site was unreasonable, the experience was not good, and finally announced defeat[8]. In 2016, China's smart phone users reached 680 million, more than 70% of national consumption using mobile payment. At the same time, with the help of advanced positioning technology, no-pile-sharing cycle without piles has gained rapid development in China.

Studies show that China's private car growth rate is more than 18%, and is the world's largest car consumer [9]. At the same time, increased travel time, increased road congestion, longer queuing length. The latest research findings show that sharing bicycles, less than a year into the city, has become the car, bus, subway outside the fourth major modes of travel. Bikes-haring schemes are associated with environmental benefits through the diminished usage of motor vehicles and the associated reduction in fuel use and traffic congestion. Shared bicycles represent the perfect combination of "Internet +" and the real economy.

3. Selection of evaluation indicators

Too many evaluation indicators not only increase the amount of calculation, but also may lead to duplication, resulting in a large proportion of an evaluation index. In this paper, according to the characteristics of shared bicycles and related literatures selected four quantitative indicators and two non-quantitative indicators[10].

1. Fuel consumption: Refers to the energy consumption of the travel vehicle under the premise of full load, different transportation vehicles, each hundred people travel a kilometre, the standard coal consumption statistics [11].

2. Economic cost: Indicate the sum of the costs of traveling through various means of transportation throughout the process.

3. Total time spent: The time spent from the departure to the destination of the passenger. It usually includes waiting time, ride time and transfer time. The Total time spent takes the following functional form:

$$T_n = \frac{L}{S} + t_w + t_b \quad (1)$$

Where L is the on the vehicle, S is the vehicle speed, t_w is the waiting time, t_b is the transfer time.

4. Emission of carbon dioxide: The carbon dioxide produced by taking the vehicle. Table 1 is the different vehicles' emission of carbon dioxide in per kilometre [12].

Table 1 Beijing passenger transport carbon dioxide emissions statistics

Travel mode	Carbon dioxide emissions g/km
bicycle	0
car	50.5
bus	13.14
subway	10.7

5. Degree of comfort andDegree of convenience: From the competent feelings of passengers.

4. Methodology

The evaluation of objective things is the advantage of DEA. It can effectively avoid the potential errors caused by human factors[13]. However, many factors in real life are difficult to represent with specific data[14]. In order to solve such problems, we can make use of fuzzy comprehensive evaluation. The research methods are as follows: Using the DEA analysed quantitative indexes. And the take the result as the input and Fuzzy. At last, take fuzzysyntheticevaluationwith qualitative indicators. We can finally obtained the evaluation results.

Assuming the comment set is $V = (v_0, v_1, \cdots v_{p-1})$, where p is the degree,and then the output of DEA can be interpreted as the membership grade of V.Assuming the $r = (r_0, r_1, \cdots r_{p-1})$ is the membership grade, which can be expressed as

$$r_j = \begin{cases} \frac{x-(j-1)\frac{1}{p-1}}{\frac{1}{p-1}}, (j-1)\frac{1}{p-1} \le x < j\frac{1}{p-1} \\ \frac{(j+1)\frac{1}{p-1}-x}{\frac{1}{p-1}} - x, j\frac{1}{p-1} \le x < (j+1)\frac{1}{p-1} \\ 0, \quad \text{others} \end{cases}, \quad r_j \in [0,1], j = 0,1,2 \cdots p-1 \quad (2)$$

Where x is the result of DEA; j is decision unit; p is the degree.

At the same time, the fuzzy results of quantitative indexes and the fuzzy evaluation results of qualitative indexes are obtained. And then make a comprehensive evaluation. As the following equation:

$$B = A^{\circ}R \rightarrow B_j = (a_{j1}, a_{j2}, \cdots a_{jk})^{\circ} \begin{bmatrix} B_{j1} \\ B_{j2} \\ \cdots \\ B_{jk} \end{bmatrix} = (b_{j1}, b_{j2}, \cdots, b_{jp}), j = 1,2,\cdots,n, \quad (3)$$

Where A is weight coefficient of the quantitative indexes and qualitative indexes; R is a matrix consist of the result of DEA and Fuzzy; k is number of the branch.

5. Modelling results

This paper takes Beijing city as an example. According to the average travel distance of 13.2km people in Beijing, the authors carried out statistical data as Table 2. The statistical methods have been mentioned in the Part 2.

Table 2 Evaluation indicators value

Level 1 indicators	Social consumption				Service management	
Weights	0.6				0.4	
Level 2 indicators	Energy consumption (kg) (input)	Total cost of economy / Yuan (input)	Time total cost per second (output)	Co2 total discharge / kga (output)	Comfort	Convenience
Weights	DEA adjustment	DEA adjustment	DEA adjustment	DEA adjustment	0.4	0.6
B+B+B	0.112	4	1/2263	1/0.173	(0.5,0.3,0.2)	(0.5,0.2,0.3)
P	0.452	6	1/1670	1/0.667	(1,0,0)	(0.7,0.2,0.1)
B	0	1.5	1/3223	0	(0.6,0.2,0.2)	(0.6,0.2,0.2)
W+S+W	0.032	5	1/1542	1/0.141	(0.7,0.2,0.1)	(0.5,0.2,0.3)
B+S+B	0.032	6	1/1129	1/0.141	(0.6,0.3,0.1)	(0.4,0.5,0.1)
W+B+W	0.112	3	1/2396	1/0.173	(0.6,0.2,0.2)	(0.5,0.3,0.2)

[a]In DEA, the higher the conversion efficiency is, the better the result is. In this paper, we hope to get less solution and less carbon dioxide emissions. So take the reciprocal.

In this paper, the comment set is $V = (good, \text{medium}, \text{poor})$. Using DEA、Fuzzy and equation (2) are used to analyse the date from Table 3. And the results are shown in the following Table 3.

Table 3 Quantitative and qualitative indexes processing results

Travel modes	Social consumption performance				Service management performance		
	DEA results	DEA score fuzzy membership (Bj1)			Fuzzy membership (Bj2)		
		good	Medium	Poor	good	Medium	Poor
B+B+B	0.8149	0.37	0.63	0.00	0.5	0.24	0.26

P	0.5429	0.09	0.91	0.00	0.82	0.12	0.06
B	1	1.00	0.00	0.00	0.60	0.2	0.2
W+S+W	1	1.00	0.00	0.00	0.58	0.20	0.22
B+S+B	1	1.00	0.00	0.00	0.48	0.42	0.1
W+B+W	1	1.00	0.00	0.00	0.54	0.26	0.20

According to the web- surveys, this paper definite Social consumption and Service management weight vector is $A = (0.6,0.2)$.Though the equation (3). Table 4 can be obtained:

Table 4 Comprehensive review matrix and evaluation results

Travel modes	Comprehensive evaluation matrix (T)						Comprehensive evaluation results membership degree			Evaluation results
	community resource			Service management						
	excellent	good	poor	excellent	good	poor	good	Medium	poor	
B+B+B	0.37	0.63	0.00	0.5	0.24	0.26	0.42	0.47	0.11	Medium
P	0.09	0.91	0.00	0.82	0.12	0.06	0.38	0.59	0.03	Medium
B	1.00	0.00	0.00	0.66	0.2	0.14	0.84	0.08	0.08	good
W+S+W	1.00	0.00	0.00	0.58	0.20	0.22	0.83	0.08	0.09	good
B+S+B	1.00	0.00	0.00	0.48	0.36	0.16	0.79	0.17	0.04	good
W+B+W	1.00	0.00	0.00	0.54	0.26	0.20	0.82	0.10	0.08	good

Finally, making the results Uniformization (Assuming good = 100, medium = 60, poor = 10), Table 5 is the comprehensive evaluation result.

Table.5. Ranking of travel modes

Travel modes	B+B+B	P	B	W+S+W	B+S+B	W+B+W
Mark	71.3	73.7	89.6	88.7	89.6	88.8
Ranking	5	4	1	3	1	2

6.Conclusion

Although sharing bike is a new way of solving the problem of first and 'last mile solution.' However, there is a lack of researches on the utility of travel combination patterns involving shared bicycles. This paper studies that issue taking Beijing as an example. The author focuses on the benefit study of various travel modes of Beijing residents traveling by public transport. The results of the mathematical model showed that the B+S+B and B in whole travel have good benefits. Considering the traveling purposes of residents,theauthor recommend a kind of travel combination pattern the B+S+B in the utilitarian trips.

These findings have important policy implications, the government departments and enterprises to formulate policies and operational plans.And this paper has a guiding significance for short and medium distance travel of residents.

To extend the research, the authors need to strengthen the studies of following dimensions:(1).Increase the evaluation factors. Because the past studies have shown that the environment has a great impact on the travel mode choices of residents.(2) Then we need to further study the role of shared bicycles to solve traffic congestion.

Acknowledgement

This research was supported by the Key Research and Development Program of Shandong province (No.2016GGX105008) and Scientific Research Foundation of Shandong University of Science and Technology for Recruited Talents (No.2015RCJJ035 and No.2015RCJJ036).

References

[1] Mobike, Beijing Tsinghua TongHeng Planning and Design Institute. Shared bicycle and urban development white paper in 2017[R].2017.4
[2] Kjartan Scelensminde.Analysis about walking and bicycle lanes network's cost benefit based on safety, health effects and external costs of motorized traffic[J]. International Urban Planning, 2012, 27(5): 18-25.
[3] Börjesson M., Eliasson J. The value of time and external benefits in bicycle cost-benefit analysis [J] Working Papers in Transport Economics, 2011, 46(4):673-683
[4] Liu Chen, Shang Li-Yuan Zhu Yun. Comprehensive evaluation of public bicycle system based on data envelopment method [J]. Journal of Transportation Engineering and Information, 2016, (03):119-128
[5] Wang Wenjing, Ma Shifeng.Analysis on the evaluation model of public bicycle service system. [J] Value Engineering, 2015, (08):127-128
[6] Tang Wengqian, Li Wenquan, Chen Qian. Efficiency analysis of the public bicycle free-service system [J].Journal of Transportation Engineering and Information, 2014, (03):87-93.
[7] Ofo, Ministry of Transport and Science Research Institute. 2017 Major cities in China Cycling Report [R] .2017.4
[8] Liu, Z., Jia, X., Cheng, W., 2012. Solving the last mile problem: ensure the success of public bicycle system in Beijing. Procardia – Soc. Behav. Sci. 43, 73–78.
[9] Wang, Y., Teter, J., Sperling, D., 2011. China's soaring vehicle population: even greater than forecasted? Energy Pol. 39 (6), 3296–3306.
[10] BigDate-Research. China SharingBicycle Market Research Report [R] 2017.2
[11] Zhang TieYing.Study on different urban transport modes' energy consumption [D].Beijing Jiaotong University, 2010.
[12] Zhang Xiuyuan,Yang Xinmiao,Yan yan.Statistical estimation method for energy consumption and carbon emissions by urban transport[J].China Soft Science, 2014, (06): 142-150.
[13] Charnes A,Cooper W W.Rhodes E. Measuring the efficiency of decision making units .European Journal of operational Reasearch,1978,2:429~444
[14] Chen Shuili, Li Jinggong, Wang Xianggong.Application of Fuzzy Set Theory [M]. Beijing: Science Press, 2005.

Impact of livestock Scale on Rice Production in Battambang of Cambodia

D SIEK[1,2], S W XU[1*], WYU[1], A-G AHMED[1]

[1] Agricultural Information Institute, Chinese Academy of Agricultural Sciences, Beijing, China 100081

[2] Regional Polytechnic Institute TechoSenBattambang, Ministry of Labor and Vocational Training, Battambang, Cambodia

*Corresponding author, E-mail: darith9398@yahoo.com, xushiwei@caas.cn

Abstract. Increasing the awareness of environmental protection especially in the rural regions is important as most the farmers reside in that region. Crop-livestock proudciton has proven in many ways to encourage environmental protection. This study analyzes among other factors the impacto of livestock scale on rice production. Two regressions: Ordinary Least Square (OLS) and stepwise regression was applied to investigate these interrelationship. The result stress of three factors encouraging livestock production namely size of farmland, scale of livestock and income acquired from other jobs. The study further provides recommends to the government based on the findings of the study.

Key word: livestock; rice; environment; Battambang; Cambodia

1. Introduction

The world's population growth is projected to reach around 9.1 billion by 2050 which pushing up the value of food consumption would need to rise by 70% and 3 billion tonnes (up from today's nearly 2.1 billion tonnes) of cereals for both food and animal feed; and food production in the developing country would need to almost double(FAO, 2009).Cambodia is one of developing country with the population estimate in 2016 is about 18 million (World Bank, 2016) and is growing rapidly about 1.83% per annum(CIPS 2013). In Cambodia today, Agricultural sector has been playing a crucial role in developing food security, economic growth and employment especially in the rural community (MAFF, 2015). Approximately 80 percent of the population resides in rural areas and 71 percent are estimated to be solely dependent on agriculture (largely rice cultivation) for their livelihoods (USDA, 2010). Rice is the overwhelmingly predominant food crop, being grown on an estimated 3.3 million hectares or nearly 85.3 percent of the country's total cultivated area (CAC, 2015). The rice cultivation grown are incredibly diverse, with crop scientists estimating there are over 3000 varieties being cultivated in the country which is dissected by several major rivers, including the Mekong and Tonle Sap, and is covered by ancient, highly-weathered and chemical fertilizer use is extremely low and native soils are often very infertile (USDA, 2010). The United Nations Food and Agricultural Organization (FAO) estimates that Cambodia has the lowest rate of fertilizer use for rice in Southeast

Asia, with only about 30 percent of total area receiving minimal application, and Cambodia rice was declared the world's best in its taste at the Competitive Rice Tasting event held at the 2013 Rice Traders World Rice Conference held in Hong Kong and awarded top prize at the Rice Trader World Rice Conference in Bali, Indonesia. As at 2016, the annual paddy rice production is about 9.5 million MT (5.7 million MT, milled basis) with an increase of 200,000 MT from year 2015 (FAO, 2016), and expected to reach10.85 million MT in 2018 (NSDP, 2015). According to MAFF, the average rice yield achieved in 2013 was at 3.163 MT/ha, around 11.52% increase over 2009 (rice yield in 2009 reached by 2.836 MT/ha); and comparing to it neighbouring countries, its yield is relatively low; Laos (4.2 MT/ha), Vietnam (5.8 MT/ha) and Thailand (3.1 MT/ha).

The low use of fertilizer by the Cambodian farmers is on one hand partially due to it unavailability or high price and on the other hand, due to the increase use of livestock manure as substitute or as a means of taking advantage of the high market prices of organic rice. The increasing engagement of livestock in rice production practices by the rice farmers is overwhelming. This study attempt to investigate whether the changes in livestock scale has adverse effect on the total rice output of the farmers in the rural region. Other factors such as the demographic status and socioeconomics of the farmers are as well. Applying Ordinary Least Square (OLS) and stepwise regression, the study will analyse the factors influencing rice output and furthermore, the study will provide recommendationsbased on the findings.

2. Methodology

2.1. Data resource

The data used in this study were conducted in Battambang province of Cambodiain 2015, involving 18 villages, 12 communes and 6 districts. 204 standard questionnaires were collected after the face to face interview with household farmers. Key informants were selected to be interview gathering useful and confidential information such as: demography, population, total cultivation area and agrarian crop-livestock farming system in target area. Data collected were processed and analyzed in Agricultural Information Institute (AII) of Chinese Academy of Agricultural Sciences (CAAS) Beijing.

The selected province Battambang is situated in the north-western of Cambodia which is about 291km from Phnom Penh city. With the well-known of vast land and long history of conventional rice farming, Battambang is one the province in the Tonle Sap reservoir and the second largest city in Cambodia with total population 1,121,001 and 1.79% for annual growth rate of population during 2008-2013 (CIPS, 2013). This region experiences two seasons: wet season (May to October) and dry season (November to April) with an estimated rice yield of 2.8 and 3.8 MT/ha in wet and dry season respectively (MAFF, 2015).

2.2. Data descriptive& statistics

As shown in Table 1, majority of the farmers (72%) operate rice area less than 3 hectares. About 63.2% of the farmers own 8 or less livestock unit (lu) ratio. The "lu" was defined byChilonda and Otte (2006) to principally represent an exchange unit among different species of livestock. This study used lu converting other livestock (pig and chicken) to cattle applying the standard cattle coefficients of about 0.4 pig and 0.02 chicken.More than half of the farmers in the region are aged farmers about 45 years. More so, the farmers are less educated. About 78.4% of the household are small have small family member equal or less than 5 persons and with lesser members are above 18 years to be considered as able labor force. Despite this, the younger children are still engaged in the farming activities. Off-farm jobs provided extra cash to farmers other than their farming activities.

Table 1 Descriptive statistics of mean and group variables of rice production

Variable	Definitions	Group	Obs.	Mean	Rice production (ton)
r_area	Rice area for farmer household working	> 3 ha	57	6.05	14.777
		≤ 3 ha	147	1.82	4.988
lu	Livestock Unit: livestock standard in cattle unit coefficients	> 8	75	13.05	8.872
		≤ 8	129	4.95	7.055
size_labor	Labour size per household	> 3	67	4.76	9.312
		≤ 3	137	2.18	6.946
size_family	Family size in household	> 5	44	6.77	8.54
		≤ 5	160	3.94	7.498
edu_hh	Education level for household farmers	> 5 grade	94	8.23	8.338
		≤ 5 grade	110	2.55	7.198
age_hh	Age for household farmers	> 45 years	113	55.96	8.346
		≤ 45 years	91	34.4	6.949
farm_off	Dummy variable: 1=Labour off-farm job in household; 0=others	1	38	--	10736
		0	166	--	7.033
_Idistrict_i	AekPhnum	*_Idistrict_1*	39	--	5.599
	Banan	*_Idistrict_2*	31	--	6.139
	MoungRuessei	*_Idistrict_3*	37	--	9.114
	RatanakMondol	*_Idistrict_4*	20	--	6.344
	Sangkae	*_Idistrict_5*	47	--	7.683
	ThmaKoul	*_Idistrict_6*	30	--	11.387

Model Specification

Initially, all information from research were entered into EpiData software for coding and grouping variables, and then transferred to STATA software for further data generation, regression and analysis In this tool, the multiple regression models using Ordinary Least Squares (OLS) were considered the relating performance on the rice production as dependent variable and a set of demographic and socioeconomic aspect of the household (Jeffrey, 2009,Hutcheson, 2011 and Stock, 2015). This variable included cultivation area for rice, livestock standard in cattle unit coefficients, size of labor, family size, household education, farmer'sage, off-farm job and other districts variables.The econometric analysis is shown in the constructed models below:

RegressionModel:

$$Ri\_total = \beta_0 + \beta_1 Lnarea + \beta_2 Lnlu + \beta_3 Lnsizlabor + \beta_4 Lnsizfami + \beta_1 Lnedu + \beta_2 Lnage + \beta_7 dum\_offarm + \beta_8 LnIdistrict + \varepsilon_i \quad (1)$$

Theconstantsβ_0,β_1....$\beta_8$are the coefficients of parameters of the econometric model that demonstrated the direction and strengths of the relationship between Ri_total and the demographic and socioeconomic factors of the household farmers while the$\varepsilon_i$is explain the error term assumed to be distributed logistic. The *"ln"* after the coefficients is the logarithms of the variables. There are two type of regression model. The first model is the direct running of the regression (equation 1), while second model is the stepwise OLS regression of equation (1).

3. Results and Discussions

The study applied both OLS and stepwise regression to analyse the rice output of Cambodian farmers via thesocioeconomicinfluencers including the rice cultivated area, livestock owned, and the economic and demography of the farmers. The definition and descriptive statistics of these influencers are shown in Table 1 above where the total rice outputs of different groups of each influencing factors. The table below (Table 2) on the other hand is the OLS and stepwise result. The result explains that three major factors stand out to be significantly influencing the total rice output in the two regressions.Although at different magnitude and level of significance especially on the impact of lu on rice output. As shown in Table 2, the factors include size of farmland, number of livestock which is put in unit "lu" (see Table 1) and off-farm jobs. All the three factors are positively related to total rice output of the farmers. The relation between rice output and farmland is of course obvious. As farm land increase so would bethe rice output given all other factors remain the same.

Table 2. The regression result of rice production

Variable	Definitions	OLS model			Stepwise		
		Coef.	T-value	P>T	Coef.	T-value	P>T
lnr_area	Logarithm of land area	6.340	13.92	0.000	6.359	15.07	0.000
lnlu	Logarithm of livestock standard in cattle unit coefficients	0.952	2.08	0.039	0.091	1.75	0.027
lnsize_labor	Logarithm of size labour	0.063	0.17	0.862	--	--	--
lnsize_family	Logarithm of size family	-1.235	-1.17	0.245	--	--	--
lnedu	Logarithm of household's education	0.053	0.64	0.520	--	--	--
lnage	Logarithm of household's age	0.754	0.68	0.496	--	--	--
Off-farm	Dummy variable: 1=Labour off-farm job; 0=others	3.037	3.72	0.000	2.665	3.69	0.000
_Idistrict_2	Banan	-0.226	-0.22	0.828	--	--	--
_Idistrict_3	MoungRuessei	-0.366	-0.37	0.714	--	--	--
_Idistrict_4	RatanakMondol	-0.536	-0.44	0.657	--	--	--
_Idistrict_5	Sangkae	0.019	0.02	0.982	--	--	--
_Idistrict_6	ThmaKoul	1.786	1.73	0.086	1.906	2.38	0.017
_cons	_cons	-0.774	-0.2	0.845	1.279	2.23	0.027
No.of obs.	Number of obs		202			203	
R^2	R-squared		0.60			0.59	
Adj R^2	Adj R-squared		0.58			0.58	

The interrelationship of livestock and rice output is that of biodiversity explaining that with more livestock, there are better chance of increasing rice output taking into account the advantages of crop-livestock production. In other words, livestock production encourages rice production. The manure of livestock serves as one of the major source of rice fertilizer in Cambodia especially during periods when fertilizer becomes too expensive or unavailable to the farmers and in turn, the rice straw, rice bran and in some cases broken rice are used as sources for the livestock feed. The crop-livestock practice not only provides cheap feed for the livestock or cheap organic fertilizer for the rice plant, it also reduces the chemical print in the environment. In Cambodia, more farmers are beginning to practice the use of organic manure. This development is pushed by the high prices of organic rice in the market. The third factor encouraging rice production is the income from off-farm jobs. With increase in income from other jobs, the farmer can invest more on factors which increases their rice output e.g. leasing more land or increasing direct beneficiary livestock such as cattle, sheep, goat pig and other larger livestock. With increase in income, the farmer will be able to afford essential inputs.

4. Conclusions and Recommendation

The study analysed the socioeconomic factors of the farmers, livestock scale, and income from other sources to determine which factor(s) have greater impact to increasing rice output in Battambang of Cambodia. Two regressions: Ordinary Least Square (OLS) and stepwise regression was applied to investigate the interrelationship and impact of the factors to total rice output. Three major factors stand out to encourage rice output in the region and they include the size of the farmland, the scale of livestock and income generated from other jobs. The high market price of organic rice has given a boost to the number of farmers involving livestock to their rice production for rich and cheap manure. Given these opportunity, the Cambodian government should continue to encourage and guide farmer on crop-livestock production, increase the awareness of environmental protection especially in the rural regions.

References

[1] CAC (2015) Census of Agriculture in Cambodia 2013, National Report on Final Census Results. National Institute ofStatistics, Ministry of Planning in collaboration with the Ministry of Agriculture, Forestry and Fisheries, Cambodia

[2] Chilonda P and Otte J 2006:Indicators to monitor trends in livestock production at national, regional and international levels. *Livestock Research for Rural Development. Volume 18, Article #117.* Retrieved December 20, 2016, http://www.lrrd.org/lrrd18/8/chil18117.htm

[3] FAO 2009, Global agriculture towards 2050, Food and Agriculture Organization of the United Nations, paper prepared for How to feed the World in 2050: High-level expert forum , Rome. www.fao.org/fileadmin/templates/wsfs/docs/expert_paper/How_to_Feed_the_World_in_2050.pdf.

[4] FAO (2016) FAO Rice Market Monitor (RMM), Trade and Markets Division, Food and Agriculture Organization of the United Nations, Volume XIX issue No. 3http://www.fao.org/economic/RMM.

[5] FAOSTAT(2016) Online Database, Food and Agriculture Organization of the United Nation (accessed December 18, 2016)http://www.fao.org/faostat/en/#data/QC

[6] Hutcheson, G. D. (2011). Ordinary Least-Squares Regression. In L. Moutinho and G. D.Hutcheson, The SAGE Dictionary of Quantitative Management Research. Pages 224-228.

[7] Jeffrey M. Wooldridge (2009), Introductory Econometrics: A Modern Approach, 4th ed., Cengage Learning. Chapters 3 - 7.

[8] Moul Thon (2016) **Municipality** and Province Investment Information, Battambang province.http://www.cambodiainvestment.gov.kh/content/uploads/2014/03/Battambang-Province_eng.pdf

[9] MAFF (2015) Ministry of Agriculture, Forestry and Fisheries, Agricultural Sector Strategic Development Plan 2014-2018, Kingdom of Cambodia.

[10] NIPS (2013) Cambodia Inter-censal Population Survey 2013, National Institute of Statistics , Ministry of Planning, Phnom Penh, Cambodia; Sponsored by United Nation Population Fund

Japan International Cooperation Agency.

[11] NSDP (2015). National Strategic development Plan 2014-2018, Royal Government of Cambodia. Phnom Penh, Cambodia.

[12] Stock, James H. and Mark W. Watson (2015), Introduction to Econometrics,3rd ed., Pearson Addison-Wesley. Chapters 4 - 9.

[13] USDA, (2010) CAMBODIA: Future Growth Rate of Rice Production Uncertain, Commodity Interlligence Report, United State Department of Agriculture, Foreign Agricultural Servicehttps://www.pecad.fas.usda.gov/highlights/2010/01/cambodia/

[14] WorldBank (2016). World Development Indicators (accessed December 17, 2016) http://databank.worldbank.org/data/reports.aspx?source=2&series=NY.GDP.MKTP.CD&country=KHM

Study Progress of Physiological Responses in High Temperature Environment

K Li, G Z Zheng*, W T Bu, Y J Wang and Y Z Lu
School of Energy Power and Mechanical Engineering, North China Electric Power University, Baoding 071003, China

Abstract: Certain workers are exposed to high temperatures for a long time. Heat stress will result in a series of physiological responses, and cause adverse effects on the health and safety of workers. This paper summarizes the physiological changes of cardiovascular system, core temperature, skin temperature, water-electrolyte metabolism, alimentary system, neuroendocrine system, reaction time and thermal fatigue in high temperature environments. It can provide a theoretical guidance for labor safety in high temperature environment.

1. Introduction

According to the relationship between environment temperature and body thermal balance, the living environment higher than 35℃ and the production environment higher than 32℃ are considered as high temperature environment [1]. The high temperature environment can be divided into natural high temperature environment and industrial high temperature environment. The natural high temperature environment is caused by solar radiation, while the industrial high temperature environment is caused by fuel combustion and mechanical rotation friction [2]. In the context of global warming, the extreme high temperature weather occurs more frequently. When humans are exposed to high temperature environment for a long time, they would suffer from a series of physiological abnormality. When the body thermal balance is destroyed, physical health will be threatened. To better understand the affects of the high temperature on human body, this paper summarizes the related studies of physiological responses in high temperature environment.

2. The physiological responses in high temperature environment

When exposed to high temperatures for a long time, people may suffer from a series of physiological abnormalities. This section introduces the various human body physiological responses of the cardiovascular system, the core temperature, the mean skin temperature, the alimentary system, the neuroendocrine system, the reaction time and the thermal fatigue.

2.1. Cardiovascular system

In high temperature environment, the human body temperature is lower than ambient temperature. Human surface skin vascular network is expanded to quicken the blood circulation to dissipate heat, and the blood flowing into internal organs decreased accordingly. In order to maintain the normal activities of human body, human heart rate increases to increase the cardiac output. Zhou [3] found that the skin blood flow increased from 1.4L/min to 2~4L/min or even 6~11L/min while the average heart rate increased by 20%~40% when people were in high temperature environment. Huang et al. [4] applied the method of Meta and compared the electrocardiogram between the high temperature group and the control group. Then they found that the left ventricular high voltage, sinus bradycardia and ST-T segment were significantly affected by high temperature. Therefore, above indexes are sensitive indicators of monitoring the cardiovascular disease of the staff in high temperature. Chen et al. [5] applied the western blotting to measure the heat shock protein 70 antibodies of the experimental group and the control group. They found that the abnormal proportion of hypertension and electrocardiogram in the experimental group was significantly higher than the control group. And they summarized that the incidence of cardiovascular diseases may be related to the high temperature. Zhang and Wu [6] randomly selected some hypertensive patients as a case group and some healthy workers as a control group. Based on physical examination and questionnaire investigation, the chi-square test indicated there may be a certain degree of correlation between high temperature and hypertension.

2.2. Core temperature

The core temperature usually refers to the temperature of the brain, liver and other important internal organs. When people are in a rest state, the core temperature is 37.2℃ to 37.6℃ [7]. Commonly, the armpit temperature, rectal temperature, tympanic temperature and oral temperature are adopted to represent core temperature. Bai [8] proposed to use the oral temperature to replace the measure of rectal temperature. The core temperature directly affects human metabolic activity. Nybo et al. [9] found that when the core temperature increased by 1.5℃, the basal metabolic rate increased by 23%. After sweating profusely, the blood concentrated, thus the muscle blood flow increased and the dermal blood flow decreased. As the temperature regulation was restricted, and the thermal balance was destroyed, the core temperature began to increase [10]. When the core temperature approach a critical value, it will result in the athletic ability decrease and make people feel tired [11]. Johnson [12] pointed out that when core temperature raises to about 40℃, people felt weary.

2.3. Mean skin temperature

The mean skin temperature can be calculated by the weighted average of several body surface temperatures [13]. In normal condition, the skin temperature is about 33.5℃. In high temperature environment, to maintain the human body heat balance, sweat volume and skin blood flow increases, thereby body heat loss increases. In hot environment, the temperature gradient between skin and environment is reverse. Thus the heat loss by convection and radiation is greatly restricted, and the skin temperature gradually increases. When the skin temperature is close to the core temperature, the human body gradually reaches the physiological limit. In addition, the personal safety would be threatened if there were no protective measures [14]. In artificial climate chamber, Liu et al. [15] simulated the heated process of atmospheric temperature. The temperature was set as 23℃, 28℃, 33℃ and 35℃, respectively. By measuring the subject's skin temperature, they found that the mean

IOP Conf. Series: Earth and Environmental Science **86** (2017) 012020 doi:10.1088/1755-1315/86/1/012020

skin temperature was 30.51℃, 32.39℃, 34.34℃ and 35.09℃, individually. And the local deviation of skin temperature from head to foot decreased. The local deviation of the chest was 0℃, therefore the chest temperature could be approximately considered as the mean skin temperature. Li [13] set environment chamber temperature at 23℃, 30℃, 35℃ and 40℃, individually. Subjects were required to read quietly, walk at the speed of 3.5km/h and run at the speed of 5.5km/h in the chamber. By measuring their mean skin temperature, she found that when the environment temperature increased, the rising rate of the mean skin temperature accelerated. When the ambient temperature was 35℃, the mean skin temperature were 36.5℃, 37.5℃ and 38℃ under different labor intensities respectively.

2.4. Water-electrolyte metabolism

Under the high temperature environment, sweat is accelerated to regulate body temperature. The sweat contains water, NaCl, inorganic salts and water-soluble vitamin C, B, etc. With the lack of timely supplement, the loss of the inorganic salt will result in the imbalance of water and salt. In normal condition, when the body fluid loses more than 1%, it can be regard as dehydration. Naghii [16] indicates that when the loss of water exceeds 2% of the body weight, the body athletic ability decreases. Therefore, 2% of the body weight is considered as the safe bottom line for body water loss. The body physiological responses in different water losses are shown in Table 1 [17].

Table 1. The body physiological responses in different water losses.

The loss of weight (%)	The loss of water (L)	Symptom
1	0.75	Slight (When the loss of weight is 1.5%, it can be considered as dehydration).
2	1.5	Body reaches endurance limit, and people starts feeling thirsty.
3	2.25	Body physical power begins to decline, and people feels moderately uncomfortable.
4	3	Body suffers from jerkiness and headache, and people feel extremely uncomfortable.
5~6	3.5~4	Body suffers from nausea and dizziness, and people are exhausted.
7	>5	Body loses consciousness.

2.5. Alimentary system

In high temperature environment, human body need to lose more heat, and it results in the heat stress. Firstly, the blood capillary and skin vascular become expansive, thus the amount of blood flowing into the alimentary canal decreases, and it causes the gastrointestinal dysfunction. Secondly, the profuse sweating leads to the loss of NaCl and other mineral salts. As the Cl ion reduces, the gastric acid decreases. And the gastric juice is diluted as a result of large water drink, thus it leads to poor appetite and other gastrointestinal diseases. In addition, as gastric emptying accelerates, the food cannot be adequately digested [18]. Zou [19] chose some bank staffs without a history of diabetes as the control group, and some steel workers as the observation group. The hemoglobin and blood glucose in whole blood were measured, and it was indicated that the differences of hemoglobin and blood glucose between the control group and the observation group were significant, and the average level of the observation group was higher than that of the control group. It summarized that the high temperature working environment could cause the increase of hemoglobin and blood glucose. Wang et al. [20] compared the physical conditions of the workers between a high temperature group and a control group. They found that the fasting blood glucose level of the worker was also influenced by the work type, work age, body mass index and blood pressure.

2.6. Neuroendocrine system

Under the high temperature environment, the excitability of human sympathetico-adrenomedullary system increases. Then the renin angiotensin II, antidiuretic hormone and aldosterone concentrations significantly increase, therefore the body oxygen consumption and heat production increase. Knochel et al. [21] let subjects have a sauna, and they found that their renin, angiotensin II and aldosterone all increased. However, due to the negative induction, the central nervous system function was depressed. Therefore, the body temperature regulation function weakened, and the impulse frequency of nervous system decreased. In addition, as the excitability of neuromuscular descends, the accuracy and coordination of action reduce and accidents easily occur. Sun et al. [22] selected two groups of workers from a steel tube plant. One group of them was exposed to high temperature, and the other group was exposed to non high temperature for comparison. The field investigation indicated that the incidence of neurasthenia syndrome in the high temperature group was significantly higher than the control group. And the incidence increased as the increase of the work age.

2.7. Reaction time

Reaction time can be divided into the simple reaction time and the complex reaction time [23]. The simple reaction time refers to the fastest average time the brain responds to a certain stimulus. It can be defined as the time from being stimulated and starting action. The complex reaction is the latency between a variable stimulus and a respectively variable response. It refers to the response of the body to several stimuli at the same time. The shorter the reaction of the brain, the less the accidents the workers suffer from. The factors influence the reaction time are age, sex, noise, fatigue, illness and intelligence, etc [24, 25]. Xu and Chen [26] indicated that the heat stress would shorten the simple reaction time, while it would increase the complex reaction time. And the severer the heat stress is, the more obvious the effect is. Ma et al. [1] simulated the high temperature and high humidity environment in a chamber. The simple reaction time of the subjects were measured in different working conditions and working intensities. The results indicated that changing the work type and adjusting the work intensity in definite time are beneficial to decrease the reaction time. Cheng et al. [27] measured the indexes of physiological, psychological and action stability in different temperatures and different work intensities. The results showed that in the fatigue condition, the attention, reactivity, action stability all reduced at different degrees. And the high temperature caused the body fatigue while it causes the increase of the error ratio of the action.

2.8. Thermal fatigue

The high temperature environments usually cause the mental and physiological fatigue of workers.

The cardiovascular load index is defined to estimate the fatigue degree of human body by measuring the human heart rate changes. Xing et al. [28] measured the working intensity and heart rate of the underground workers and proposed that the working hours of the miners should be lower than 8 hours. Hao et al. [29] selected some healthy college students as subjects. Subjects doing exercise in high temperature condition was set as the control group, and subjects doing exercise in normal temperature condition was set as the observation group. The body temperature, heart rate, blood pressure, superoxide dismutase (SOD) and malonaldehyde (MDA) contents in blood serum after exercise were measured. The results indicated that doing exercise in the high temperature environment would cause the fatigue and injury degree of the body cell increased, namely the SOD and MDA increased.

3. Conclusion

When exposed to high temperatures for a long time, people may suffer from a series of physiological abnormalities. This paper summarizes the various human body physiological responses of the cardiovascular system, the core temperature, the mean skin temperature, the water-electrolyte metabolism,the alimentary system, the neuroendocrine system, the reaction time and the thermal fatigue. It can provide a theoretical guidance for labor safety in high temperature environment.

Acknowledgements

This research has been supported by National Natural Science Foundation of China (Grant No. 51408220).

Reference

[1] Ma H, Yuan X X and Zhang C 2014 *Journal of North China Institute of Science and Technology* **11 (6)** 68-72.
[2] Liu Y H 2006 *Journal of Liaoning Teachers College* **8(2)** 66-67.
[3] Zhou Z Z 2011 *Neijiang Technology* **3** 29-30.
[4] Huang L Z, Yang T. F, Huang J J, et al. 2011 *Chinese Journal of Evidence- Based Medicine* **11 (2)** 174-180.
[5] Chen Z P, Su W G, and Wang S M 2010 *Chinese Medical Journal of Metallurgical Industry* **27 (5)** 521-523.
[6] Zhang Q, and Wu J B 2014 *Chinese Health Industry* **16** 126-127.
[7] Rutty G N. http://link. Springer. com/book/10. 1007/978- 0- 85729-519.
[8] Bai YK. 2015 *Xi'an Technology University.*
[9] Nybo L, Moller K, Volianitis S, Nielsen B, and Secher NH 2002 *J. Appl Physiol* **93(1)** 58-64.
[10] Peng ZH 2016 *Liaoning Sport Science and Technology* **38 (1)** 191-193.
[11] Nybo L 2008 *J Appl Physiol* **104(3)** 871-878.
[12] Johnson J M 2010 *Med Sci Sports.* **20 (13)** 29-39.
[13] Li N 2011 *Tianjin University.*
[14] Li G J 2008. *Tianjin University.*
[15] Liu YF, Zhou X and Wang LJ 2014 *Journal of Environment and Health* **31(1)** 68-70.
[16] Naghii M R 2000 *Nutr Health* **14 (2)** 127-132

[17] Guyton A C and Hall J E 2006 *Elsevier Saunders.*
[18] Huang S H and Shen Y Q 2016 *Chinese Journal of Medical Science* **6(4)** 221-228.
[19] Zou P. 2014 *Journal of Sichuan Metallurgy* **36 (3)** 66-67.
[20] Wang J, Sun J, Wang C X and Chen CG 2014 *Prevntive Medicine Tribune* **20(9)** 675-676.
[21] Knochel JP, Dotin LN and Hamburger RJ 1972 *Journal of Clinical Investigation t* **51(2)** 242-255
[22] Sun P, Yang Y L and Wang R Y 2004 *Modern Preventive Medicine* **31 (1)** 21-23.
[23] Wang B 2003 *Journal of Capital Institute of Physical Education* **15 (4)** 110-113.
[24] Li K D 2005 *China Mining News* **5 (12)** 25-26.
[25] Chen R and Tang TB 2002 *Chinese Journal of School Health* **23 (3)** 277-278.
[26] Xu XD and Chen XR 1994 *Chinese Journal of Safety Science* **4 (2)** 42-45.
[27] Cheng C, Zhang Q, Xu AM and Ding Z 2014 *Journal of North China Institute of Science and Technology* **11(6)** 79-82.
[28] Xing J J, Liu W D and Wang XL 1998 *Labor Protection Science and Technology* **18 (4)** 54-57.
[29] Hao Z J, Liu H P, Li GH and Sun Y N 2013 *Journal of Ningxia Medical College* **35 (3)** 199-200.

Statistical Analysis of Meteorological Disasters in the Jiangsu Power Grid and Prevention Methods

Y L Lu[1,a], F B Tao[1,b], C B Hu[1,c], Z C Zhou [1,d] and L Sa[1,e]

[1]Electric Power Research Institute of Jiangsu Province Electric Power Company, Nanjing, China

15105182955@163.com

Abstract. With the obvious global climate change and the rapid development of an industrial economy in Jiangsu, the number of failures in power equipment caused by meteorological disasters has been increasing every year, especially contamination flashover faults and wind deflection faults. Based on extensive investigation and analysis, this paper conducts a statistical analysis of the temporal and spatial distribution of the typical meteorological conditions in the Jiangsu area. This paper also analyses the effects of the main meteorological disasters and defensive measures that have been taken to prevent them in recent years, and the paper summarizes the characteristics of the faults in a typical case. Finally, the paper gives some overall advice for preventative measures from the perspective of design planning, applications of new technology, disaster warnings and so on. These recommendations may provide an important reference for the design, maintenance and accident prevention for power grids that experience similar meteorological disasters.

1.Introduction

Jiangsu Province is located in a subtropical and warm temperate climate transition zone. Under the combined effects of solar radiation, atmospheric circulation and its specific geographical location, this province experiences a variety of meteorological disasters. The meteorological disasters that have the greatest affect on power grids include wind, lightning, thunderstorms, fog, haze, floods, ice and snow. Wind, fog and hazy weather often cause disasters. Due to global warming, Jiangsu's climate has changed significantly in recent years, and network failures caused by meteorological disasters are also increasing every year. Therefore, collecting meteorological information and planning for disaster prevention are essential for the safe and stable operation of power grid.

2.Main Meteorological Disasters and Defense Situations

2.1 Wind Disasters

2.1.1 Influence of Wind Disasters on Jiangsu Power Grid. According to statistics, winds may occur each month in Jiangsu, and they mainly occur in March, April, July and August. Regional winds occur an average of 22.4 days per year. The year with the most wind was 1965, with 78 days, and the years with the least were 1995 and 1999, with continuous six days.

On August 9, 2013, the 500kV East-III-line in Jiangsu was successively tripped due to winds. The instantaneous squalling of the wind was at level 12, with speeds of 33.0 m/s, which is the most powerful wind Jiangsu has had in recent years. The strong wind exceeded the designed maximum wind speed of the tower (30m/s) and resulted in the discharge of an insulator string with an "I" shape on the tower body, as shown in Fig. 1. On August 10, 2013, severe convective weather with strong winds and thunderstorms appeared in Wuxi, and the maximum wind speed was 29 m/s, which was greater than the tower's designed wind speed of 25m/s, and caused the No. 25, No. 26 and No. 27 towers in the 110kV Xi-Ting-line to fall down.

Historical data shows other incidences of wind damage to 500kV transmission towers in Jiangsu; for example, the collapse of several 500 kV towers in the Ren-Shang- 5237-line, which is an important line for the area's West-to-East transmission channel. This accident caused 10 transmission towers to collapse in a row, causing power failure across a very large area[1].

Fig.1. Discharge traces of wind deflection fault on the East -III- line

In addition to transmission line tower collapse caused by tornadoes and squall line winds, damage caused by typhoons has also occurred on the grid. In recent years, when a typhoon has passed by, winds from level 7 to level 10 and heavy rains have occurred in Jiangsu and caused much damage to the distribution network equipment [2].

2.1.2 Wind Disaster Prevention.(1) Consider the local environment when designing transmission equipment to ensure the design standards conform to wind prevention standards.

(2) Carry out wind observations at different heights and topographical features and discuss the selection of the meteorological conditions that were considered during the design process.

(3)Comprehensively check the air gap of AC and DC lines after adjusting insulation, or use long-type insulators according to wind zone figures and weather conditions.

(4) Define technical parameters such as the maximum instantaneous wind speed, the uneven coefficient of wind pressure and the line's trajectory in strong winds by studying the meteorological parameters of transmission towers and the online wind-deflection monitoring system of the transmission lines.

2.2 Thunderstorm Disasters

2.2.1 Influence of Thunder Disasters on Jiangsu Power Grid. Affected by environment and location, lightning activities in Jiangsu province are very frequent [3]. Areas with an unusual amount of lightning activities are Jiangsu's north, coastal areas and a part of Jiangsu's south. Lightning activities mainly appear in June and August each year. Thunder activities increased greatly in 2011 and were the highest in recent years, as shown in Fig. 2. According to spatial distribution data, lightning activities in the areas of Huai'an, Suzhou, Yancheng and Nantong are higher than average.

IOP Conf. Series: Earth and Environmental Science **86** (2017) 012021 doi:10.1088/1755-1315/86/1/012021

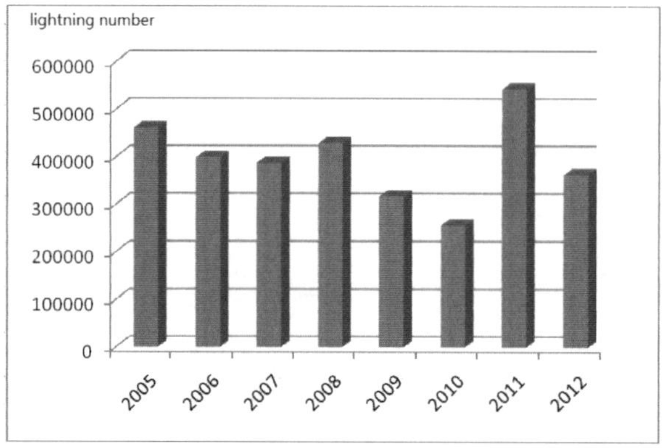

Fig.2. Temporal distribution of lightning in a recent 8-year-period

According to statistics, lightning-hit faults on transmission lines of 220kV and above occurred 304 times from 2005 to 2013, including 67 times on 500kVlines and 237 times on 220kV lines. Lightning-hit faults on 35kV and 110kV lines occurred 408 times and the percentage is 93.36.The number of lightning-hit faults is much higher in mountainous areas, areas with high soil resistivity and complex terrain. The variation of lightning hit faults in 220kV and above transmission lines in Jiangsu in recent years is shown in Fig. 3.

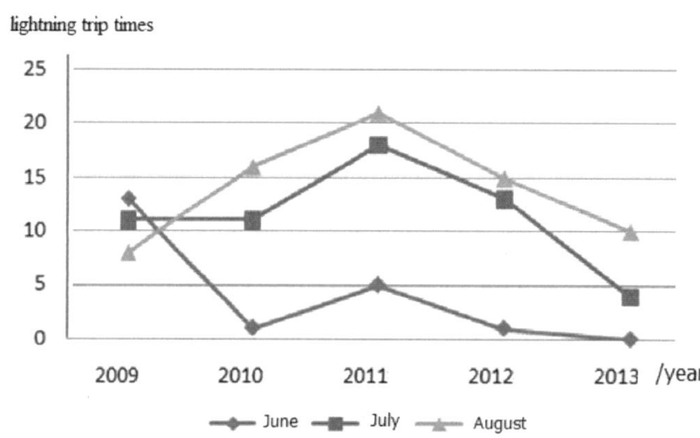

Fig.3. Changes in lightning trip times in 220kV and above lines in a recent 5-year period

2.2.2 Thunderstorm Disaster Prevention. (1) Collect meteorological data near transmission lines and make systematic analyses during the design phase to avoid areas with more lightning activity.

(2) Strengthen inspection of protection facilities and grounding devices in areas with more lightning activity in the operation phase.

(3) Install line-type lightning arresters on important 500kV and above transmission lines, including on backbone grids, important transmission channels for large power transfers, and high towers in mountain areas, which are vulnerable to lightning.

(4) Set up coupling ground wires for 35kV-110kV lines which flashover frequently because of lightning-hits.

(5) Use lightning locating technology and warning systems to locate lightning-faults, and to identify and warn about lightning-faults quickly.

2.3 Haze Weather

2.3.1 Influence of Haze Weather on Jiangsu Power Equipment. There are two main aspects of hazy weather that affect power equipment. First, humidity dampens the external insulation surface of equipment and causes flashovers. Second, hazy grime increases the contamination of the insulation;

when coupled with high humidity, this may also cause flashovers[4].Haze is most frequent in autumn and winter in Jiangsu, and its conditions are quite different in different areas because of the different locations, terrain and surroundings.

2.3.2 Haze Disaster Prevention. (1)Adjust the insulation distance of partially polluted lines, replace porcelain insulators with composite insulators or increase the number of insulators according to tower size.

(2)Strengthen statistical analysis of various meteorological information to provide a basis for design, operation and maintenance. For example, try to avoid pollution sources and strengthen the inspection of the equipment.

(3)Establish an updated mechanism to identify pollution sources and properly shorten the revision period of polluted area maps. A management unit should adjust the anti-polluting-flash strategy according to the newest polluted area map.

(4)Establish a grid pollution warning system and analyze the current situation of pollution flashover.

2.4 Ice

2.4.1 Influence of Ice Disaster on Jiangsu Power Grid Equipment. There are few ice disasters in Jiangsu and the ice disaster in 2008 did not affect the Jiangsu power grid. But in 2009, an ice disaster occurred in a small area of Nanjing. According to fault locations, most fault lines were in vertical alignment and discharge points were in the spans where the line was high and exposed. According to investigations, the terrain was close to the Yangtze-River and located in place with a tuyere, so the ice coating was formed in a micro-landform.

2.4.2 Ice Disaster Prevention. (1) Adopt reasonable planning and designing methods.The planning and design department should select the line path appropriately to avoid areas where ice is easily created, such as near rivers and lakes. If it is impossible to avoid such areas, take necessary measures to improve ice disaster prevention, such as increasing the ice-overload capacity of tension tower[5].

(2)Adopt effective measures of melting ice and deicing.Deal with ice-coated equipment quickly using current-melting technology, mechanical deicing and passive deicing methods [6].

(3)Take measures to prevent iced-lines waving.Install inter-phase anti-wave devices such as interval rods, double waving suppressors and clamp-rotary-type interval rods to inhibit uneven ice-coated waving and jumping when ice falls off.

(4)Take effective measures of ice monitoring and observation.

2.5 Other Disasters

In addition to the main meteorological disasters discussed above, there are have also been snow storms, heat, hail, rainstorms and other disasters in Jiangsu caused by abnormal climate conditions in recent years. The bad weather has also influenced power grids [7][8].For example, in November of 2009, Nanjing's lines were hit by a blizzard and several 35kV and above transmission lines broke.

3.Meteorological Disasters Prevention Advice

3.1 Strengthen Meteorological Survey and Information Collection; Raise Disaster Relief Design Level of Equipment Scientifically and Reasonably. Collect meteorological information on elements such as wind, lightning, ice and temperature from meteorological stations near to electric grids and consider wind zone figures, ice figures, polluted area maps, lightning location systems and operating experience comprehensively. Define designed wind speed, designed ice-thickness, insulation creep distance, thunderstorm days and temperature, and eliminate various meteorological factors.

*3.2 Deepen Research on Disasters; Explore new Technology and new Devices to Actively Improve the Disaster Prevention Ability of Power Grid.*Make further study of disasters such as lightning, typhoons and ice, and build fault probability models to assess the effects of disasters on power grid equipment.

Cooperate with meteorological departments and deploy new wind resistance technologiesto prevent collapses accidents [9].

3.3 Strengthen On-line Monitoring of Grid Disasters; Establish Warning, Emergency and Rapid Recovery Mechanisms. Carry out accurate monitoring systems, collect basic data and make risk assessments. Research typhoons, thunderstorms, freezing conditions, fog and other disasters that may cause large-area faults. Combine geographical data, attribute data and results of evaluation with GIS technology, establish an early warning system for meteorological disaster accidents [10],[11],[13] ,set up emergency mechanisms and post-disaster quick recovery and reconstruction mechanisms.

4.Acknowledgements

This work is supported by Natural Science Foundation of China (Grants No. 51407026)

References

[1] Xie Q, Zhang Y and Li J2006Investigation on Tower Collapses of 500kV Ren-Shang 5237 Transmission Line Caused by Downburst. *Power System Technology.***10**59-63.
[2] Peng X Y, Huang Z W 2010 Analysis on the Cause of Distribution Line's Damage during Typhoon and Counteract Measures. *Southern Power System Technology Practice & Experience.***4** 99-102.
[3] Liu Z Y, Gu L and Ye H2008Analysis of Lightning Flashover for Overhead Transmission Lines from 2004 to 2006 in Jiangsu Power Grid. *Jiangsu Electrical Engineering.***2**42-45.
[4] Yuan G Z, Luo S B and Luo Q H 2011 Meteorological Research on Contamination Flashover of Guangdong Power Grid. *Yunnan Electric Power***4**82-83.
[5] Li Q 2008 Analysis of the Freezing Rain and Frost Disaster of Year 2008 and Its Significance to Power Grid. *Electric Power Construction.***6**18-21.
[6] Li Z H, Bai X M,Zhou Z G and etc 2008 Prevention and Treatment Methods of Ice Coating in Power Networks and Its Recent Study. *Power System Technology.***4**8-14.
[7] Wang Z K 2012 The Comprehensive Assessment of the Power Loss and the Economic Impact of the Power Outages under the Flooding.*Hunan University.***6**4-6.
[8] Zhang S L 2011 Influences of High Temperature on Transmission Equipment and Countermeasures. *Shanxi Electric power.***3**35-37.
[9] Cong R G 2013 Impact of Natural Disasters on Electric Power Systems of China. a Literature Review.*Journal of Xihua University Natural Science.***32** 105-112.
[10] Zhou W, Miao S, Qu J T and etc 2008 Research on precise warning of meteorological disasters in electric network system. *Journal of Yunnan University.***30** 286-290.
[11] Xie Q 2006 Current situation of natural disaster in electric power system and countermeasures.*Journal of Natural Disasters.***15**126-131.
[12] BAO B, Cheng R L, Xiong X F, Weng S J and Wang J 2014 A typhoon risk early warning method for power transmission line considering micro-terrain correction. *Power System Protection and Control***42**79-86.
[13] Xiong X F, Weng S J,Wang J, LI Z and Liang Y 2015 An online early warning method for windage yaw discharge of jumper towards JG type strain tower considering corrected by rainfall.*Power System Protection and Control.***43** 136-143.

Feature-based Approach in Product Design with Energy Efficiency Consideration

D D Li [1] and Y J Zhang*[2]

[1]Faculty of Humanities and Foreign Languages, Xi'an University of Technology, Xi'an, China.
[2]School of Mechanical Engineering, Xi'an Jiaotong University, Xi'an, Shaanxi, P. R. China.

E-mail:*yjzhang@mail.xjtu.edu.cn

Abstract. In this paper, a method to measure the energy efficiency and ecological footprint metrics of features is proposed for product design. First the energy consumption models of various manufacturing features, like cutting feature, welding feature, etc. are studied. Then, the total energy consumption of a product is modeled and estimated according to its features. Finally, feature chains that combined by several sequence features based on the producing operation orders are defined and analyzed to calculate global optimal solution. The corresponding assessment model is also proposed to estimate their energy efficiency and ecological footprint. Finally, an example is given to validate the proposed approach in the improvement of sustainability.

1.Introduction

As known all, the energy efficiency and environmentally friendliness of a product is very much depending on its usages, production and distribution only towards the end of the lifecycle. In fact, the energy efficiency and ecological footprint during usages, distribution and dispose stages have been studied for many authors [1-4]. Although the concept of features and the concept of design for manufacturing has been introduced for many decades, the influence of design on the energy consumption and carbon emission in the manufacturing stage have already been overlooked for many decades. From the process planning point of view, the allocation of machining stock to each machining process is very crucial in terms of energy consumption. Optimization of process plans and process parameters can effectively reduce the production energy [5-7]. For manufacturing companies, the abilities of manufacturing to respond changes in the market rely heavily on the process plan. The process plan is vitally important in providing a competitive edge to manufacturing [8]. The process sequence is fixed in the numerical control code on numerically controlled equipment. The operation instructions are described according to the relative motion between cutting tool and work piece. The drawback of using the fixed process sequence is the lack of activity to the unexpected events that occur at shop-floor level during executing phase [9].

To achieve the objective of energy efficiency in manufacturing stage, all available manufacturing ways should be evaluated in product design stage for optimum. To address this problem, in this paper, a manufacturing oriented product design approach is proposed. In the following sections, Section 2 describes the definition, classification and energy estimation modelling of features. Section 3 presents the energy estimation model of products. Section 4 is a case study, which is following by the section with the conclusions and indicates some directions for further study.

2.Classification and Energy Modeling of Features

From the viewpoint of product manufacturers, product features can be classified as cutting, welding and forming operations, etc. In some cases, these generation approaches are alternative if the requirements and constraints could be guaranteed, but their required energies are mostly different. To help to select adaptive features, techniques for the estimation of energy consumption and environmental impact of these features should be studied. Although the cutting feature and welding feature are only covered in this discussion the technique can similarly be transferred to other types of features.

Cutting features are generated by machine tools, and the types and power required of machines are usually concerned. Although there are different in energy consumption for all kinds of machine tool, they are very less than those in other generating approaches. According to the energy model proposed by Gutowski et al[10], the overall power required of a machining process depends on the base power demand and the specific power required of the cutting operations. The idle power is the base power demand of equipment features that support the machine tools, such as, power to start-up the computer, motor, coolant pump, and so on. Furthermore, the power consumption in generate a given feature can be deduced by considering its process time, as follows.

$$P_f = (P_0 + k \cdot MRR) \cdot t \quad (1)$$

where t is the required process time for creation of the feature. From equation (1), we can be clearly seen that P_0 dominates direct power consumption, and depends on the machine tool. Therefore the selection of machine tools has significant impact on the direct power consumption in creation of the feature.

(a)

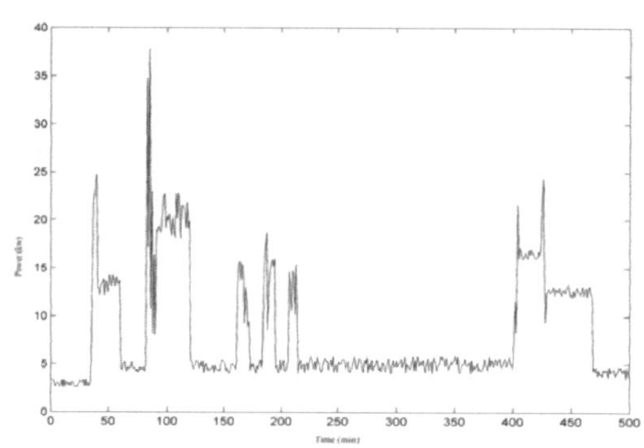

(b)

Fig. 1 Energy profile of total welding operation process: (a) workpiece; (b) load curve

Unlike the cutting features, a welding process not only concerns with the process equipment but several geometry features and actual joining needs. To analyze electrical power consumption of the welding processes, voltage and current are fundamental process parameters, which are supplied by the welding power source.

As shown in Fig.1, there are several events in the energy profile, i.e. the load curve of the welding process, but we can extract a specific the welding process by analyzing it in detail. Then the total energy consumption in the welding process can be estimated from this load curve segment.

Practically, this load curve describes total energy consumption and states of all relative equipment used in actual welding operations. Those include preparation, setup and post processing of machine equipment, etc.

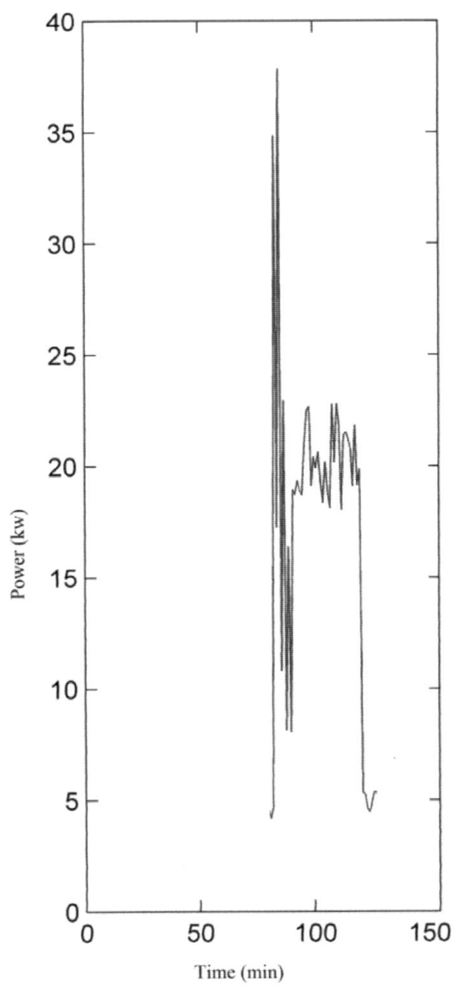

Fig. 2 Load profile of the welding operation

By analyzing the load curve of a welding feature, as shown in Fig.2, the relation between the size of welding seam and the time taken on this welding process is determined. The energy consumption of unit length of the welding seam with given welding approach can be estimated. Also the energy consumption in other stages, such as startup, warming, waiting for operation can be taken into considered. Since the load curve still is not steady as a strict straight line in the actual welding operation stage, averaging is required in the estimation of energy consumption of the welding seam with unit length. The electrical power consumption of a welding feature can be computed as

$$P_f = P_s + P_m + P_m' + P_e \quad (2)$$

Where P_s represents the electrical power consumption at start up stage, P_m the consumption of warming up, P_m' the consumption of waiting, P_w the consumption during welding operation. After the total consumption is determined in the welding period, assume that L represents the welding seam length of the welding feature, the electrical power consumption in welding seam with unit length can be computed as

$$\xi = \frac{P_w}{L} \quad (3)$$

Once P_s, P_m, P_m', ξ are obtained for a welding feature they can be stored and used as the parameters of the feature. Then the electrical power consumption for a specific welding feature ξ with seam length L_ξ can be estimated as

$$P(\xi) = P_s + P_m + P_m' + \xi \cdot P_w \quad (4)$$

3.Energy Consumption Model from the Viewpoint of Product

Although the score of a specific feature in the energy and environmental impact can be determined from Section 2, from product viewpoint, all possible schemes with different features should be checked or measured during the design stage and throughout the manufacturing process. To solve this problem, we proposed a general model to capture total energy footprint and ecological footprint. Suppose that N features of a product need to be generated, and all these possible generating approaches are represented by $\xi(s,k)$. Where s is the feature number, which is used for the identification of features $(s = 1,2,\ldots,N)$, and k denotes a specific approach for the generation of features $(k = 1,2,\ldots,K)$ .The energy efficiency and ecological footprint metrics can be measured as an assignment of a generation approach to a corresponding feature of the part, in such a way, the primary assessment function of energy efficiency and ecological footprint for a part can be defined as follows:

$$E[p] = \sum_{i=1,N} \xi(f(i), m(j)) \quad (5)$$

Where $\xi(f, m)$ represents the total energy consumption and ecological footprint of the feature f by using the specific approach m .In addition, the energy footprint and ecological footprint f of the features are defined and used as a two dimensional vector so that the energy and ecological footprint measures could be estimated separately.

4.Experiments

To demonstrate the feasibility of proposed approach, a simple product that consisted of two parts is used. Assume that there are two alternative structures of the product, as shown in Fig.3.

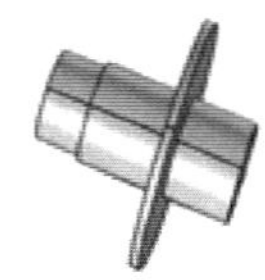

(a)

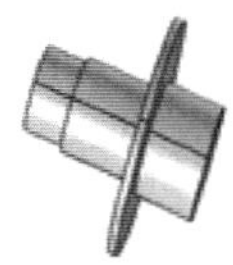

(b)

Fig.3 Two kinds of alternative structures of a product

In (a), the disk shape part should be assembled by welding operation. In this case, no shoulder is required in designed for the rotation axis. Against, if cutting features were considered in the assembly of the disk shape part a shoulder would have to be planned, as shown in (b), to locate the disk in the assembly operation. Sometime, a shoulder and rings for shoulder also are required. In this experiment, bar-stocks are used as blanks for the generation of the rotational components. To generate this product with structure (a), the bar-stock is chosen with diameter 100 and length 300. For the structure (b), bar-stock with the diameter 120 and length 300 is used. In this test, the machine tool (MAHO 60) is selected to perform the machining operation, and the welding equipment (HB350) is used to carry out the welding operation. The material removal volume and welding seam length on the components can be determined, further the total energy consumption in generating the two components can be obtained. The results are given in Table 1.

Table1Energy required of total manufacturing processes

products	**Material removal Volume** ($\times 10^5 mm^3$)	**Welding seam Length (mm)**	**Required Total Energy (J)**
Fig.3 (a)	7.037	282.7433	5895.4
Fig.3 (b)	17.719	0	6723.8

5.Conclusion

In this paper, the product design with energy efficiency consideration of manufacturing stage is studied. The concept of manufacturing features is introduced by grouping and definition of various production processes. Moreover, the estimation for the energy consumption of features in the production process is modelled for the evaluation purpose. Although the cutting features and welding features have been only considered and analyzed other features can be modelled by the same way.

However, considering that each of manufacturing feature type has its own characters, it has to be modelled and analyzed in practice. On the other hand, the models for energy consumption are very rough

IOP Conf. Series: Earth and Environmental Science **86** (2017) 012022 doi:10.1088/1755-1315/86/1/012022

in current study. Further work should be done to improve the evaluation. It still is a challenge for future research.

6.Acknowledgements

This work is supported by National Natural Science Foundation of China (Grant No. 51375377)

References

[1] ChristianM and NilsW 2015 Process chain evaluation for an over all optimization of energy efficiency in manufacturing-The welding case. *Robot Comput-Integr Manuf.* 34.
[2] FilippiAD andIppolitoR 1981 NC machine tools as electrical energy users. *Ann of the CIRP* 30.
[3] AvramOI and XirouchakisP 2011 Evaluating the use phase energy requirements of a machine tool system. *J Clean Prod.* 19.
[4] ByrneG and ScholtaE 1993 Environmentally clean machining processes. *A strategic approach,Ann of the CIRP*.42.
[5] YansongG, JoostR D and BertL 2014 Energy-based optimization of the material stock allowance for turning-grinding process sequence. *Int J AdvManuf Technol.* 75.
[6] XinyuS, XinyuL, LiangG and ChaoyongZ 2009 Integration of process planning and scheduling-a modified genetic algorithm based approach. *ComputOper Res.* 36.
[7] ChenQ and KhoshnevisB 1990 Scheduling with flexible process plans. *Prod Plan Control* 4.
[8] BhaskaranK 1990 Process plan selection. *Int J Prod Res* 28.
[9] PellegrinelliS and TolioT 2013 Pallet operation sequencing based on network part program logic. *Robot ComputIntegrManuf*29.
[10] GutowskiT, Dahmus J and ThiriezA 2006 Electrical energy requirements for manufacturing processes, Proceedings of the 13th CIRP Conference on Life Cycle Engineering. *Leuven, Belgium.*

Based on Artificial Neural Network to Realize K-Parameter Analysis of Vehicle Air Spring System

San-Shan Hung, Chia-Ning Hsu, Chang-Chou Hwang and Wen-Jan Chen

No.100, Wenhua Rd., Xitun Dist., Taichung City 407, Taiwan (R.O.C.),Taichung, Taiwan

Email:*anita74987498@gmail.com

Abstract. In recent years, because of the air-spring control technique is more mature, that air-spring suspension systems already can be used to replace the classical vehicle suspension system.Depend on internal pre-ssure variation of the air-spring,thestiffnessand the damping factor can be adjusted.

Because of air-spring has highly nonlinear characteristic, therefore it isn't easy to construct the classical controller to control the air-spring effectively. The paper based on Artificial Neural Network to propose a feasible control strategy. By using offline way for the neural network design and learning to the air-spring in different initial pressures and different loads, offline method through, predict air-spring stiffness parameter to establish a model. Finally, through adjusting air-spring internal pressure to change the K-parameter of the air-spring, realize the well dynamic control performance of air-spring suspension.

1.Introduction

The suspension system is the structure between auto-mobile body and wheel. With more and more wide-spread use of air-spring, there are many car manu-factures uses for the pneumatic suspension system. Using compressed air inflated or deflated by a spring inside the bellow shape capsule, changes in air pressure inside the spring in order to achieve the adjustment of the spring K parameter and car chassis height adapt to different road conditions.
This research will aim at the different initial pressure and load capacity, is forecast the air suspension soft and hard degree and internal pressure for the kind neural network. chosen the driving operator schema by the user, the operator schema had decided the suspension system parameters (damping factor and spring constant) (most greatly longitudinal changed and stable time with the system dynamic characteristic highly), but by the change air suspension internal pressure, according to the driver change of demand spring constant, achieved the user to anticipate the pressure scope that by the pressure dependent control opening solenoid valve air admission or the exhaust , realized the goal of spring K value adjustment.

2.Methodology

2.1.Air-spring Introduction and Analysis. In this research, the air spring is a dual winding box with bellow, metal wires are inlaid in the rubber body in order to improve the ability to withstand the pre-ssure of the cavity. Since the air spring K value non-linear characteristics, we can learn that the air spring not only determined by the material of the air spring but also the internal pressure. The coefficient K of the air spring is shown in Eq. (1).

$$K = K_0 + \Delta K \tag{1}$$

Where, $K_0$is determined by the air spring material, further initial pressure P_0, the amount of stretching of the spring and internal pressure changes will produce different ΔK increments. Therefore, the air springK value is nonlinear. In this research, we focus of the relevant parameters of the spring K value variable as the neural network input and output, thereby we can predict the current K value of the air spring and internal pressure [1].

2.2.Particle swarm optimization method combined with neural network learning. A multi-layer feed-forward networks architecture was used in this research[2][3], when the input transformed by the connection weight and activation function, the prediction output of neural network compared with the real target output, by adjusting the weights, so that the prediction output more close to the real output. Its neural training input and output architecture were shown in Fig. 1:

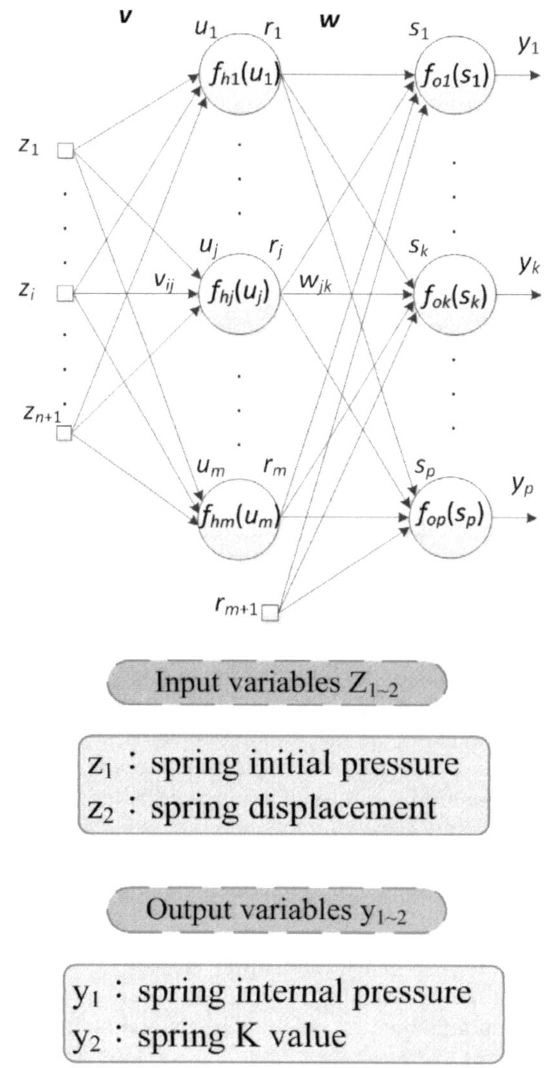

Fig. 1 Air springs neural training input-output chart

Among them, the neural network training input vari-able for the initial internal pressure of the air spring and compression spring, a total of two dimensions, as input variable of $Z_{1\sim2}$ in Fig. 1 The output variable were internal pressure of the air spring and spring coeffi-cients two dimensions, as input variable $y_{1\sim2}$ in Fig. 1 The neural network connection weight, connecting the input layer to the hidden layer weights number v_{ij} and the connecting hidden layer to the output layer weights number w_{jk}, constituting each particle component of the PSO algorithm, the sum of the number of re-ANN rights, such as Eq. (2):

$$\text{number of } v_{ij} \text{ weights}: (n+1)\times m \tag{2.a}$$

$$\text{number of } w_{jk} \text{ weights}: (m+1)\times p \tag{2.b}$$

$$\text{number of weights}: (n+1)\times m + (m+1)\times p \tag{2.c}$$

Among them, n is the input node, m is the number of hidden layer node, p is the output node. Using the Eq. (2) to obtain particle dimension of PSO, by constantly revised weights, selecting weights v_{ij} and w_{jk} to minimize the sum of squared error which represents Eq. (3) below:

$$E_{total} := \frac{1}{2}\sum_{q=1}^{l}\sum_{k=1}^{p} e_{qk}^{2} \quad , q \in \underline{l},\ k \in \underline{p} \tag{3}$$

The objective function values calculated by PSO algorithm, change the particle component is minimized so that the objective function[4]. Mean Square Error and Mean Absolute Deviation said that if the Eq. (4) and (5) below, as machine learning to assess the effectiveness indicators.

$$MSE = \frac{\sum_{q=1}^{l}\sum_{k=1}^{p} e_{qk}^{2}}{l \cdot p} \tag{4}$$

$$MAD = \frac{\sum_{q=1}^{l}\sum_{k=1}^{p} \left| e_{qk} \right|}{l \cdot p} \tag{5}$$

In the Eq. (4) and (5), l is the total number of data, p is the number of output nodes.

2.3. Controller design. This research will be adjusted the air spring and setting up the rational K value range of the air spring by drivers, according to road conditions in accordance with the driver to air spring parameter K value. Upgrade the intersection the type of input and parameter of neural network, stopping predict the intersection of k value and reasonable range, until predict output the inter-section of K and value according with driver. Pressure forecast output pressure becomes the target K value desired internal air spring arrival value. Using pressure dependent control, through the solenoid valve's on-off, changing the supply of compressed air to the air pressure inside the spring or the spring as part of the internal air emissions, to achieve the appropriate inter-nal pressure, thus achieving the purpose of adjusting spring K value. The controller of this research can be divided into four blocks, it contains offline training the neural network weights, the neural network to predict the output of the operator to adjust the driving pattern and pressure regulation system controller design block diagram, shown as Fig. 2 :

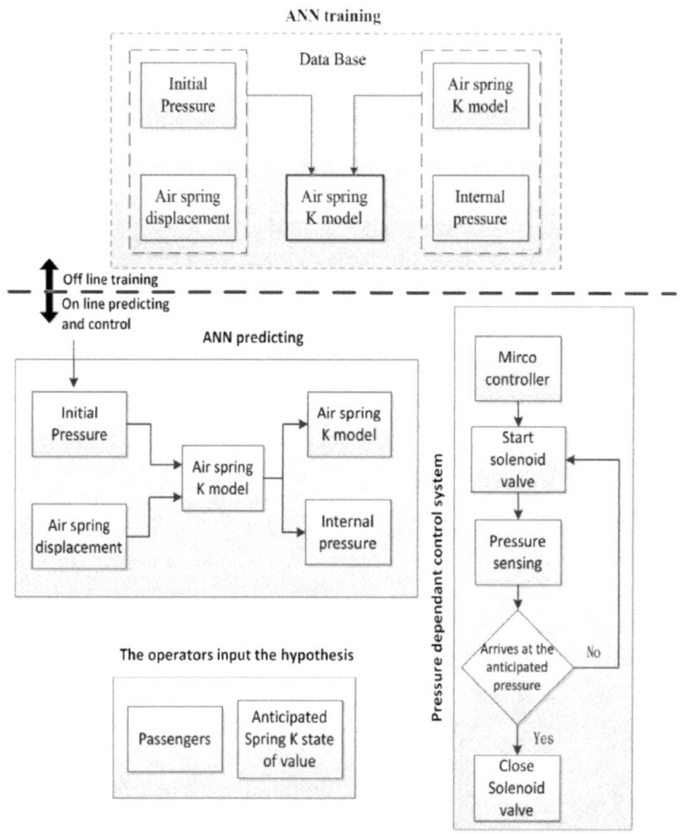

Fig. 2 controller design block diagram.

3.Resultstructure

*3.1.Training the neural network forecasting results.*This research experimental stuff is 1,200, is divided into the training material 900 and test material 300, by hideaway level node integer (m)4~6 forecast that the training material and has not participated in the training flow the test data analyzing. Will conform to the hideaway level node integer and weight value of this research uses the weight value as the forecast. The training and predicting input/output parameter, as shown in Table 1:

Table 1The training and predicting input/output parameters

ANN input parameters		Actual output		ANN predicted output parameters	
Z1	Z2	d1	d2	y1	y2
Air -spring initial pressure	Spring displacement	Spring internal pressure	Spring K values	Spring internal pressure	Spring K values

When the kind neural network hideaway level node integer (m) hypothesis is 4~6, obtains the best weight value after the training, the test data by the kind neural network forecast output result, and will compare with the actual output, and may know the forecast output result good or bad degree. Because the hypothesis neural network output integer is 2, in order to demonstrate good or bad degree of its predict that will take the abscissa hypothesis as actual output d1 and d2, the ordinate outputs y1 and y2 for the forecast presents the forecast good or bad degree, shown in Fig. 3 to Fig. 5.

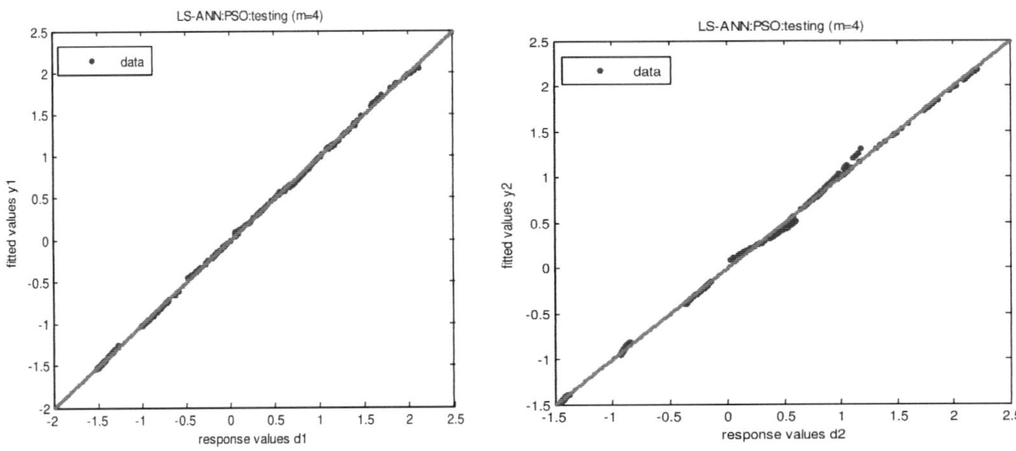

Fig. 3Test data actual value and predicted value material compare（m=4）

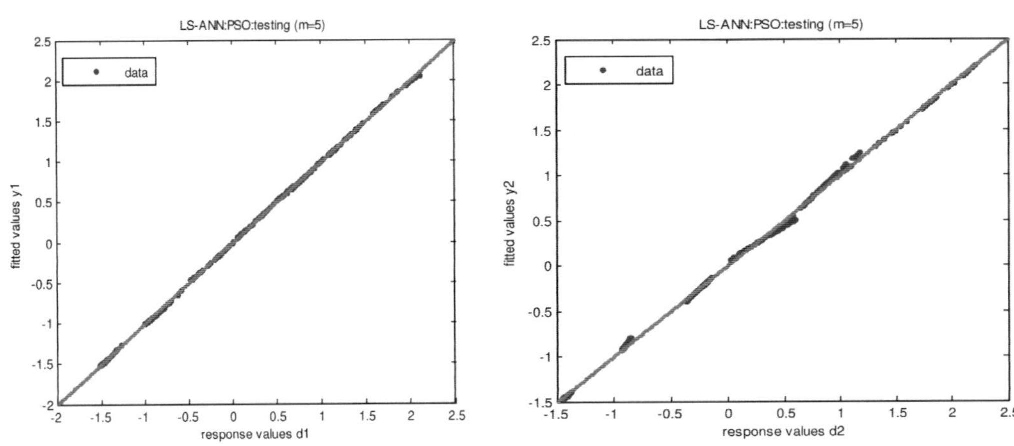

Fig. 4Test data actual value and predicted value material compare（m=5）

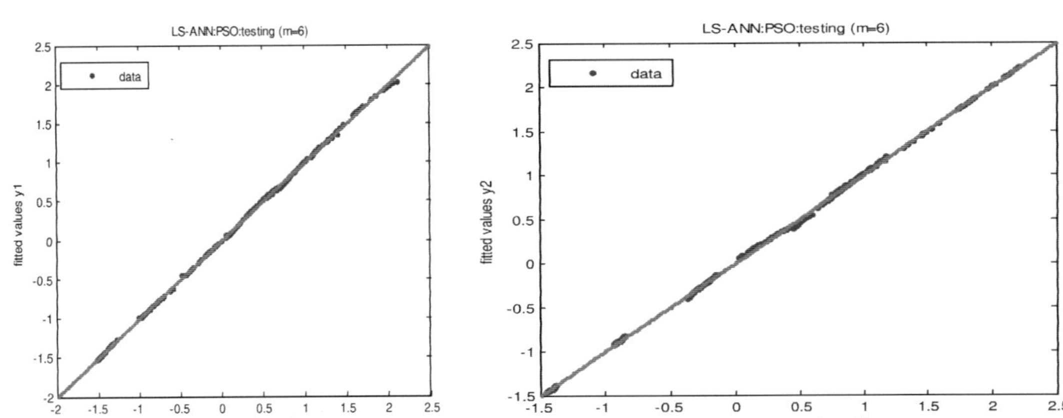

Fig. 5 Test data actual value and predicted value material compare（m=6）

In cross-validation method to verify the accuracy of their training and testing [5], and depending on the number of hidden nodes (m= 4~6) interactive validation process of training data and test data of MSE(Mean Square Error, MSE) and MAD(Mean Absolute Deviation, MAD). Results in 10 tests were averaged (Mean) and the number (Median), the number of hidden nodes chosen for this study. The results are shown in Table 2.

Table 2 Different hidden node (m) test data validation results.

m			4	5	6
Testing data (120 pen data)	MSE	Mean	0.0007	0.0004	0.0003
		Median	0.0007	0.0004	0.0003
	MAD	Mean	0.0191	0.0156	0.0135
		Median	0.0195	0.0162	0.0133

3.2.Analysis of experimental results. From the results of cross-validation method, the selection (m = 4) hidden nodes as a prediction of the control of the use, the predicted results comply with the requirements of this study and neural network training parameters less than other hidden nodes, so that the training speed increases. For different initial pressures, depicting theoretical, compares actual and forecast Figure shown in Fig. 6. Table 3 is in the different initial pressure, the maximum percentage of predicted and actual neural network regression curve of relative error.

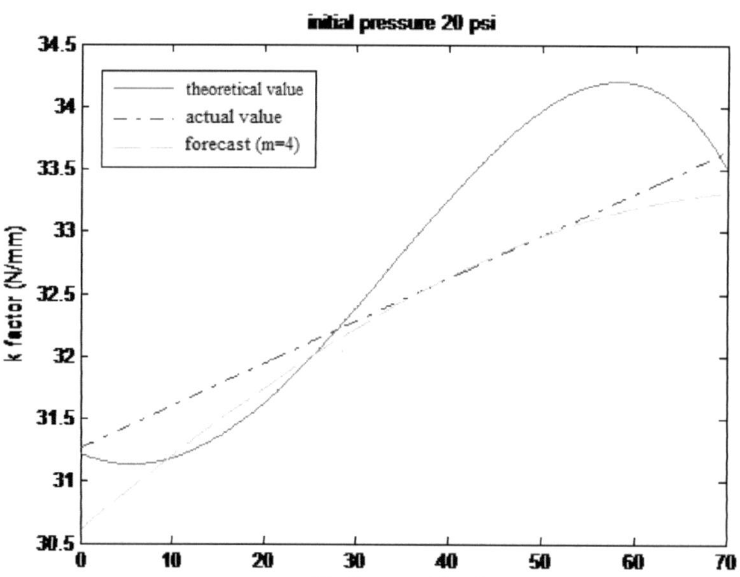

Initial pressure 20 psi

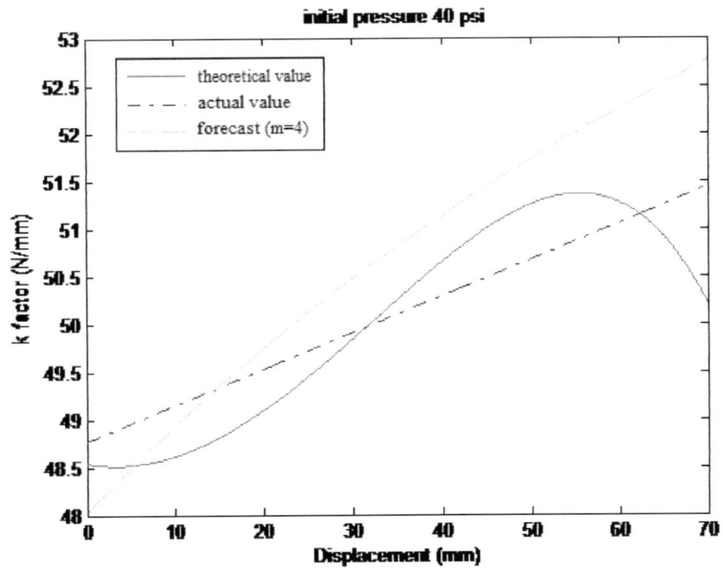

Initial pressure 40 psi

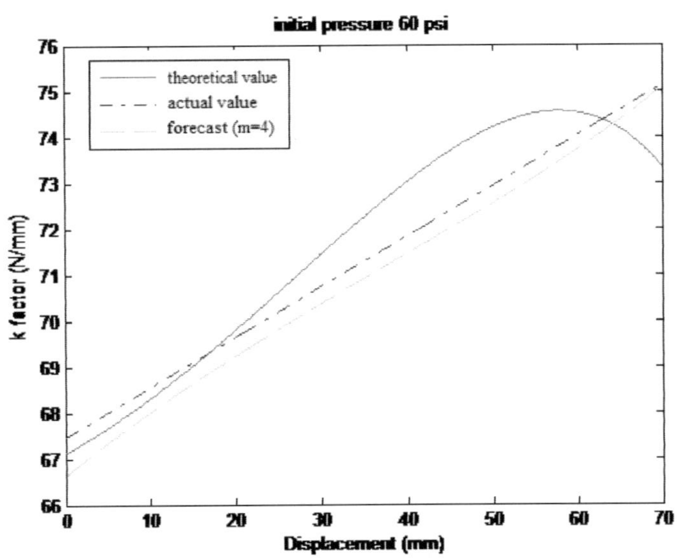

Initial pressure 60 psi

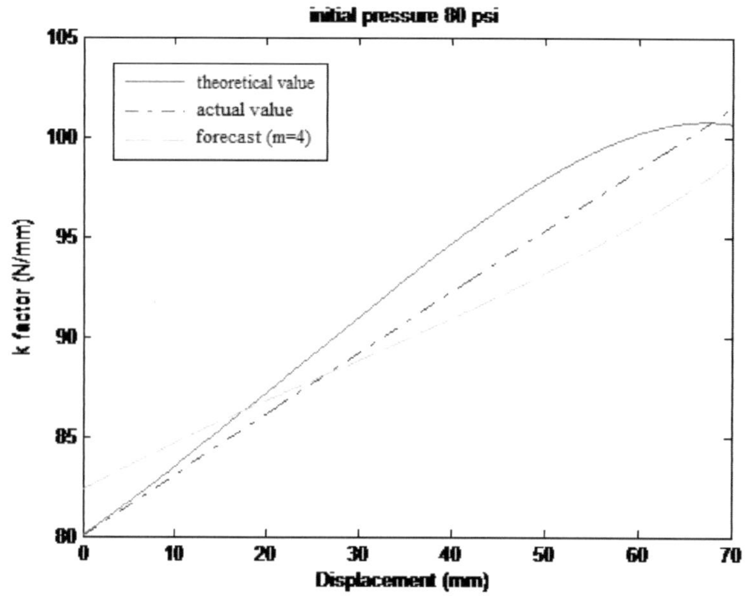

Initial pressure 80 psi

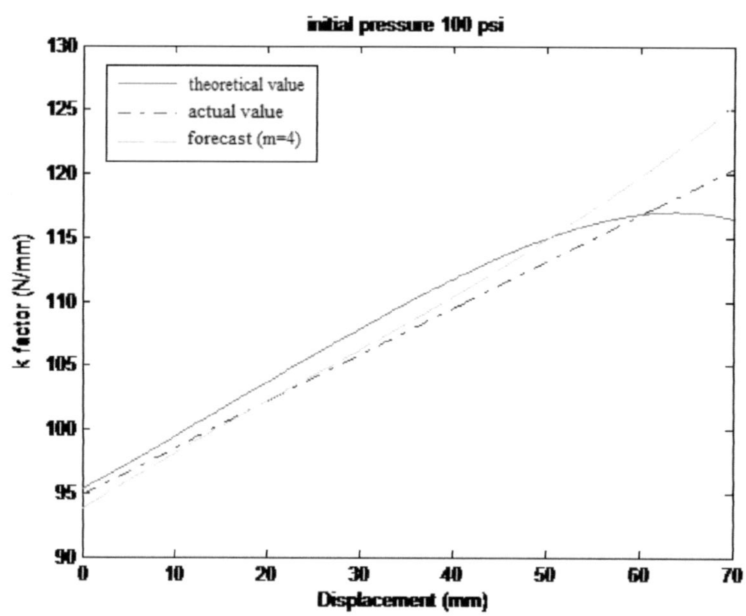

Initial pressure 100 psi

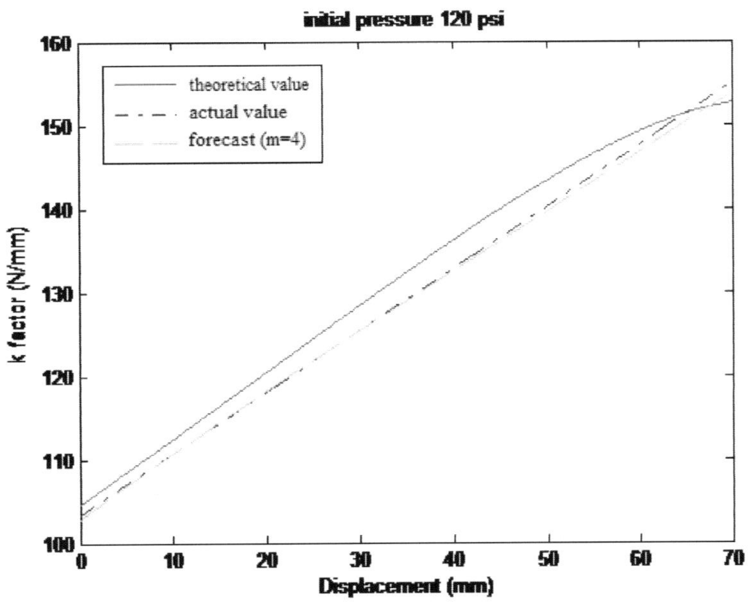

Initial pressure 120 psi

Fig.6 Different initial pressure springs K parameter for theoretical, actual and forecast comparison chart.

Table 3 Different initial pressure, the maximum spring coefficient predicted value and the actual value of the relative percentage error.

initial pressure（psi）	20	40	60	80	100	120
relative percentage error（%）	2.1	2.5	1.2	2.9	4.1	0.7

From Fig. 6 and Table 3, at low initial pressure, it predicted a big gap between the actual value, possible reasons for the internal pressure and K values of the two data points with the amount of compression on the more obvious changes, resulting in poor forecasts. Conversely, at high initial pressure, internal pressure and K values vary with the amount of compression is more obvious. Therefore, at high initial pressure, the predicted results are more accurate..

4.Conclusion

The parameters of suspension system decide the main key of riding comfort and handling. For riding on the rough road, the definition of air springs' shock peak and settling time are operational factors of riding comfort and concussion change time of tire and road. Pneumatic suspension system is subjected to the characteristic of hardware, it cannot and don't need to react immediately for those small changing road (ex: non slip nosing). The research proposes an effective control policy to the air spring of pneumatic suspension system, according to the driver's requirement that changes the internal pressure of the air spring make corresponding adjustments, realize good suspension system dynamic control.

The research uses neural network combine with PSO algorithms to learn the characteristic curve of air springs offline, depend on the initial pressure, the maximum percentage error of learning value K and the actual value are less than 5%, therefore it proving the feasibility of neural network predictive value K. With the trained weights, we can predict the spring value K of compression in air spring by different internal pressure. Thus, construct the air spring controller, then predicting the numerical difference between the value K of air spring and driver's expected value, by pressure-dependent change inside pressure of air spring to reach the perfect spring parameter K..

5.Acknowledgements

This research was supported by the Ministry of Science and Technology, R.O.C. We appreciate their tremendous support and sponsorship.

References

[1] PresthusM 2002 Derivation of Air Spring Model Parameters for Train Simulation. Lulea University of Technology, M.S. Thesis.
[2] HsiehJG, JengJH and LinY L 2015 Pathways to Machine Learning and Soft Computing.
[3] HaganMT, Demuth HBandBealeM 1996 Neural Network Design, Thomson Publishing Inc.
[4] KennedyJ and EberhartR C 1995 Particle swarm optimizationin: Proc. IEEE Int. Conference on Neural Networks, Perth, Australia, Vol. 4pp1942-48.
[5] KohaviR 1995 A Study of Cross-Validation and Bootstrap for Accuracy Estimation and Model Selection. Appears in the International Joint Conference on Artificial Intelligence (IJCAI), San Francisco, USA, Vol.2, pp1137-43.

IOP Conf. Series: Earth and Environmental Science **86** (2017) 012024 doi:10.1088/1755-1315/86/1/012024

Research on Power Calculation Method of High Speed Rotary Device under Wind Loads Crystals

M S Ji[1], Y Xue[2] and N Wu[3]

[1]School of Mechatronics Engineering, Harbin Institute of Technology,Harbin, China;

Email: *jimingshi19@163.com

Abstract. The wind load has a great influence on the power of large rotary devices working outdoors. In the power calculation formula of the rotary devices, the static air pressure is often used as the wind resistance of the whole device. But in fact, the rotating device bears the dynamic wind pressure during the rotation. This method of calculation will lead to large deviation. Based on this, this paper emphatically studied the dynamic wind load of the rotating device under rotation, and gave a more accurate formula for the calculation of the rotating power. This formula solves the problem of power calculation of the rotating device in high speed rotation. It can be widely used in all kinds of rotating devices.

1.Introduction

The rotating mechanism is widely used in all kinds of mechanical devices, such as tower crane, port machinery, rotating antenna and so on[1]. When calculating the power of rotating devices working in door or the devices with small integral size and slow speed, the wind load change is not considered. So the traditional power calculation formula of the rotary mechanism only considers the static wind load. With the development of modern society, the rotating devices are required to be worked under strong wind, and the rotating speed is also required to be faster. At this time, the wind load under the rotational state cannot be ignored.It becomes the main resistance of the rotating mechanism.

In the calculation formula applied in engineering machinery, the stationary wind load is taken as the wind load of rotary device[2]. But in fact, the rotating devices bear dynamic wind pressure in the process of rotation. So the calculated result obtained by this method are less than that in the true condition, when the speed is fast, the error is large. In the calculation formula applied in antenna, dynamic windresistance

resistance is obtained through the method that static wind resistance is multiplied by the dimensionless coefficients[3], and the results are often larger than the measured data. When the static wind resistance is zero, this result is not consistent with the actual situation.

In order to solve this problem, this paper considered the static and dynamic wind load conditions and gave the accurate calculation formula of dynamic wind load. Finally, the accuracy of the formula is verified by an example.

2.The mathematical model of the power calculation of rotary device

Calculating the driving force is the first step to calculate the power of the rotary device. The composition of rotary resistance moment M_{sw} is as follows:

$$M_{SW} = M_f + M_S + M_W + M_P \quad (1)$$

M_{sw}—rotary resistance moment; M_f—frictional resistance moment; M_S—the resistance moment caused by the incline of the rotary platform; M_W—the resistance moment caused by wind pressure; M_P—the resistance moment caused by inertia.

Power calculation of rotary mechanism:

$W = M_{SW} w$ (2)

W—rotary drive power; w—angular velocity.

2.1. The rotation resistance torque without wind load. The frictional resistance moment of the rotary devices can be determined by the following formula, and the force diagram is shown as Figure 1.

$M_f = 0.5\mu k(N_1 + N_2)D_0$ (3)

The resistance model is shown in Figure 2. According to the model, the calculation formula is shown as follow:

$M_S = (P_Q R + G_b r - G_1 l_1)\sin\gamma \sin\varphi$ (4)

G_1—Weight load of rotary platform; G_b—Weight load of the upper part of the rotary

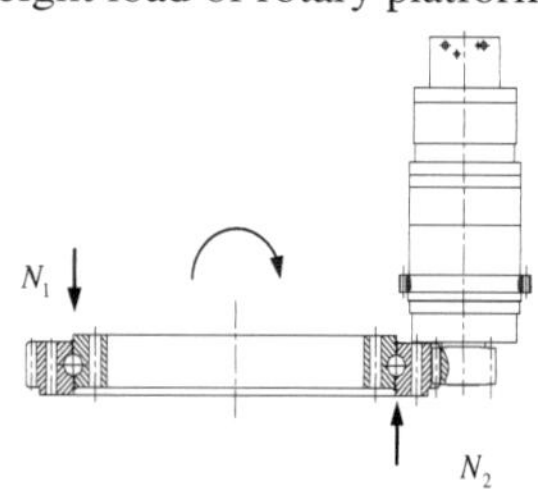

Fig 1 friction resistance

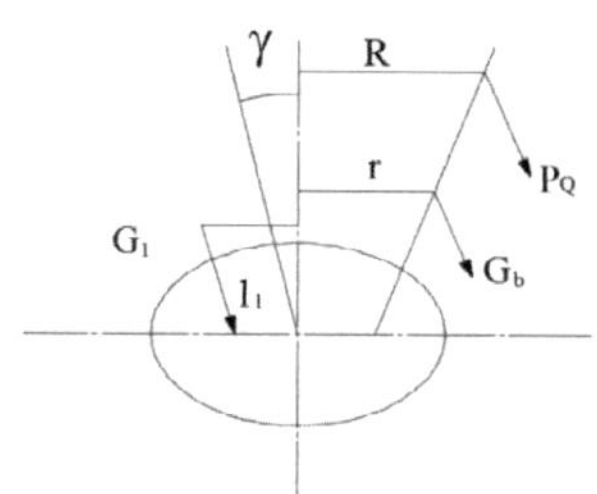

Fig 2inclination resistance

The rotating torque caused by inertia is composed of two parts[4], the calculation formula is shown as follow:

$$M_P = \frac{n}{9.55t}\left[\frac{G_b R^2}{3g} + \frac{4G_1 l_1^2}{3g} + i^2 K J_1 \eta\right] (5)$$

2.2. The resistance torque caused by wind pressure. The wind load include the wind pressure while the device is at rest and the wind pressure while the device is rotating.

When considering the wind pressure under the rotating state, the resistance is caused by external wind load and by the rotating of rotary device under windless condition. As the rotating devices are generally symmetrical about the central axis of the rotation $e \approx 0$, the coupling relationship between the two kinds of resistance can be ignored[6].In order to facilitate the calculation, the two resistances can be treated independently.

$$q = 0.613v^2;$$

$$p = CK_n q;$$

$$M_{Wz} = qC(A\sin\alpha\sin\beta)\cdot e\cos\beta \ (6)$$

q—wind pressure; A—the area of upwind surface; r—the distance from centroid to the center of rotation; C—wind force coefficient;e—eccentricity

Equivalent wind resistance moment:

$$M_{W1} = \frac{2\int_0^{\frac{\pi}{2}} M_{Wz} d\varphi}{\pi} = \frac{qC\,\mathrm{Aesin}\,\alpha \sin 2\beta}{2} \quad (7)$$

In the windless condition, due to the rotation of the rotating device, the wind load which is opposite to the rotation direction. The front view of rotating device is shown as figure 3. The force diagram of arbitrary cross section is shown as figure 4.

According to the location of the center of rotation, the force condition can be divided into two kinds of situations: $x_0 < h_0 \cos\alpha$ and $x_0 > h_0 \cos\alpha$

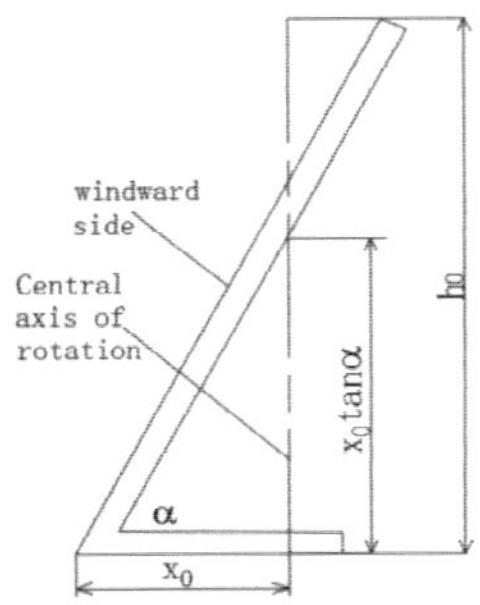

Fig.3 front view of rotating device

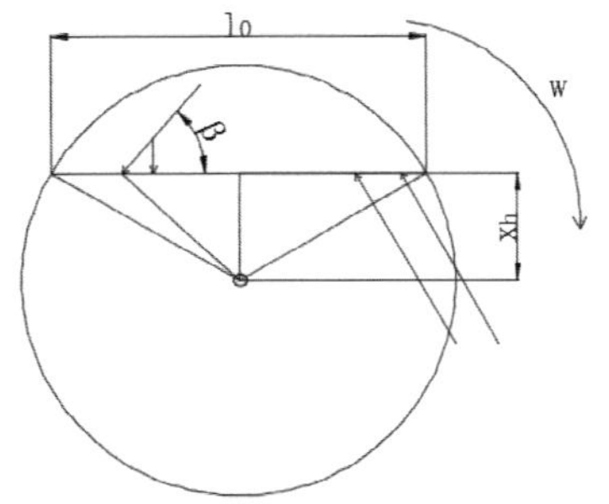

Fig.4 vertical view of rotating device

x_0 is the distance between the gyration center and the base of rotation device

a). When the gyration center $x_0 < h_0 \cos\alpha$, the wind load is divided into two kinds of situations: $0 < h < x_0 \tan\alpha$ and $x_0 \tan\alpha < h < h_0$

While $0 < h < x_0 \tan\alpha$:

$$R_h = \sqrt{(x_0 - h\cot\alpha)^2 + (\frac{l_0}{2})^2} \quad (8)$$

Because the relative wind load is symmetry in any cross section while rotating without wind outside. $0 < l < 0.5 l_0$ At a height of h of arbitrary cross section, the distance from the windward side to the midpoint of the windward side is l , the distance of any point in the cross section to the rotation center: $R_l = \sqrt{l^2 + x_h^{\ 2}}$, the relative speed of any point in the cross section:

$$v_l = wR_l ;$$

$$\sin\beta = \frac{l}{\sqrt{l^2 + x_h^{\ 2}}} \quad (9)$$

Wind pressure at this point:

$$q_l = 0.613 w^2 (\mathrm{l}^2 + \mathrm{x}_h^2) \quad (10)$$

The torque of wind resistance without wind:

$$M_1 = 2\int_0^{x_0 \tan\alpha} \int_0^{\frac{l_0}{2}} p_l \cdot R_l \cdot \sin\beta \cdot dhdl \quad (11)$$

While $h > x_0 \tan\alpha$:

$$x_h = h\cot\alpha - x_0 ;$$

$$M_2 = 1.226 C K_n w^2 \int_{x_0 \tan\alpha}^{h_0} \int_0^{\frac{l_0}{2}} (l^2 + x_h^{\ 2}) \cdot l \cdot dhdl \quad (12)$$

The total torque of wind resistance without wind:

$$M_{W2} = M_1 + M_2$$

$$= N(\frac{1}{24}h_0^3 l_0^2 \cot^2\alpha - \frac{1}{8}x_0 h_0^2 l_0^2 \cot\alpha + \frac{1}{8}x_0^2 h_0 l_0^2 + \frac{1}{64}h_0 l_0^4) \quad (13)$$

While $N = 1.226CK_n w^2$

b). When the gyration center $x_0 > h_0 \cos\alpha$:

$$M_{W2} = 2\int_0^{h_0}\int_0^{\frac{l_0}{2}} p_l \cdot R_l \cdot \sin\beta \cdot dhdl$$

$$= N(\frac{l_0^3}{24} + \frac{x_0^2 l_0^2}{8} - \frac{l_0^2 x h_0 \cot\alpha}{4} + \frac{l_0^2 h_0^2 \cot^2\alpha}{8}) \quad (14)$$

The total torque of wind resistance:

$$M_W = M_{W1} + M_{W2} \quad (15)$$

3.Application examples

The design parameters of a large scale mobile turntable are as follows:

$\alpha = 70^\circ, w = 6\text{r}/\min, \text{l}_0 = 15.18m, h_0 = 16.352m, \text{G} = 80t,$

$D = 2m, e = 0m$

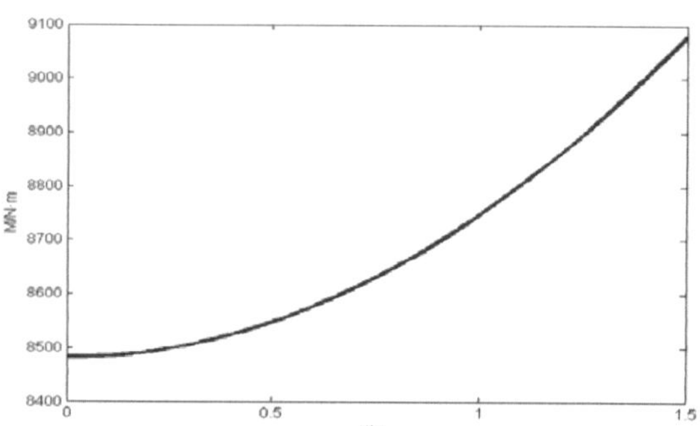

Fig 5 Wind load resistance torque at different distances

All design parameters are calculated in the formula (1):

$M_f = 9049N \cdot m, M_s = 1360N \cdot m, M_p = 3200N \cdot m$

The wind load is depended by the distance between the central axis of rotation and the centroid axis of the windward side. The wind load of different distance torque as shown in figure 5.From Figure 5, the minimum value of the wind load resistance torque is: $M_W = 8500N \cdot m, M_{SW} = 22KN \cdot m, P = 13.8Kw$ Wind tunnel test results from the literature [5] shows that the dynamic wind torque is $8250N \cdot m$. When this data is calculated in the formula (17), the rotary power of the device is $13.5Kw$. Using the calculation formula of literature [1], the rotary power of the device deserve to be $9.1Kw$. While using the calculation formula of literature [7], the rotary power of the device deserve to be $17.4Kw$. Compared with other computational methods, the formula of wind resistance, given in this article, is much more accurate.

4.Conclusion

(1)In this paper, the condition of the wind load of the rotating device in the rotating state was studied. Static and dynamic wind pressure were taken into account. By introducing the wind load, under windless situation, caused by the rotation of the rotary device, this paper deduced the formula of dynamic wind resistance torque.
(2)This formula solve the problem of power calculation of rotating device at high speed, and can be widely applicable to all kinds of rotary device.
(3)Finally, it is proved that the formula has higher accuracy through Citing an application example.

5.Acknowledgements

In this paper, the research was sponsored by:
（1）DIY project of Natural Science Foundation of Heilongjiang Province（E201433）
（2）The fundamental research funds for the Central Universities（HIT.KLOF.2010054）

References

[1] Gu D M 1995 Construction Crane. China Building Industry Press.12.
[2] GuoQ J and Han H L 2009 Approximate Calculation of Tower Wind Load. *Energy and Environment.***04** 26-29.
[3] Li D W 2004 Crane Rotary Mechanism Design. Construction Machinery05.
[4] Zhang X X, Ai Y, Lin X, Zhao Z P and Zhang Y F 2014 Wind Load Mode of Tower Crane Boom under Non-working State. *Shenyang Construction University.***01**86–91.
[5] Chen F B 2011 Wind Tunnel Test and Theoretical Analysis of Large Span Structural Wind Effects. *Journal of Human University of Technology*.9.
[6] Wu Y P and Xiao P 2011 Research on Wind Load Characteristics of Vehicle Borne Radar Antenna and Its Platform. *Journal of Wuhan University of Technology*,9.
[7] Ye S H and Li Z G 1986 Antenna structure design. Northwest Telecommunication Engineering Press.

IOP Conf. Series: Earth and Environmental Science **86** (2017) 012025 doi:10.1088/1755-1315/86/1/012025

Technical Study on Improvement of Endurance Capability of Limit Short-circuit Current of Charge Control SMART Meter

W W Li[1,2], Z Z Du [3], R m Yuan[12] , D Z Xiong[4], E W Shi[5],G N Lu[12] ,Z Y Dai[3], X Q Chen[4], Z Y Jiang[12] and Y G Lv[12]

[1]North China Electric Power Research Institute (NCEPRI), China.

[2]Electric Power Institute of State Grid Jibei Electric Power Company, Beijing, China.

[3]Henan Xuji Instrument Co., Ltd. Xuchang, China.

[4]Metering Center of Hunan Electric Power Company

[5]China Electric Power Research Institute

*821328633@qq.com

Abstract. Smart meter represents the development direction of energy-saving smart grid in the future. The load switch, one of the core parts of smart meter, should be of high reliability, safety and endurance capability of limit short-circuit current. For this reason, this paper discusses the quick simulation of relationship between attraction and counterforce of load switch without iteration, establishes dual response surface model of attraction and counterforce and optimizes the design scheme of load switch for charge control smart meter, thus increasing electromagnetic attraction and spring counterforce. In this way, this paper puts forward a method to improve the withstand capacity of limit short-circuit current.

1.Introduction

As the terminal equipment of smart power grid, smart meter is no longer a traditional product with only basic function of electric energy measurement, but the one with multiple functions, such as storage of electricity consumption information, two-way data communication. Therefore, smart meter represents the development of energy-saving smart grid in the future. The load switch, one of the core parts of smart meter, can automatically cut off users' power when their electric charge is not enough and switch on the power after recharging.So the quality of load switch will directly influence the safety and reliability of power supply and consumption. In addition, the quality is also related to interests of thousands of households and the corporate image of the State Grid.

At present, magnetic latching relay is mainly adopted for built-in load switch for smart meter. For its low energy consumption and small size, the magnetic-latching load switch has been widely used. However, during real operation of the load switch for smart meter, fault rate of built-in load switch is high. The failure causes include burn-out of smart meter due to disconnection at short circuit, temperature rise of contactdue to increasing of contact resistance and accelerated aging of plastic parts because of overheating. In addition, the design of load switchinsome domestic manufacturers mainly depends on imitation and experience. Aside from that, great fluctuation of materials' distributed parameters and different technical levels lead to relatively large dispersion of key characteristic parametersof load switch, poor consistency of quality and short service life. Therefore, it's necessary to research the optimization design method to improve endurance capability of limit short-circuit current of load switch for charge control smart meter. The research can urge manufacturers to ensure product quality, improve technical level in design phase,enhance power system's safety and reliability, reduce failure rate and prevent potential safety hazards.

2.Analysis on Endurance Capacity of Short-circuit CurrentofLoad Switch

When there is current in a load circuit, electro-dynamic force on dynamic contactor mainly includes Lorentz force and Holm force. Out-of-phase current takes place due to Lorentz force at circuit. According to electro-magnetic induction, the conclusion that out-of-phase can produce repulsive force can be drawn. While, Holm force takes place because real contact area is far less than apparent contact area, which will lead to reduction of current before and after flowing through real contact points. In this way, out-of-phase current will appear, resulting in repulsive force (see Fig.1 and Fig.2).

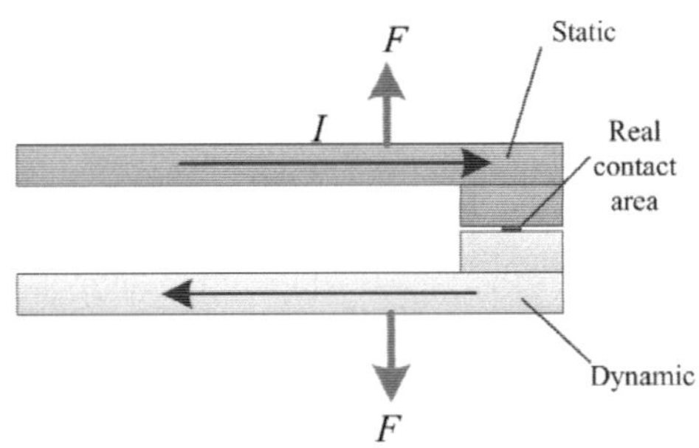

Fig.1 Schematic Diagram of Lorentz Force

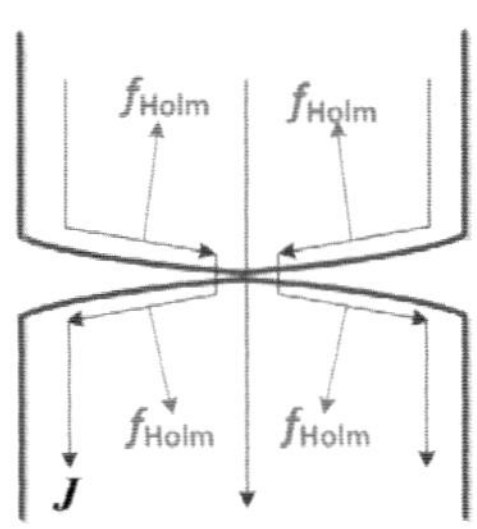

Fig.2 Schematic Diagram of Holm Force

As for the load switch for electric meter in Fig.3, the foldable and moveable spring makes the Lorentz force, resulting from short-circuit current, compress contactors easily, which in turn increases their pre-compression and keeps contact status. However, Holm force, as the force to separate contactors, is the main reason leading to abnormal disconnection of contacts at short circuit.

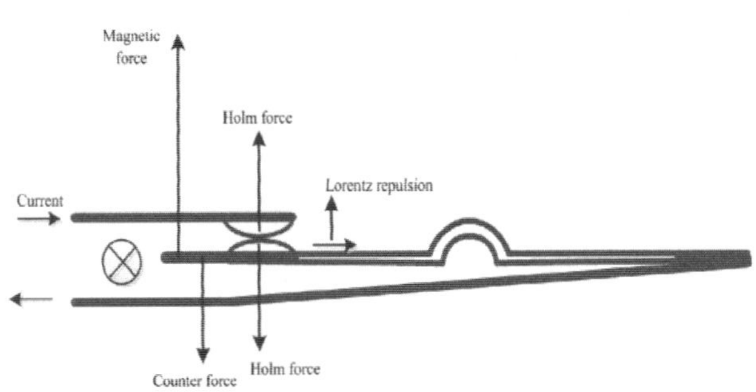

Fig.3 Force Analysis on Load Switch for Charge Control Smart Meter

Due to dispersed processing, some products have relatively small retention after closing of contacts; herein, the retention is the difference between permanent magnetic attraction and spring counterforce. When the load circuit incurs short circuit, several thousand amperes will flow through contacts. If Holm force, resulting from contraction of current line, is larger than contact retention, off-contact and electric arc of short-circuit current will take place, and lead to electric meter burnout. As a result, contact closure retention needs to be increased to improve resistance of contacts to short-circuit current.Increase of counterforce can lead to rise of contact pressure at contacts and their contact area, thus reducing Holm force.

3.Design Technology on Resistance Improvementto Short-circuit Currentof Load Switch

3.1 Fitting design of dual response surface model. Response surface methodology (RSM) is a statistical treatment technology used for multivariable modeling and analysis. In most designs, the relationship between response surface and independent variable is unknown. Accordingly, the first step of response surface method is to determine an appropriate approximation of f(x) by adapting the least square method. By using the least square method, error sum of squares between the calculated data and real data can be the smallest. Generally, lower-degree polynomial can be used for approximation within independent variables in limited range. For example, use the independent variables in linear function(e.g. function of first degree) in formula(1); or use higher-degree polynomial (e.g. quadratic polynomial) in formula (2).

$$y = \beta_0 + \beta_1 x_1 + \beta_2 x_2 + \cdots + \beta_n x_n + \varepsilon \quad (1)$$

$$y = \beta_0 + \sum_{i=1}^{n} \beta_i x_i + \sum_{i=1}^{n} \beta_{ii} x_{ii}^2 + \sum\sum_{i<j} \beta_{ij} x_i x_j + \varepsilon \quad (2)$$

Combination of Taguchi method and response surface method will get dual response surface method, as is shown in formula (3) and (4), where ε_u and ε_σ are error terms of mean value model and variance model respectively. Similarly, response surface models of mean value and variance with third and fourth orders or even higher can be constructed. Performance and fluctuation characteristic of products can be reflected more intuitively through building response surface models of mean value and variance.

$$Y_u = b_0 + \sum_{i=1}^{k} b_i x_i + \sum_{i=1}^{k} b_{ii} x_{ii}^2 + \sum\sum_{i<j}^{k} b_{ij} x_i x_j + \varepsilon_u \quad (3)$$

$$Y_\sigma = c_0 + \sum_{i=1}^{k} c_i x_i + \sum_{i=1}^{k} c_{ii} x_{ii}^2 + \sum\sum_{i<j}^{k} c_{ij} x_i x_j + \varepsilon_\sigma \quad (4)$$

Unknown coefficients are determined through the least square method. Specifically, unknown coefficient $B=[b_0, b_1, b_2,\dots, b_L]^T$ can be determined by formula (5) and calculation of $C=[c_0, c_1, c_2,\dots, c_L]^T$ is similar.

$$\begin{cases} E(\varepsilon) = \sum_{j=1}^{p} \{[Y(\tilde{X}_j) - \sum_{i=0}^{L} b_i \phi_i(\tilde{X}_j)]^2\} \\ \left.\dfrac{\partial E(\varepsilon)}{\partial b_i}\right|_{b_i^*} = -2\boldsymbol{X}^T\boldsymbol{Y} + 2\boldsymbol{X}^T\boldsymbol{X}\boldsymbol{B} = 0 \end{cases} \rightarrow \boldsymbol{B} = (\boldsymbol{X}^T\boldsymbol{X})^{-1}\boldsymbol{X}^T\boldsymbol{Y} \quad (5)$$

Where, $\boldsymbol{Y}=[Y(\tilde{X}_1),\ Y(\tilde{X}_2),\dots,Y(\tilde{X}_p)]^T$ is the response vector at P (P>L) test points; X refers to primary function matrix, as is shown in formula (6).

$$\boldsymbol{X} = \begin{bmatrix} \phi_1(\tilde{\boldsymbol{X}}_1) & \cdots & \phi_L(\tilde{\boldsymbol{X}}_1) \\ \vdots & & \vdots \\ \phi_1(\tilde{\boldsymbol{X}}_P) & \cdots & \phi_L(\tilde{\boldsymbol{X}}_P) \end{bmatrix} \quad (6)$$

Carry out error estimation to models from dual response surface method; namely, verify fitting of response surfaces and sample points by using determination coefficient $R^2$and adjusting determination coefficient R^2_{ad}. See formula (7) and (8) for definition.

$$R^2 = \frac{\sum_{i=1}^{P}(\hat{y}_i - \bar{y}_i)^2}{\sum_{i=1}^{P}(y_i - \bar{y}_i)^2} \quad (7) \qquad R^2_{adj} = 1 - \frac{\sum_{i=1}^{P}(y_i - \hat{y}_i)^2(P-1)}{\sum_{i=1}^{P}(y_i - \bar{y}_i)^2(P-k-1)} \quad (8)$$

Where, P refers to the number of selected test points; adjustment factor minus **1** is k, degree of freedom; y_i and $\bar{y}_i$ mean measured value and average value of response variable; $\hat{y}_i$is predicted value of response variable; whenR^2 and $\mathbf{R}^2_{\text{adj}}$are closer to 1, fitting degree of response surface model is better.

Based on U-type design, uniform design is a design method to pursue high quality and disperses test points evenly into input parameter space. And the method can be applied to various models and has robustness on model change. Uniform design is to arrange tests through the uniform design table, without randomness. The design is aimed at finding the minimum deviation between grand average of output variable and real total average among test points.

According to the analysis, the key structural design parameters ofelectromagnetic system with a significant influence on attraction and counterforce are respectively right half-length of long-pole face, left half-length of long-pole face, left half-length of short-pole face, and right half-length of short-pole face. Key design parameters of contact springsystem are respectively offset at Ω initial position, thickness of spring of two lower layers, thickness of spring of upper layers and width of spring. There are two determined key output characteristic parameters, namely F_1, electromagnetic attraction at pick-up position under0Vand F_2, corresponding counterforce at the pick-up position.

Uniform design sampling table will be used in uniform design of the sample. In parameter fitting of dual response surface model, expressions ofF_1 (electromagnetic attraction at pick-up position under 0V) andF_2 (the maximum counterforce at corresponding pick-up position) are defined as follows:

$$\begin{aligned} F_1 &= 1537.75 + 102.84 \cdot x_1 - 235.76 \cdot x_2 - 292.57 \cdot x_3 + 77.79 \cdot x_4 + 8.79 \cdot x_1 x_2 \\ &\quad + 8.75 \cdot x_1 x_3 - 34.55 \cdot x_1 x_4 + 14.06 \cdot x_2 x_3 + 7.072 \cdot x_2 x_4 + 15.17 \cdot x_3 x_4 \\ F_2 &= 4744.77 + 5956.13 \cdot x_5 + 28829.53 \cdot x_6 - 11631.49 \cdot x_7 - 554.97 \cdot x_8 - 7748.27 \cdot x_5 x_6 \\ &\quad + 5472.32 \cdot x_5 x_7 - 525.36 \cdot x_5 x_8 + 11906.20 \cdot x_6 x_7 - 3258.89 \cdot x_6 x_8 + 2180.72 \cdot x_7 x_8 \end{aligned} \quad (9)$$

Where, $x_1$refers to long right armature length,$x_2$long left armature length, $x_3$short right armature length,$x_4$short left armature length, x_5 Ω offset, $x_6$thickness of spring of two lower layers, x_7 thickness of spring of upper layers and x_8 width of spring.

3.2.Optimization design of multi-objective particle swarm. Constraint conditions:

(1) Feasible domain constraints: particles are confined within feasible region, with optimization parameters being subject to the upper and lower bounds. Given real assembly links and the feasibility in real production, dimensions of relevant parts cannot be infinitely increased and reduced, but limited within a certain range. Assuming dimension X, then $X \in [X_{\min}, X_{\max}]$.

(2) Other index constraints: retention (difference betweenF_1 and F_2at pick-up position under 0V) at pick-up position under 0V should be greater thanF_{Holm}.

According to standard particle swarm optimization (PSO), the key step is to find out local optimal position of individuals and global optimal position of swarms. An ordering strategy of niche fitness has been introduced based on general multi-objective particle swarm optimization (MOPSO), so a niche sorting multi-objective particle swarm optimization (NSMOPSO) can be provided to solve multi-objective optimization of load switch's electromagnetic system. Key strategies to improve MOPSO calculation are as follows:

(1) Selection of historical local optimal parameters of individuals and global optimal parameters of swarms

Particles' individual historical local optimal parameters can be selected according to Pareto dominance relation. If parameters of particle under current iteration dominate the historical local optimal parameters of particle (P), then P will be adopted as current parameters; if P dominates the parameters of particle under current iteration, then P will not be adopted; If the two cannot be compared, one of them will be selected randomly with probability of 50%.Particle swarm's global optimal parameter (Pg) is sorted according to all particles' fitness values of historical global optimal parameters under current iteration. The particle with the best fitness value will be selected as the swarm's global optimal parameters.

(2) External space of particle of particle

To improve computational efficiency of the calculation, external space of particle with capacity of N will be established to store Pareto optimal solutions. After iteration, all the new historical local optimal parameters of particles shall be put into the external space. Meanwhile, sort fitness according to descending order in the external space and delete dominated and repeated solutions. At this moment, if the number of particles in external space exceeds the standard capacity N, the tail end should be deleted to ensure validity of optimal parameters in the external space.

(3) Niche strategy

Niche sharing strategy is adopted as the calculation to sort fitness, with calculation formula defining fitness of particle Xi in the external space being as follows:

$$F_i = \frac{1}{S_i},\ i = 1, 2, ..., N_S \quad (10)$$

Where, NS refers to the number of particles in the external space;

S_i sharing degree of particle X_i.

$$S_i = \sum_{j=1}^{N_S} f_{sh}(d_{ij}),\ j = 1, 2, ..., N_S \quad (11)$$

Where,$f_{sh}(d_{ij})$ is a sharing function between $X_i$and X_j, representing osculation of particles in the external space.

$$f_{sh}(d_{ij}) = \begin{cases} 1 - (\frac{d_{ij}}{\sigma_{share}})^{\alpha} & , 0 \le d_{ij} \le \sigma_{share} \\ 0, d_{ij} \ge \sigma_{share} \end{cases} \quad (12)$$

Where, α (usually 1 or 2) refers to power of sharing distance;

σ_{share} value of predetermined sharing distance (usually constant);

d_{ij} distance between$X_i$andX_j.

$$d_{ij} = \|X_i - X_j\| = \sqrt{\sum_{k=1}^{n} (x_{ik} - x_{jk})^2} \quad (13)$$

In the external space, the denser the particles at someplace are, their sharing function will be larger and the fitness will be smaller; on the contrary, the more sparsely particle distribution is, the fitness of particle will be larger. Through fitness sorting in external space, the niche strategy has avoided densely-distributed particles and enhanced the exploitation ability of calculation.

(4) Gaussian mutation
Gaussian mutation mechanism has been added into MOPSO to initialize again those particles inferior to P with some probability. Meanwhile, random initialization will be conducted to particles (out-of-range particles) going beyond the feasible region so as to improve availability of particles.See Fig.4 for NSMOPSO solution flow of optimization model of structure parameters.

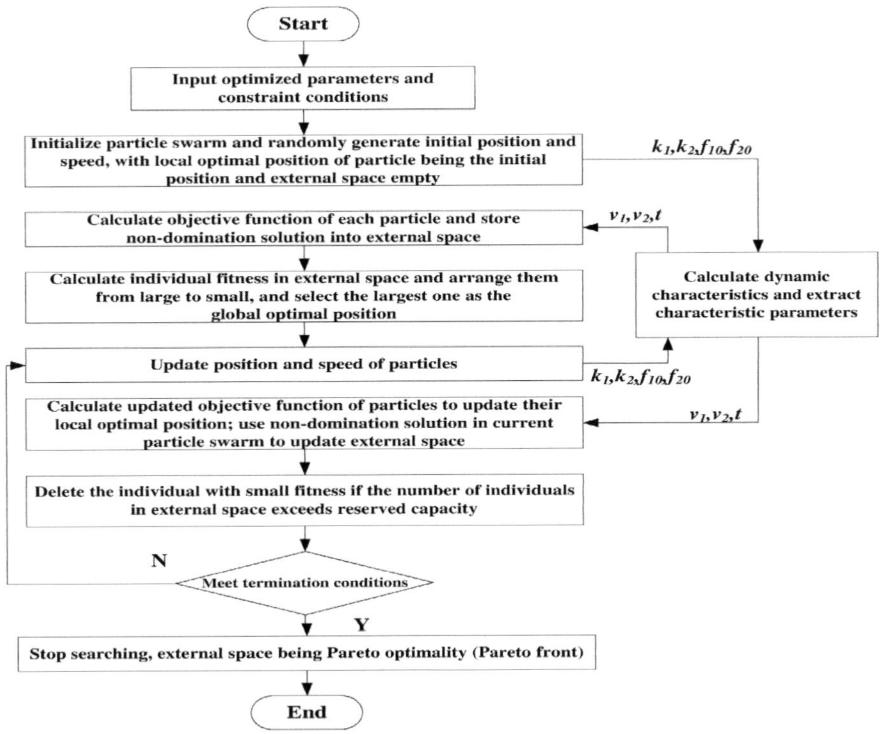

Fig.4 NSMOPSO Flow of Multi-objective Optimization of Structure Parameters

4.Design Results on Improvement of Load Switch's Endurance Capability of Limit Short-circuit Current

According to the aforesaid analysis, the mathematical model of multi-objective optimization of load switch is determined below:

$$\min y = f(x) = (f_1(x), f_2(x))$$
$$s.t. \begin{cases} \dfrac{1}{f_1(x)} - \dfrac{1}{f_2(x)} > F_{\text{Holm}} \\ x \in R^8 \end{cases} \tag{14}$$

Where, $x = x_1, x_2, \dots, x_8 \in R^8$ are 8 optimized parameters;$f(x)$ refers to objective function, including$f_1(x)$ and $f_2(x)$, of which$f_1(x)$ is reciprocal of attractionF_1at pick-up position under 0V and $f_2(x)$ reciprocal of the maximum counterforce F_2at corresponding pick-up position;$g_i$is the constraint of$i$inequality and $h_i$the constraint of $j$equality.
In this paper, we take a common load switch as an example, with its optimized parameter boundary and constraints shown in Table 1.Parameter setting of NSMOPSO calculation:
inertia weight $w_{start} = 0.9, w_{end} = 0.4$, accelerated factor$c_1 = c_2 = 2$, niche radius$\sigma_{share} = 1$, controls parameter $\alpha = 1$, capacity of external space$N$ is 200, the number of particles in a swam 30 and the maximum

number of iterations 40. See Fig.5 for Pareto front.

Table 1. Optimized Parameter Boundary

	Long Right Armature Length	Long left Armature Length	Short Right Armature Length	Short Left Armature Length
min	7.6mm	8.1mm	8.1mm	7.1mm
max	7.9mm	8.4mm	8.4mm	7.4mm
	Ω offset	Thickness of spring of two upper layers	Thickness of spring of lower layers	Width of spring
min	-1.65mm	0.205mm	0.255mm	10.85mm
max	-1.35mm	0.235mm	0.283mm	11.15mm

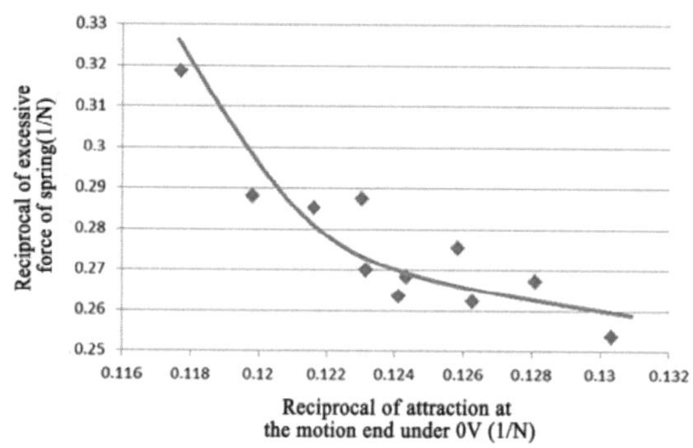

Fig.5Optimization Results of Structure Parameters

According to PSO and processing technology, final optimal combination of parameters is shown in the following table.

Table 2.System Parameters of Electromagnetic System and Spring after Optimization (Unit: mm)

Parameters	Long Right Armature Length	Long left Armature Length	Short Right Armature Length	Short Left Armature Length
Before optimization	8.25	8.25	7.25	7.25
After optimization	7.74	8.57	8.21	7.32
Parameters	Ω offset	Thickness of spring of two upper layers	Thickness of spring of lower layers	Width of spring
Before optimization	0	0.2	0.25	11
After optimization	-1.42	0.21	0.27	11

5.Effect Verification

In Fig.3, contact pre-compression is about 5.5-6.5N, proposed short-circuit current 1k-5kA, surface material of contactors $A_gS_nO_2$ and BH 80-100N/mm^2.Holm force is calculated under these conditions. See formula (15) for empirical calculating formula of Holm force.

$$F_h = \frac{\mu_0}{4\pi} I^2 \ln\frac{R_0}{R_c} \quad (15)$$

Where, $F_h$refers to Holm force, μ_0 magnetic conductivity in vacuum, I short-circuit current, R_0contactor radius and $R_c$real contact area. Among them, real contact area can be calculated according to formula (16).

$$R_c = \sqrt{\frac{F}{\varepsilon \pi H}} \qquad (16)$$

Where, Fmeans contact pre-compression, ε correction factor generally related to status (surface roughness) of contact surface and the value ranges between 0.3 and 0.6, the typical value is generally 0.45 and H refers to BH of contactor material.

First, calculate contact radius$R_c$and the ratio (R_c/R_0) of contact radius to contactor radius under different correction factors according to formula (16), where contactor pre-compression is 6N and correction factors range between 0.3 and 0.6. See Fig.12 for calculation results. The ratio of contact radius to contactor radius fluctuates between 0.05 and 1 and a median of 0.45 is generally taken as the correction factor and then the ratio reaches 0.064.

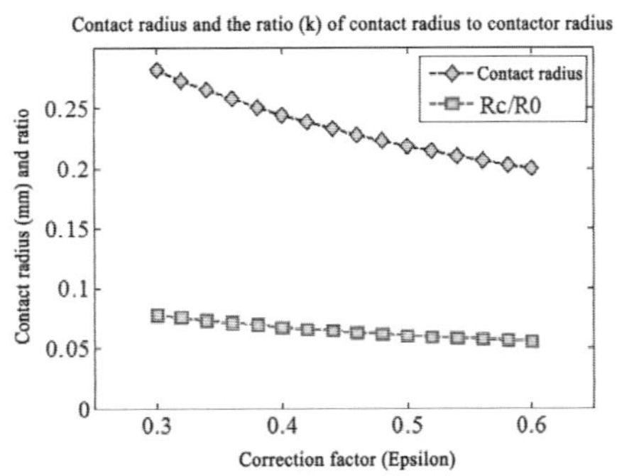

Fig.6. Calculated Value of Contact Radius (contact force: 6N)

When the correction factor is 0.45, contact pre-compression ranges between 5 and 7N. See Fig.17 for Holm forces under different short-circuit currents. Holm force will increase rapidly according to the square of current when short-circuit current increases. When the contact pre-compression varies, it has little influence on Holm force due to little change of real contract area.

When short-circuit currentapproaches 6kA, Holm force is closed to 5N, which means it nears contact pressure (i.e. contact retention). At this moment, contactors may be separated by short-circuit current, leading to contact failure.

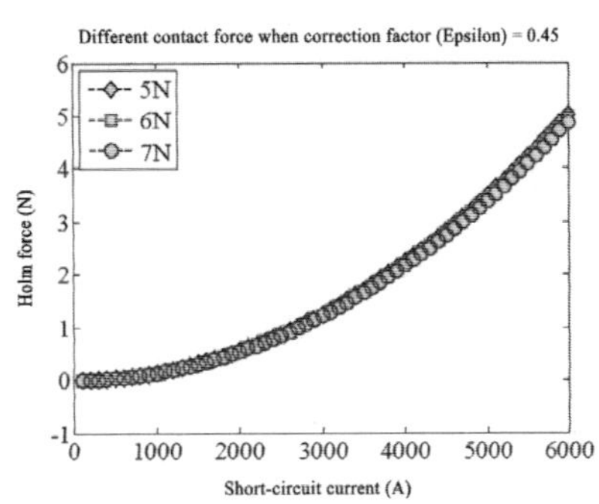

Fig.7. Holm Force under Different Pre-compressions

To understand the influence of different contact pre-compressions on Holm force more clearly, we also calculate the change rule of Holm force under different ratios of real contact area to contactor radius. See

Fig.8 for the results.

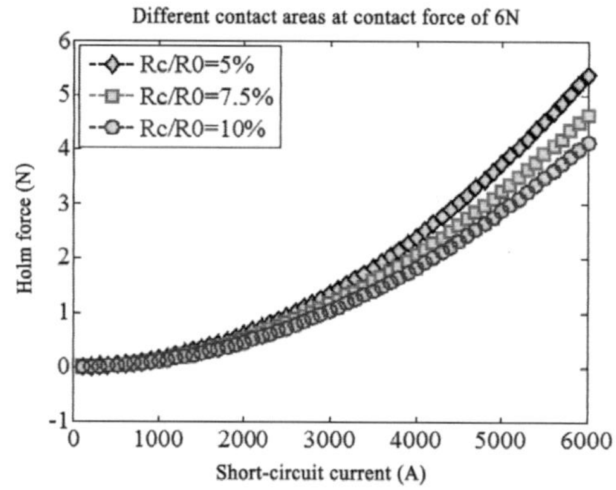

Fig.8. Influenceof Contact Area on Holm Force

According to the results, the real contact area in this range has a certain influence on Holm force, but the effect is small. That means estimation formula of real contact area and results have little influence on calculation results of Holm force, which is of relatively credibility. In addition, short-circuit current is the decisive factor of Holm force.

If the load switch is to be subjected to a short-circuit current of 6 kA, then the retention must be at least over 6N.

In summary, short-circuit current is the decisive factor of Holm force and the retention must be at least over 6N to ensure the load switch can endure a short-circuit current of 6kA.Holm force can also be calculated precisely by modeling through simulation software such as ANSYS. Literature shows that the difference between the results and empirical formulas is less than 5%.

According to the aforesaid analysis, the resistance of contact to short-circuit current can be improved to 6kA after the contract pressure having been increased from 4.5N and 6.3N, ensuring no abnormal disconnection between contacts when a short-circuit current of 6kA takes place in the loading circuit.

6.Conclusion

Based on general MOPSO calculation, this paper introduces ordering strategy of niche fitness and puts forward an improved multi-objective particle swarm calculation (Niche Sorting NSMOPSO) to solve multi-objective optimization of load switches. This calculation adopts niche sorting to select global optimal parameters and Gaussian mutation mechanism to repeatedly utilize particles beyond the boundary, making Pareto solution distribute more widely and evenly. The optimization design scheme on load switch for charge control smart meter from study of this paper can ensure that the endurance capacity of limit short-circuit current can be improved to 6kA, which effectively guarantees safe operation and reliability of smart meter.

References

[1] Clarke S M, Griebsch J H and Simpson T W 2005 Analysis of Support Vector Regression for Approximation of Complex Engineering Analyses.*Journal of Mechanical Design.***127**1077-87.

[2] Turner C J 2012 MetamodelingIn Product and Process Design. *ASME International Design Engineering Technical Conferences and Computers and Information in Engineering Conference.*639-651.

[3] Jin R, Chen W and Simpson T W 2001 Comparative Studies of Metamodelling Techniques under Multiple Modeling Criteria.*Structural and Multidisciplinary Optimization.***23**1-13.

[4] Coulomb J L, Kobetski A, Costa M C, Maréchal Y and Jönsson U 2003 Comparison of Radial Basis

Function Approximation Techniques. *The International Journal for Computation and Mathematics in Electrical and Electronic Engineering*.**22**616-629.

[5] Mehta D M and Pant R S 2013 Surrogate Based Design Optimization of Aerostat Envelope. *AIAA Lighter-than-Air Systems Technology (LTA) Conference*.43-50.

[6] Kim J B, Hwang K Y and Kwon B I 2011 Optimization of Two-Phase In-Wheel IPMSM for Wide Speed Range by Using the Kriging Model Based on Latin Hypercube Samplin. *IEEE Transactions on Magnetics*.**47**1078-81.

[7] Douglas C M 2014 Stu Hunter's Contributions to Experimental Design and Quality Engineering. *Quality Engineering*.**26**5-15.

[8] Hunter J S 1985 Statistical Design Applied to Product Design. *Journal of Quality Technology*.**17**210-221.

[9] Diplaris S C 2000 Sfantsikopoulos M M. Cost-Tolerance Function: A New Approach for Cost Optimum Machining Accuracy. *International Journal of Advanced Manufacturing Technology*.**16**32-38.

[10]Chi, C.T 2006 A Study of Closing Adaptive Control in Electronically Controlled Intelligent Contactor. *2006 IEEE Region 10 Conference*.82-88.

IOP Conf. Series: Earth and Environmental Science **86** (2017) 012026 doi:10.1088/1755-1315/86/1/012026

Intelligent Home Control System Based on ARM10

G X Chen [1],J Jiang[2] and L H Zhong[3]

[1]School of Electrical and Electronic Engineering, Tianhe College of Guangdong Polytechnic Normal University, Guangzhou, China

[2]Economic and Trade Department, Guangdong Polytechnic of Industry and Commerce, Guangzhou, China

Abstract. Intelligent home is becoming the hot spot of social attention in the 21st century. When it is in China, it is a really new industry. However, there is no doubt that Intelligent home will become a new economic growth point of social development; it will change the life-style of human being. To develop the intelligent home, we should keep up with the development trend of technology. This is the reason why I talk about the intelligent home control system here.

In this paper, intelligent home control system is designed for alarm and remote control on gas-leaking, fire disaster, earthquake prediction, etc., by examining environmental changes around house. When the Intelligent home control system has detected an accident occurs, the processor will communicate with the GSM module, informing the house keeper the occurrence of accident. User can receive and send the message to the system to cut the power by mobile phone. The system can get access to DCCthrough ARM10 JTAG interface, using DCC to send and receive messages. At the same time, the debugger on the host is mainly used to receive the user's command and send it to the debug component in the target system. The data that returned from the target system is received and displayed to the user in a certain format.

1.System design

The system can be divided into CPU control module, GSM communication module, stepper motor drive module, display module, sensor module, AD conversion module and calendar module and so on. The system uses ARM10 as CPU master, with TC35 module for remote communication. It uses relay module to control the main circuit that is on and off; CPU can detect sensor group to gain the data.Power module is responsible for providing the power supply to all modules. It is to read the TC35 that the systemcanreceive the contents of the message to control the corresponding external relay , or control the data acquisition system. However, it sends the corresponding data to TC35 module through the serial port, and then deals with the data.You can also press the button to control the curtain switch. The block diagramofthe system is shown inFigure 1.

IOP Conf. Series: Earth and Environmental Science **86** (2017) 012026 doi:10.1088/1755-1315/86/1/012026

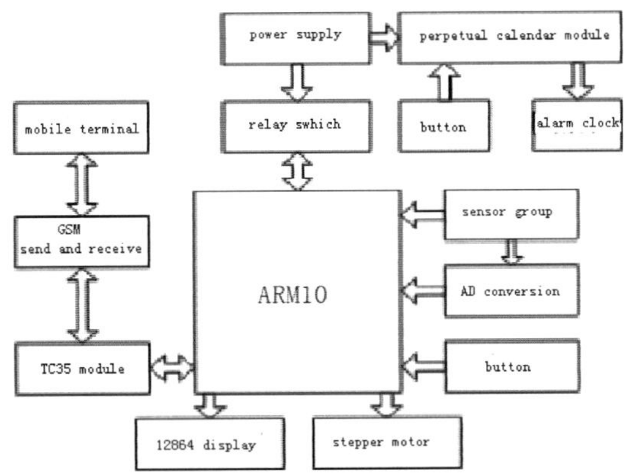

Fig. 1.remote block diagram of the system

The system requires to receive short messages, you can automatically disconnect the power supply system, so it uses the relay module. The relay module has six interfaces, such as the signal side, VCC, GND, normally open, common, normally closed. When the signal terminal is low and the common connects with the normally closed terminal, the circuit turns on to guarantee that the system has normally power supply. When the signal terminal is high and the common connects with normally open, the circuit is disconnected to stops supplying the power for the system. Power control system circuit is shown in Figure 2.

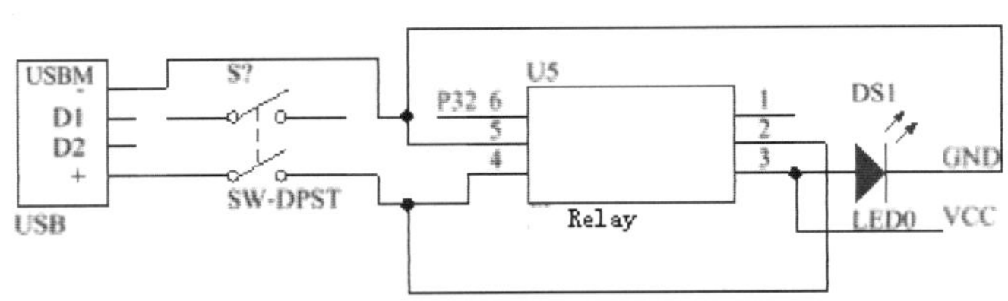

Fig. 2.power control system circuit

2.System software design

The system software design uses C language to develop it high efficiently. It is divided into the main system and an attached system. The main system is divided into main program and subprogram. Meanwhile， subprogram mainly includes temperature sensor acquisition program, PCF8591AD conversion program, LCD12864 LCD display program, GSM short message transceiver program and so on. The attached system has clock chip program, key program, display program.

2.1. The main system design. The main program is the main part of the microcontroller program, the program execution is carried out mainly through the main program.the whole system is basically completed in the functions of the main function. In this process the main program calls the subprogram and interrupting service program.The flow chart of the main function is shown in Figure 3.

IOP Conf. Series: Earth and Environmental Science **86** (2017) 012026 doi:10.1088/1755-1315/86/1/012026

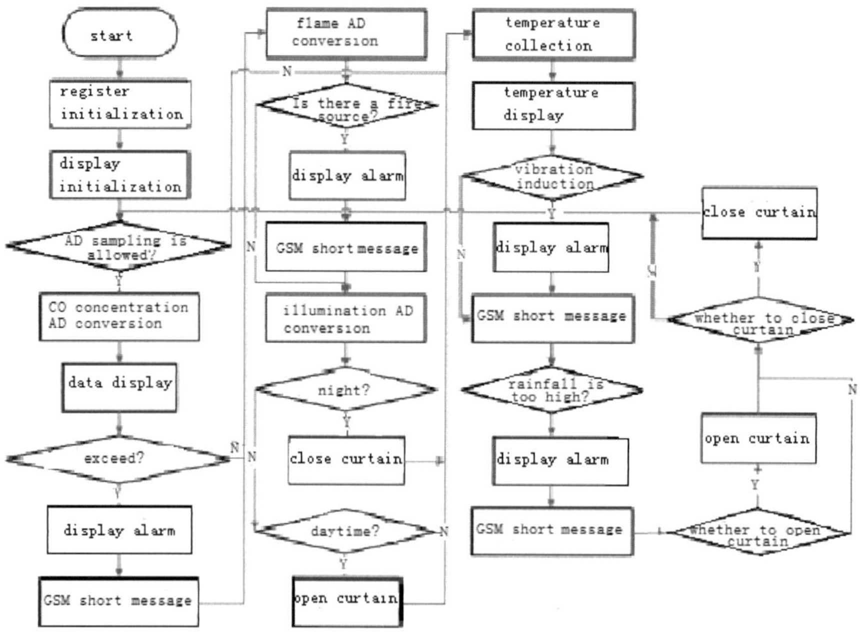

Fig. 3. The main function flow chart

2.2. GSM communication program design. In the software design ofthis system, GSM module through the serial port sends AT commands to the GSM module in order to achieve the function of sending and receiving short messages. The short messages must initialize before sending or receiving text messages. The flow chart of the relevant program is shown in Figure 4.

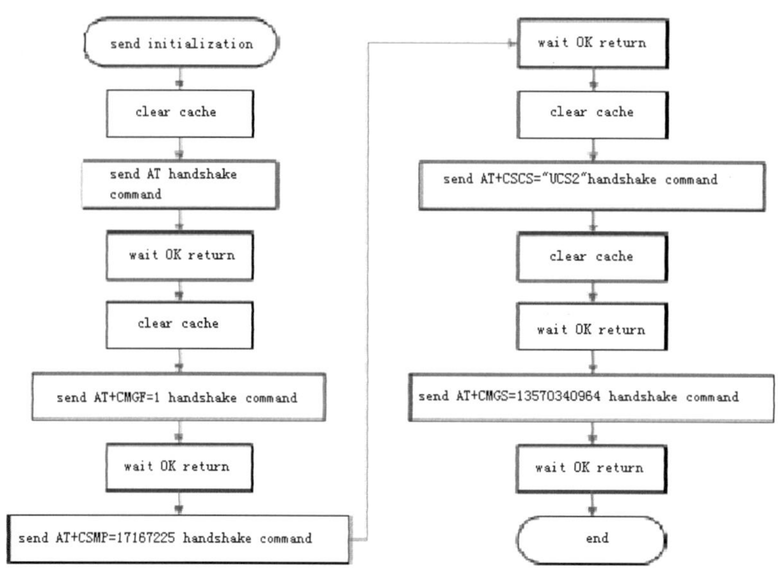

Fig. 4. Transmit Initialization

3.System debugging

The system results are debugged here.The test of the system thatis mainly from the test of several major modules starts to carry on GSM short message communication test, gas test, flame test, raindrop test, vibration test, light test, human body infrared test, temperature test,etc.

Gas Test is used to simulate gas with a combustible gas in the lighter. If the MQ-2 gas sensor detects a flammable gas leak, it will send a short message with the message "Send high CO concentration, please note property safety" to the victim's mobile phone. GSM SMS test is that the phone receives short messages, once again it sends "Close The Power" of the message to the system, the system can automatically cut off the main power supply. Flame test is used to contact the flame sensor with the flame of the lighter to see if it is displayed on the screen and sends short messages.

The rest of the test is a reason, it simulates the sensor detecting environment with a tool to see if there is a display, a short message prompt. This is the whole system diagram shown in figure 5.

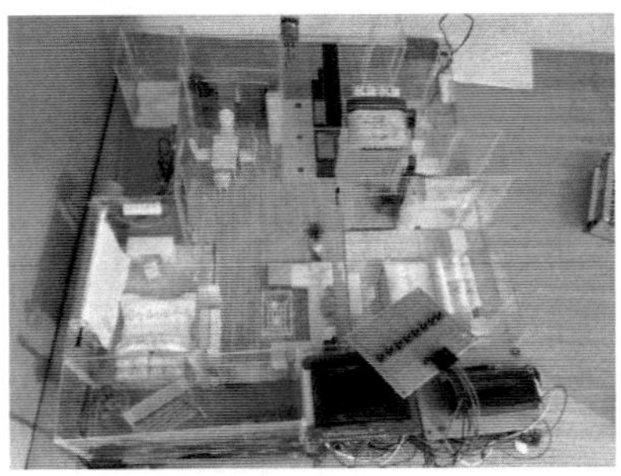

Fig. 5.system diagram

4.ARM10JTAGinterface communication

The debugger can also see the R and W bits when querying the DCC control register through ARM10 JTAG interface shown in figure 6.When reading the write registers of communication data， it can set W bits to scan and output the data. The DCC control register is read-only, which enables synchronization between the processor and the debugger.When the processor wants to send a message to the debugger,it is necessary to check whether the DCC data register is in the idle state by checking the W bit in the DCC controller.The processor reads the DCC controller to check the status of the W bit.In addition,it sends messages from the debugger to the processor.At this point,debugger can query and debug the R bit of communication control register. When the read register of communication data is free, the data is written to the JTAG interface.

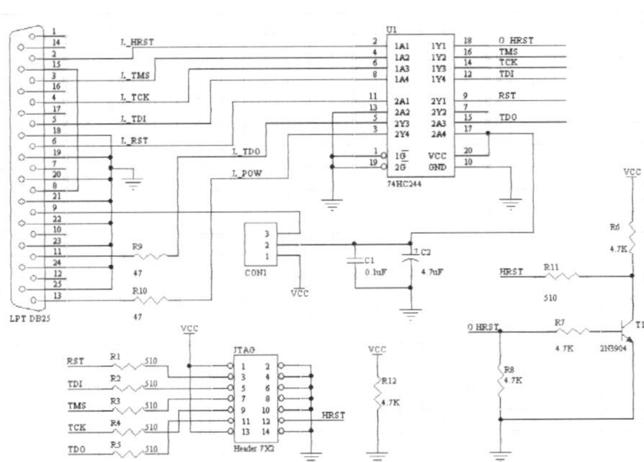

Fig. 6. ARM10 JTAG interface principle

Send data function as follows：

```
void* send（void* data）
{
int c=‘0’；
printf（“send  data\n”）；
while（STOP= =FALSE）
 {
c++；
 c%=255；
  write（fd，&c，1）；
usleep（100000）；
 }
return  NULL；
 }
```

Receive data function as follows：

```
 void* receive（void* data）
{
int c=‘0’；
printf（“read modem\n”）；
 while（STOP= =FALSE）
   {
 read（fd，&c，1）；
    write（1，&c，1）；
   }
printf（“exit from reading modem\n”）；
    return NULL；
   }
```

5.Acknowledgements

The system has the function of disaster prevention,mitigation, protection,removal,and can realize automatic detection and remote alarm.According to the information collected by each sensor ,it

IOP Conf. Series: Earth and Environmental Science **86** (2017) 012026 doi:10.1088/1755-1315/86/1/012026

determines whether there is abnormal situation.If there are unusual circumstances, it will send a text message to the victim through the GSM module, to remind them to pay attention to property safety.Users can also send text messages, remote control to cut off the power supply to prevent accidents.The system has the function of disaster prevention,mitigation, protection,removal,and can realize automatic detection and remote alarm.According to the information collected by each sensor,it determines whether there is abnormal situation.If there are unusual circumstances, it will send a text message to the victim through the GSM module, to remind them to pay attention to property safety.Users can also send text messages, remote control to cut off the power supply to prevent accidents.

This work is supported by Youth Innovative Talent Project of Guangdong Province. Item number is 2015KQNCX237.』

References

[1] Qi W and Yang S 2006 MCU C5l language tutorials and experiments. Beijing: *Beijing University of Aeronautics and Astronautics Press.*

[2] Han B J 2001 GSM principle and its network optimization. *Mechanical Industry Press.*

[3] Lin G H 2006 Data transmission and real-time control system based on GSM module. *Computer and Modernization.*__2__22-24.

[4] Gao F 2009 Design of Car Burglar Alarm System Based on GSM Network Short Message. *Hebei Industrial Science and Technology.*__26__182-184.

IOP Conf. Series: Earth and Environmental Science **86** (2017) 012027 doi:10.1088/1755-1315/86/1/012027

The Application Design of Solar Radio Spectrometer Based on FPGA

Q F Du [*], R J Chen [1], Y C Zhao [2], S W Feng[3], Y Chen [4] and Y Song[5]

[1]School of Mechanical, Electrical & Information Engineering, Shandong University ,Weihai, China

[2] School of Mechanical, Electrical & Information Engineering, Shandong University ,Weihai, China

[3]Institute of Space Sciences, Shandong University ,Weihai, China

[4]Institute of Space Sciences, Shandong University ,Weihai, China

[5]School of Mechanical, Electrical & Information Engineering, Shandong University ,Weihai, China

Email: *1793620354@qq.com

Abstract. The Solar radio spectrometer is the key instrument to observe solar radio. By programing the computer software, we control the AD signal acquisition card which is based on FPGA to get a mass of data. The data are transferred by using PCI-E port. This program has realized the function of timing data collection, finding data in specific time and controlling acquisition meter in real time. It can also map the solar radio power intensity graph. By doing the experiment, we verify the reliability of solar radio spectrum instrument, in the meanwhile, the instrument simplifies the operation in observing the sun.

1.Introduction

The solar radio dynamic spectrometer instrument originated from radio astronomy. It is a kind of radio telescope to do sun continuous observation. It works with wide range of frequency and can get the dynamic data of the outbreak of sun. When the sun break out with huge energy output, the radio frequency usually distribution between MHz and GHz. The radio power characteristic is rich in this range of frequency, so the collect frequency of the instrument is mainly between 0M and 250MHz. The spectrometer is make up with wide band antenna system, wide band receiving system and display system. Because of its characters, the performance is shown in frequency and time resolution ratio.

At present, time resolution of instrument in the same kind is generally higher than 10ms, and frequency resolution ratio cannot reach more than 60 KHz at the same time. In this design, we program a software to do the data processing in high speed by FPGA and control the data collection. This spectrometer can collect the data with the time resolution ratio of 10 milliseconds and the frequency resolution ratio of 32 KHz. The data can be get together by the software. And the software can also change the parameter and model of data collection. When it loads the mass of data, it can carry on data analysis and get the strength of the solar radio signal so that we can observe the sun's outbreak.

2.THE TOTAL DESIGN OF SOLAR RADIO DYMANATIC SPECTROMETER

The total design of solar radio dynamic spectrometer mainly contain 3 part: data collection, data saving and data processing. In the data collection part, the signals come from array antenna, then we get 2 road solar radio signals. After that signals are magnified by bandpass filter for low noise amplifier and sent to the high speed collection card. The A/D converter turn the analogue signal into digital signals and transmit it to FPGA. The high speed data stream is collected by FIFO multi model. When the FPGA get data, it can be packed with reality signal data and time data together to finish the collection. The data packages are sent to computer by PCI-E port and load in memory of the virtual hard disk. The data load in the directory which is named by time so that we can find data easily. We also analyzethe data with FFT arithmetic in specified time and get frequency radio figure. We can see the solar radio signal in this figure.

3.THE SOFTWARE AND HARDWARE OF SOLAR RADIO SPECTROMETER

3.1 The Hardware Design of Solar Radio Spectrometer. The sampling frequency is the most important parameter. During the data collection, the sampling frequency must be in the range limited by Nyquist theorem so that we can get all the original data. Because the spectrometer's output signal frequency we made should be between 0M and 250MHz, the acquisition ratio should be 500MHzsps at least.

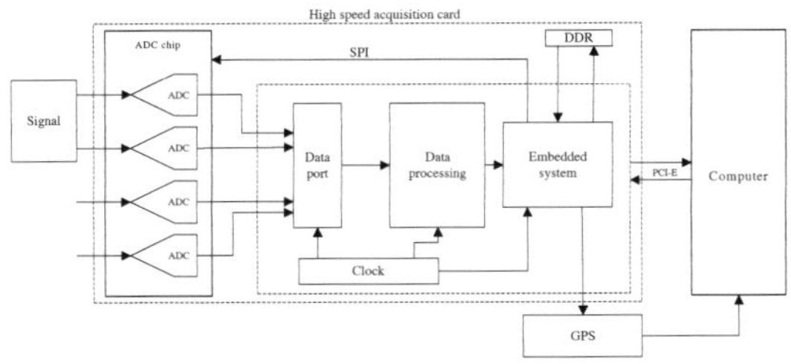

Fig.1 The total flow chart of system

In this design, we set QT1135 designed by Queentest as the high speed acquisition card. This card not only has direct-coupled amplifier but also supports wide frequency input.The speed of data transmission by PCI Express 2.0 x8 port can be 2.8GB/s. It is fast enough for transmission. The FPGA of QT1135 is Xilinx Virtex-6 FPGA which has high arithmetic speed and can be programming inside by user. The acquisition card has four collection channels and support 500Msps transmission rate. Transmission rate in direct current can reach 250MHz and fit for the solar radio collection.

When observing the solar radio, the input signals could be sent to high speed acquisition card from antenna array. After that, we can do a series of work to get data and save it in computer. Because of the mass of data input and avoiding the weak of low speed,we use high speed virtual hard disk to save and protect data.

3.2 The Software Design of Solar Radio Spectrometer. The QT1135 high speed acquisition card support programing in C. Main control program is made up with three part: PCI-E control part, API port program and main control interface.

3.2.1 PCI-E Control and API Port Program. The duty of PCI-E is to control data transmission port. There are a lot of control function inside the card and can be easily used. After compiled these files, they will become PCI-E call file. And in the process of the main program compiling, we can use these files to control PCI-E data transmission part. In the process of sampling, we need get continuous data, so the integrity of data is necessary. The fifo multi model can figure out this program.

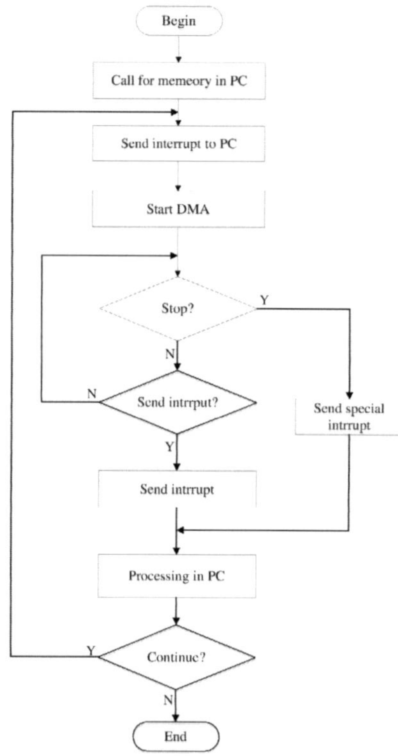

Fig. 2 The flow chart of PCI-E uploading

Fifo multi is a kind of model which can get data from part to part. It will load continuous data in FPGA firstly. When the DDR3 memory in cardis full, data will move to computer. The weakness of this job just be the speed of moving data from FPGA to computer and the data will not lose if the computer use high speed hard disk. The Fifo multi model is defined in PCI-E control program and the main program use it directly.

The card complete DMA move by PCI-E port. The specific operation process is as follows:

(1) Computer program opensup and lock a contiguous physical address memory to prevent being used;
(2) When the memory in FPGA is full, send an interrupt to computer;
(3) The computer check the interrupt and begins DMA;
(4) FPGA gets the order of starting the DMA and transfers data to computer by PCI-E port. The data package will be write to the size we set. After that, FPGA will send an interrupt to computer and finish transmission;
(5) When computer gets the finish interrupt, it will move the physics address data to the memory and the step change to step (2).

PCI-Euploading diagram is shown in “Fig. 2”.

3.2.2 User API. The user API is used to control ADC in card. Moreover, it defines some card functions to finish the work of data move. The main program must use the API functions to compile together so that complete the work of controlling card. The API projects contain FFT algorithm and it could process the data from A/D converter directly. There are also many card parameter set functions. Before the main program control the acquisition card, it should use these functions to set board parameter.

The main API functions are shown in the table 1:

TABLE 1 MAIN API FUNCTION

API function	State
QTOpenBoard	Open the board
QTResetBoard	Reset the board
QTCloseBoard	Close the board
QTRegisterWrite	Write registers
QTRegisterRead	Read registers
QTStart	Start DMA
QTWaitDMA	Wait DMA to stop
QTDMADone	Finish the DMA

The data transfer into binary style by A/D converter and load in memory for reading. The binary style is very easy to read in the later processing. The package of data contain timestamp and it is the offset from the start of the collection time. We can know the precise time of data by read the timestamp. It will make easily to find the rule of sun's outbreak.

The timestamp include 32 bit, there is different meaning for different bit. The former seven bits stand for the number of A/D, the extension is for the highest bit of last byte. It will be expanded to 2 bytes and enough to stand for the number of byte. For example, if the number of A/D is 14, then the first byte will be 0xFF and the second will be 0x3F. The sum of power is defined by the bits from 8to 15 which stand for the wholepower during the collection time. The number of 10 milliseconds is defined from 16 bit to 25 bit, it can show us the past number of time from the beginning of collection. The unit is 10 milliseconds. Similarly, the bits from 24 to 34 are also stand for time. But the unit becomesminute.

3.2.3 The Main Data Collection Program. The main data collection program is a MFC program. The duty is to set parameter, send work orders, control data collection time and set data save directory. Before the start of main program, we should build API project and PCI-E project. After that, the main program can use these function to control card. The interface of main program is shown as follow:

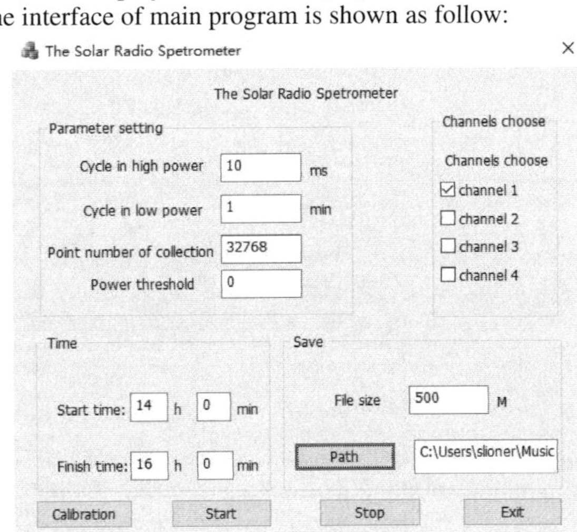

Fig. 3 The main program

The main program include five parts, there are parameters setting part, channels choose part, collection time part, file save part and control button. Because of the indiscipline of sun's outbreak, we set power threshold to control the size of data. When the power is more than the power threshold, the fast collection begin and get a lot of data. While, if the sun is in peace, the main program will not collect a lot. Moreover, the parameter setting part can also control sampling number, the more sampling number is set, the higher sampling rate it will be.

QT1135 has 4 input channels, they can be open in the channels choose part. The start and finish sampling time should be set in collection time part. Data file size will be set in file save part, the directory is also set in this part. When the program works, files will be load in directory named by time, we can find data by these directory easily. The main program uses multithreading program to make data saving speed improved.

There are calibration, begin, stop and exit button in this program. Before data collection, we should click calibration button to correct the IO delay of channels. Then, we can click begin button to start sampling. The stop button can finish sampling.

3.2.4 The Sampling and Storage Method. Data acquisition is the core of this paper. The key work is get data from PCI-E DMA to user memory and save it in binary file. The whole data collection and storage will be carried out in a "while" cycle. Before "while" cycle, the program use "UCHAR *pBuf = new UCHAR[buflen]" to call for a data memory. When the card is on, the program use "memcpy" function to copy data from DMA memory. The storage file will be named by time. If the stop button is clicked, the program finishes.

It should be noted that we should call for the user memory whose size is more than DMA memory. If it is less thanDMA memory, the program will return fault code. The user memory will be free after program stop.

The sampling and storage flow chart is shown in figure 4:

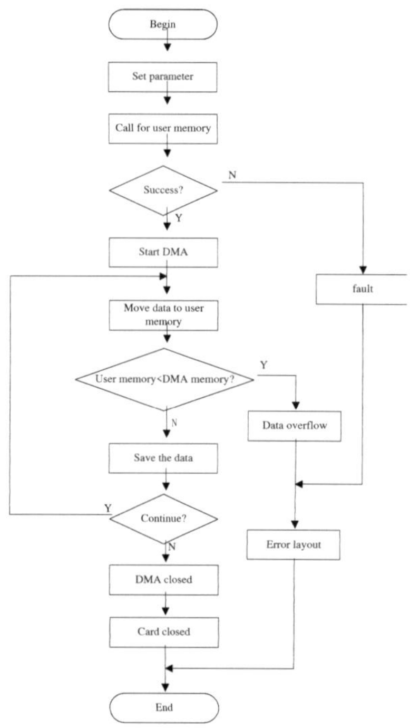

Fig. 4 The flow chart of saving data

3.2.5 DataProcessing Program. Data processing program can make the frequency intensity map. The data will be computed with FFT algorithm. A map will be shown in a cycle of analysis. As is shown in "Fig. 5", the strength is shown by the color. Color boxes shown the strength of each color. You can get your need color freely. The ordinate in left side stand for frequency. It can show the strength of frequency from 0 to 250MHz. The abscissa stand for time.

IOP Conf. Series: Earth and Environmental Science **86** (2017) 012027 doi:10.1088/1755-1315/86/1/012027

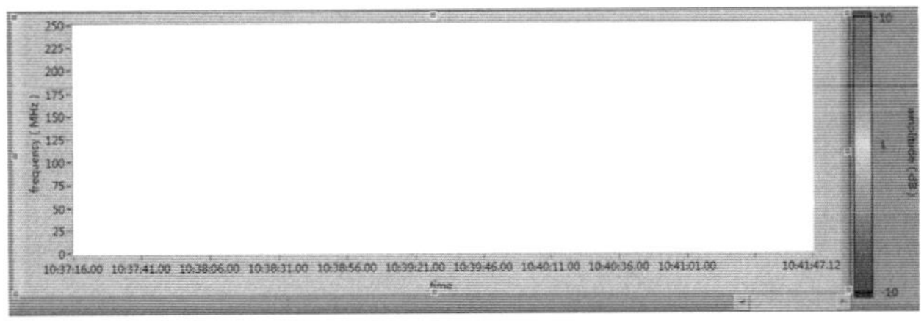

Fig. 5 Data processing program

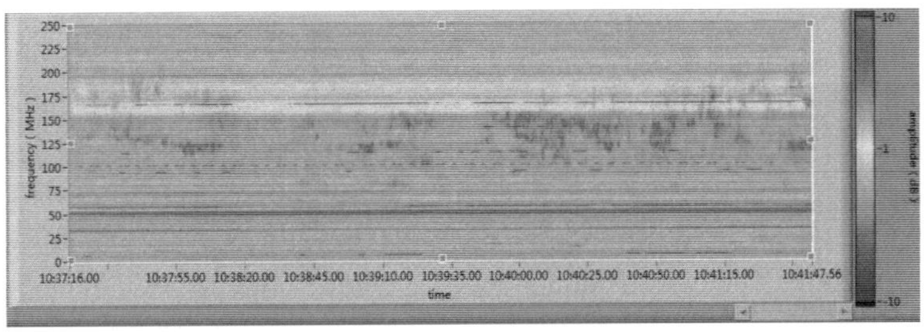

Fig.6 The result of data processing

4.DATA ANALYSIS

"Fig. 6" is a map which show the strength of signals in June 20, 2016. The time resolution is 250 milliseconds. The map one is the result of real data. The power of signals was high from 10:37 to 10:42. It stands for the outbreak of sun. The main frequency is between 0MHz and 250MHz. The map two remove the background color and we can find outbreak easier.

5.CONCLUSION

By using MFC and LabVIEW, we designed a kind of autonomous control solar radio spectrometer. It can collect data with 10 milliseconds resolution ratio and 32 KHz frequency resolution ratio tested by experiments. The software make it easy to do acquisition and save data file accurately. It will become the main tool to do the solar observation and make great sense.

REFERENCES

[1]Durante P., Neufeld N and et al 2015100 Gbps PCI-Express readout for the LHCb upgrade. *Nuclear Science IEEE Transactions on*.**62**1752-57.

[2]Fernandes A, Pereira R C, Sousa J and et al 2016 FPGA Remote Update for Nuclear Environments.*IEEE Transactions on Nuclear Science*.**63**1-1.

[3] Weinhardt M, Lang B, Thiesing F M and et al. 2015 SAccO: An implementation platform for scalable FPGA accelerators. *Microprocessors & Microsystems*.**39**543-552.

[4]Jeon S, So H, No H and et al. 2010 Development of Real-Time Software GPS Receiver using Windows Visual C++ and USB RF Front-end. *Proceedings of the International Technical Meeting of the Institute of Navigation*.713-721.

[5] Dong L., He L., Wang M. and et al. 2014 A high performance solar radio observation system in terrible RFI environments for existing telescope. *General Assembly and Scientific Symposium (URSI GASS), 2014 XXXIth URSI. IEEE*.1-4.

[6] Zhao A, Yan Y and Wang W 2015 A Subsystem Test Bed for Chinese Spectral Radioheliograph. *Publications of the Astronomical Society of Australia*..31.

IOP Conf. Series: Earth and Environmental Science **86** (2017) 012028 doi:10.1088/1755-1315/86/1/012028

Photovoltaic Cells Mppt Algorithm and Design of Controller Monitoring System

X Z Meng [1] and H B Feng [2]

[1] Qingdao University of Science & Technology , Qingdao,China

[2] Qingdao University of Science & Technology , Qingdao,China

Email: *fenghongb@163.com

Abstract. This paper combined the advantages of each maximum power point tracking (MPPT) algorithm, put forward a kind of algorithm with higher speed and higher precision, based on this algorithm designed a maximum power point tracking controller with ARM. The controller, communication technology and PC software formed a control system. Results of the simulation and experiment showed that the process of maximum power tracking was effective, and the system was stable.

1.Introduction

Solar energy as a kind of inexhaustible and low pollution energy, it has aroused widely attention and is being studied. Photovoltaic power generation is the most effective way to use and story solar energy. How to improve the efficiency of Photovoltaic power generation has become the core issue of solar power [1]. The PV cell output characteristic is non-linear, which not only affected by the light and temperature, but also by the load when the light intensity and temperature are fixed. According to its output characteristics, changing the output voltage will realize the maximum power point tracking (MPPT) [2]. In this paper, the algorithm combined the advantages of the common algorithms of Constant Voltage, Perturb & Observe and Incremental conductance, take constant voltage method first, then Perturb & Observe, the last incremental conductance to make the process of MPPT rapid and accurate. Through the simulation, the algorithm is more effective than the other algorithms used alone. The solar energy maximum power tracking controller was designed with this algorithm, improving the efficiency of PV cells. The control system consisted of the controller, communication technology and the PC software. The purpose of this system was to monitor and control processes of power generation of PV, MPPT and charge of storage battery.

2.PV Output Characteristics

PV module equivalent circuit diagram as Figure 1.

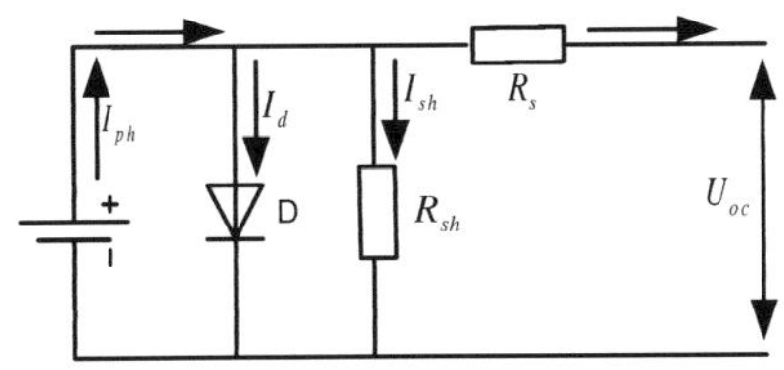

Figure 1. Equivalent circuit of a PV module

The output characteristic equation is:

$$I_L = I_{ph} - I_o\left\{\exp\left[\frac{q(U_L + I_L R_s)}{AkT}\right] - 1\right\} - \frac{U_L + I_L R_s}{R_{sh}} \quad (1)$$

In this formula, IL-photovoltaic battery output current; Iph- Photoproduction current; Io-reverse saturation current of photovoltaic cells; q-quantity of electric charge constant; k-Boltzmann constant; UL-photovoltaic battery output voltage; Rsh-bleeder resistor; Rs-the internal resistance of the photovoltaic (PV); A -ideal factor (the characteristics of the diode factor，usually between 1 to 3).

Under the standard conditions (light intensity of 1000W/m2, battery temperature Tref=25℃), selecting a photovoltaic cells with the open circuit voltage U_{oc}=*22.5V*, short circuit current I_{sc}=*2A*.The relationship between I and V, P and V as Figure 2 and Figure 3.

Curves in the figure showed the I-V characteristics of PV and P-V characteristics of PV. Point ‘a’ correspond to two curves, they are I-V characteristic curve and P-V characteristic curve of PV in a value of light intensity. Before point ‘a’, as voltage increases, current change little, power increase as voltage; after point ‘a’, as voltage increases, current and power decreases sharply. When voltage came to voltage of ‘a’, output power will come to maximum. If try to improve PV output power, take some measures to make the working point at point ‘a’, ‘a’ is the maximum power point(MPP).

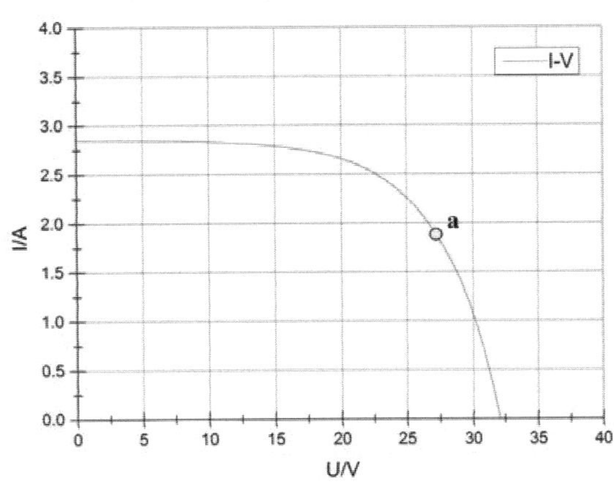

Figure 2. I-V curve of PV

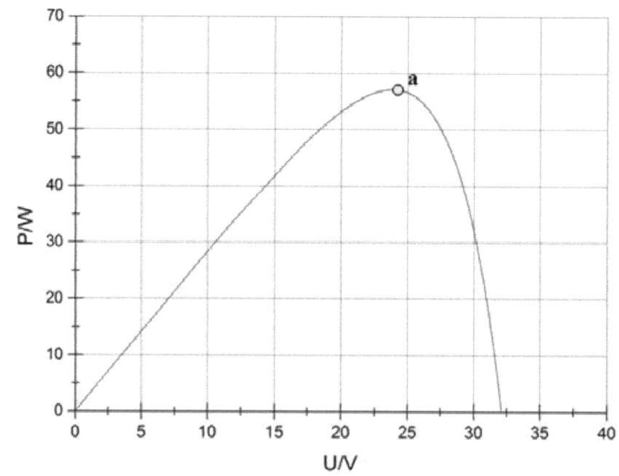

Figure 3. P-V curve of PV

3.Photovoltaic MPPT Algorithm

To achieve maximum power output of PV cell, the most practical way is continuously to acquire output current and voltage of the battery, after calculation, controlling DC-DC circuit and changing the output voltage of photovoltaic battery until to the maximum power point. There are many kinds of maximum power point tracking algorithm, common and practical algorithms include Constant Voltage(CV), Perturb & Observe(P&O), Incremental conductance(INC), etc [3].

3.1 Advantage and Disadvantage of Traditional Algorithm. Each algorithm has its feature: the CV(Figure 4(a)) is simple and easy to implement, has good reliability and stability, but PV cells I-V characteristic is influenced by environment, so CV is only applicable when environmental factor is stable; P&O (Figure 4 (b)) is simple, less accused parameters, easy to implement, hardware system need is simple, but P&O perturbation step size is not easy to be determined, too large step size cannot guarantee accuracy, too small step size lead to low speed, using P&O near the maximum power point is easy to cause misjudgment; INC(Figure 4 (d)) is one of the highest accuracy algorithm can be not affected by environmental conditions, INC algorithm is relatively complex, has large amount of calculation, the hardware requirement is high [3][4].

IOP Conf. Series: Earth and Environmental Science **86** (2017) 012028 doi:10.1088/1755-1315/86/1/012028

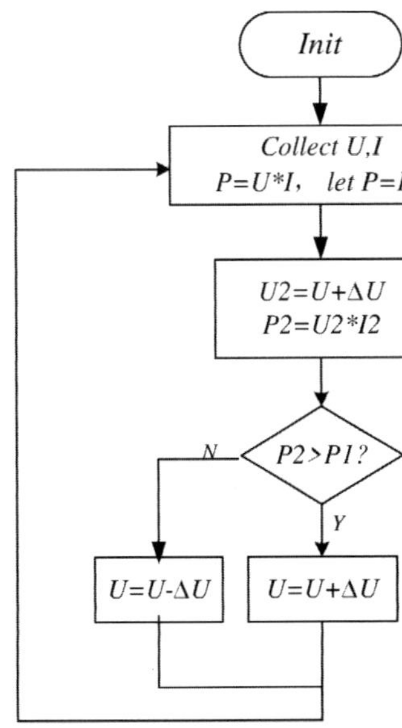

Figure 4(a). Algorithm flow chart of CV

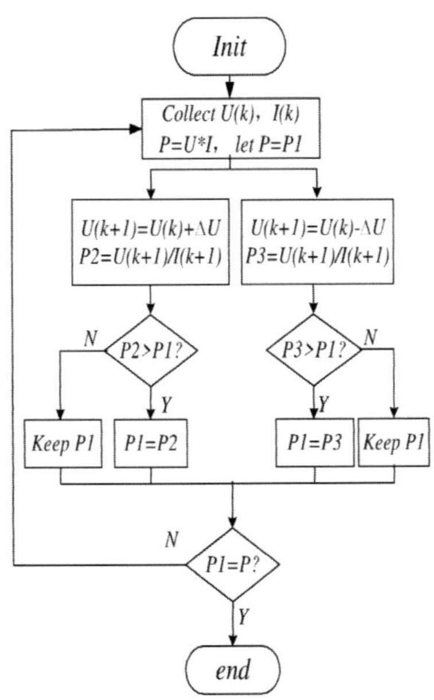

Figure 4(b). Algorithm flow chart of P&O

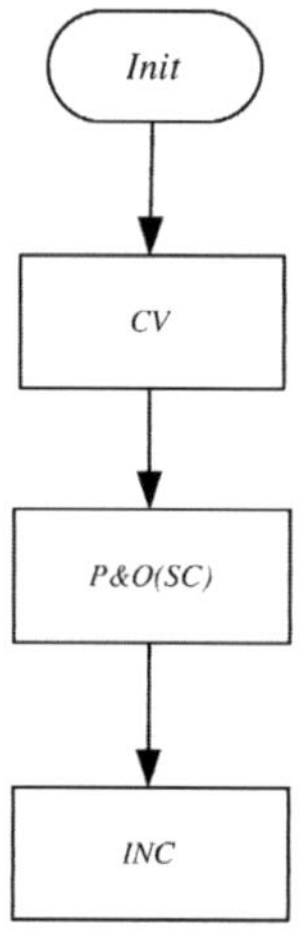

Figure 4(c). Algorithm flow chart of CV-P&O-INC

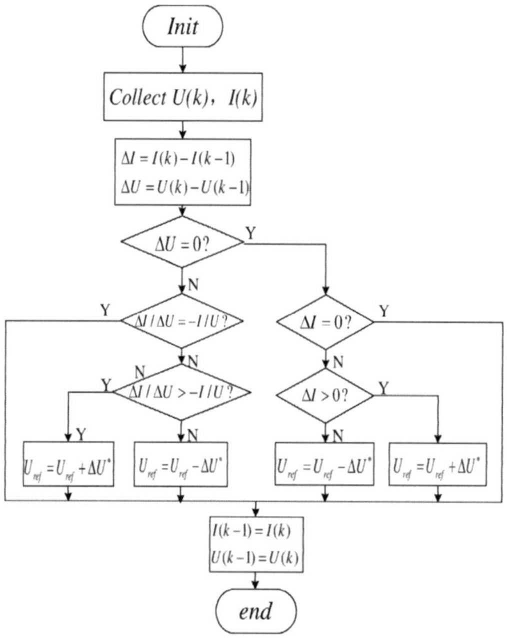

Figure 4(d). Algorithm flow chart of INC

3.2 New Algorithm of CV-P&O-INC. In algorithms, control the output voltage changes every time, we call the change of the value of voltage *ΔD* compensation factor, the compensation factor is determined by the output voltage as U_L and step length *ΔD*. In the buck and the boost circuit, adjust output voltage by changing the duty ratio D of PWM wave, every time adjust the duty ratio D, the size of the duty ratio *D* is called step length *ΔD*. During the control, long *ΔD* lead to high speed, but the precision is not easy to guarantee; small step length lead to high precision but low speed. So choose the right step is a key factor to improve the algorithm efficiency.

The algorithm used in this article combined advantages of above algorithms, bypassed the shortcomings of each algorithm, put forward a modified CV-P&O-INC(FIG.3(c)). In general, photovoltaic maximum power point voltage is 65% to 85% of the open circuit voltage U_{oc}, therefore, after initialization of algorithm, let the PV voltage fixed to 75% U_{oc},, it's easy to realize by using the DC - DC circuit. When voltage is steady, the algorithm will enter P&O, the algorithm adopted in this paper has some differences with the general P&O, in order to prevent the inaccurate judgement caused by disturbance in a single direction, the part of P&O in this algorithm (in this paper, we call it Sliding Comparison(SC)) adopts two-direction disturbance, the SC add disturbance before and after the selection point when enter the process of calculate, the way to add disturbance is to amplify or minish the output voltage, and it will get three different output voltage value point, compare the three get the biggest value point of power, it will gradually close to the maximum power point after repeated comparison. Sliding comparison in the process of the method aimed at achieving high speed to arrived point near the maximum power point so choose the larger step length ΔD=0.008, step size determines the change of voltage each time as $\Delta U=\Delta D*U_L$. When output power come to the maximum power point , algorithm turn into INC for more detailed tracking, then the step size ΔD=0.004 After calculation and the algorithm control DC - DC circuit can realize the tracking of maximum power point quickly and accurately.

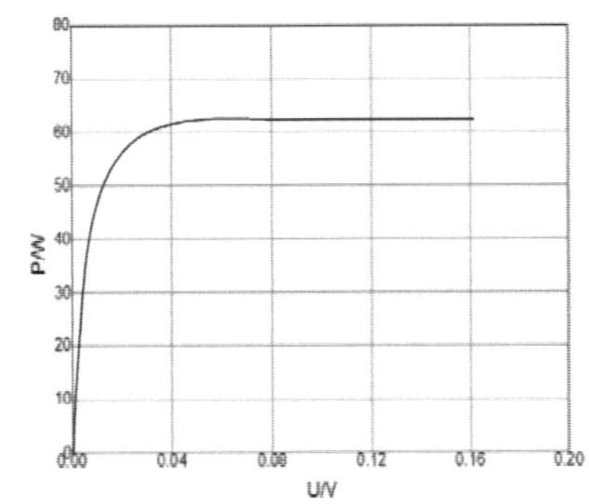

(a)

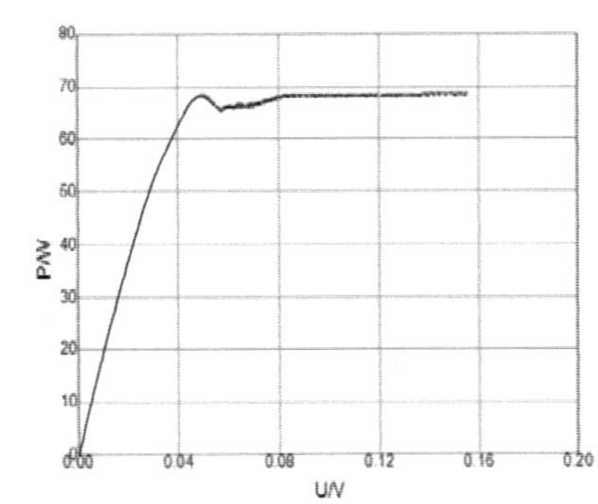

(b)

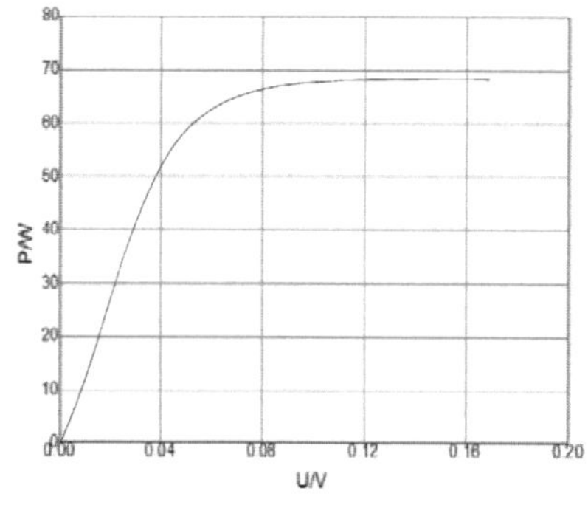

(c)

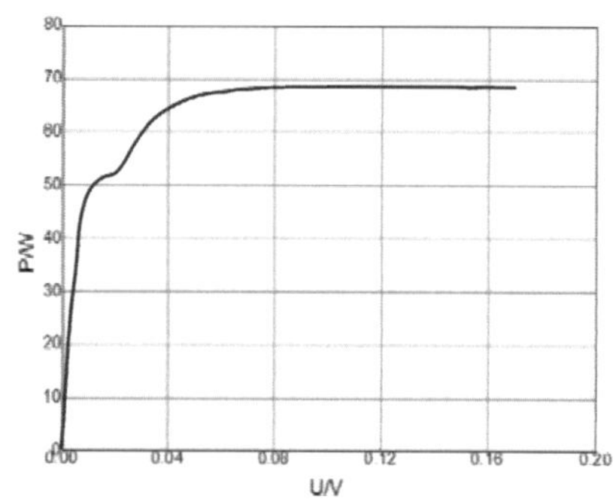

(d)

Figure 5. Simulation result of each algorithm

In Figure 5, (a) is the CV simulation result, (b) is the P&O simulation result, (c) is the INC simulation result, and (d) is the CV-P&O-INC simulation result. The results showed that CV could keep voltage to a certain value, can't reach MPPT; P&O could reach MPPT, but it was unstable and had a wave, INC was smooth but it used more time; Comprehensive comparison showed that the CV-P&O-INC was better than others. See the CV-P&O-INC curve, it could be divided into three parts. The first part of CV process, voltage fast approached 75%Uoc; once the voltage have not change obviously, algorithm entered into the second part of P&O, this part of the algorithm calculation speed was moderate, it could roughly close to the MPP; the third part of INC would keep voltage fixed to the maximum power point. Simulation results proved that the CV-P&O-INC algorithm of MPPT is fast and precise.

4.The Design of The Controller and System

Based on the research on algorithm above, we designed a controller. The controller structure was as shown in Figure 6, included DC - DC circuit, communication module, power supply module, display module, test module, protection module, etc. The ARM produced PWM wave to control DC - DC circuit to change the output voltage so the controller could change the output power. DC - DC contains buck and the boost circuit, buck circuit used for buck, its relationship between the output voltage and input voltage: $U_o=U_s*D$; voltage boost circuit used for boost, its relationship is $U_o=U_s*(1-D)$. The key of the algorithm control the controller was changing the duty ratio(D) of PWM wave. The controller also has communication module and display module. Communication module used 485 serial communication and wireless communication based on the IEEE 802.15.4, 485 communication was mainly used for the distribution of the multipoint Photovoltaic power generation, controller embedded MODBUS protocol, connect to PC, wireless communication has solved occasions the signal line cannot go directly. 12864 OLED display module show information, the module was a kind of low power that display 4 lines, used to display the input, output voltage and current of the power generation efficiency [5].

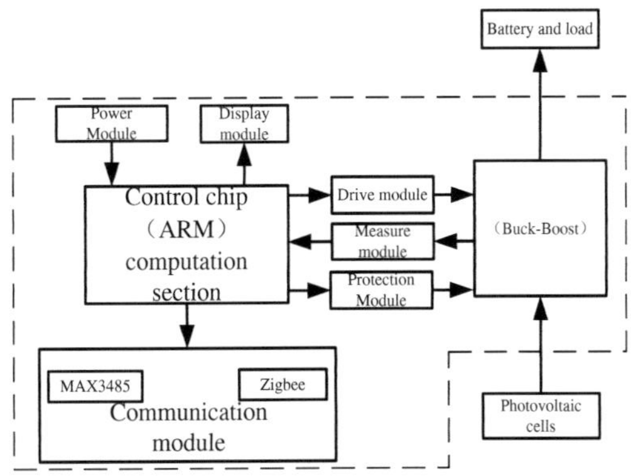

Figure 6. The structure chart of the controller

After finished the controller ,we built a system as Figure 7, this system includes solar panels, the MPPT controller, storage battery and PC and Communication module. The PV cell converted light energy into electrical energy, electric energy sent to the storage battery through the MPPT controller, the controller display module display voltage, current and efficiency, etc. Multiple MPPT controllers and PC machine and the use of 485 communication can form 485 network. PC software was based on the QT framework for C++ language was a PC monitoring analysis software, it was used for real-time monitoring the MPPT controller for solar energy network and analysis the state.

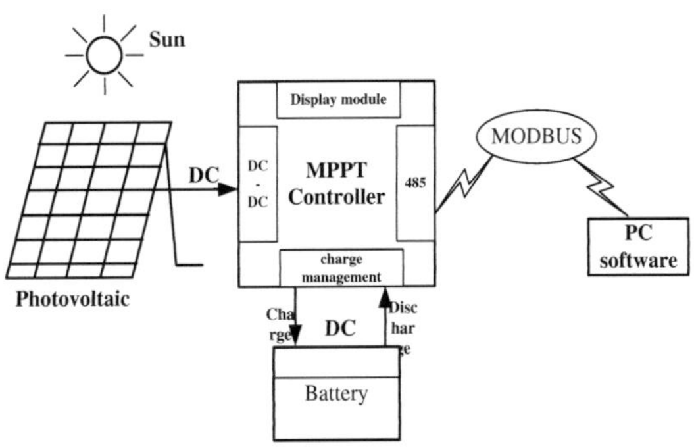

Figure 7. The structure chart of the system

5.Conclusion

This paper analyzed the advantages and disadvantages of the commonly used algorithm in photovoltaic power generation, combined the advantages of three algorithms namely CV, P&O and INC and put forward a modified algorithm. Result of simulation prove the characteristics of the algorithm was fast and accurate. The MPPT controller was designed based on the algorithm, this controller was a important part of the photovoltaic power generation system. The system would improve the PV cells efficiency and had a vital significance.

References

[1] Yan Y F, Zhang Z E, Zhang L and Dai C L 2012 Use technology and application of solar power. *Acta energiae solaris sinica*. **12** 47-54.
[2] Guan Q, Liu,X K, You Y X and Yang L 2012 Study on Maximum Power Point Tracking Techniques for Photovoltaic Power Generation System. *Shanxi Electric Power*. **40** 18-21.
[3] Wang B Z, liu W F, Fu N N and Song D F 2010 Photovoltaic power generation systems and study on MPPT algorithm. *Science technology and engineering*. **12** 8433-38.

IOP Conf. Series: Earth and Environmental Science **86** (2017) 012028 doi:10.1088/1755-1315/86/1/012028

[4] Kotti R. and Shireen W 2015 Efficient MPPT control for PV systems adaptive to fast changing irradiation and partial shading conditions. *Solar Energy*. 397-407.
[5] Tsang K M and Chan W L 2015 A single swith DC/DC converter with galvanic isolation and input current regulation for photovoltaic systems. *Solar Energy*. 203-211.

IOP Conf. Series: Earth and Environmental Science **86** (2017) 012029 doi:10.1088/1755-1315/86/1/012029

Wide-area Power System Damping Control Coordination Based on Particle Swarm Optimization with Time Delay Considered

J Y Zhang [1] and Y Jiang [2]

[1]Department of Electrical Engineering, North China Electric Power University,Baoding,China

[2]State Grid Hebei Baoding Branch, Baoding, China

Email: *jyzhang327@126.com

Abstract. To ensure satisfactory dynamic performance of controllers in time-delayed power systems, a WAMS-based control strategy is investigated in the presence of output feedback delay. An integrated approach based on Pade approximation and particle swarm optimization (PSO) is employed for parameter configuration of PSS. The coordination configuration scheme of power system controllers is achieved by a series of stability constraints at the aim of maximizing the minimum damping ratio of inter-area mode of power system. The validity of this derived PSS is verified on a prototype power system. The findings demonstrate that the proposed approach for control design could damp the inter-area oscillation and enhance the small-signal stability.

1.Introduction

The low frequency oscillation of interconnected system brings challenges to the inter-area power dispatch and transmission [1-2] and the utilization of remote signals is of vital significance to damp low frequency oscillation and boost dynamic performance. However, the centralized controller entails inputs that may arrive after a certain communication delay in WAMS. Because of its destructive influence on small-signal stability, system robustness and voltage oscillation of power grids [3-4], time delay can be hardly ignored in wide area control.

Various approaches have been taken to overcome the limitations of transmission delay in WAMS, such as robust control theory [5], fuzzy control [6] and model predictive control (MPC) [7]. In contrast to the wealth of studies tackling the time-delayed problem with classical control theories, there has been less research adopting intelligent algorithm, which is an optimization method with superior random searching strategy and faster convergence speed. In the light of this, the aim of this study is to explore a combined control strategy with intelligent algorithm and equivalent transformation.

In this paper, a combinatorial approach of power system stabilizer (PSS) design is adopted in this work to address the damping control problem with regard to time-delayed power systems. Firstly, by employing Pade rational polynomials to approximate the value of time delay, the augmented state matrix of power systems is formulated. Secondly, the coordination configuration scheme of power system controllers is achieved by a series of stability constraints and PSO, and an eigenvalue-based objective function to increase the system damping is proposed. Lastly, the validity of this method is verified on a two-area four-machine system.

2.Model of time delayed power system

The dynamic model of power system is usually illustrated by a set of differential algebraic equations, which can be linearized at an equilibrium point in the form of state space as follows:

$$\begin{cases} \dot{x}_p(t) = A_p x_p(t) + B_p u_p(t) \\ y(t) = C_p x_p(t) + D_p u_p(t) \end{cases} \quad (1)$$

With local and wide-area supplementary damping controller installed on the power system (1) shown in Fig.1, transmission delay is introduced.

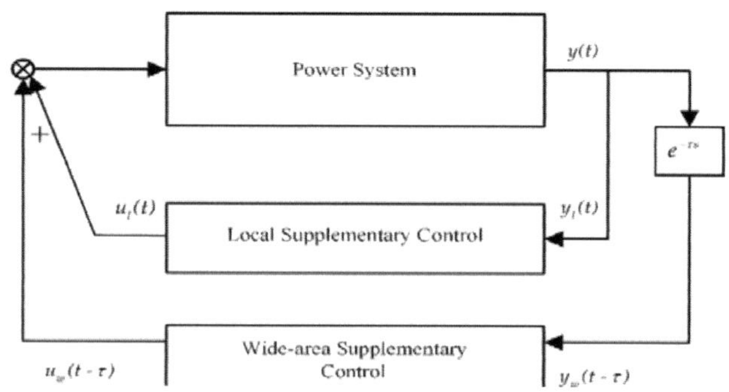

Figure I. The Structure of power system

Therefore, the time-delayed power system is

$$\dot{x}(t) = A_c x(t) + A_d x(t-\tau) \quad (2)$$

where A_c and A_d are matrix containing both system and control parameters.

3 Controller design

The aim of this paper is to design supplementary controllers to damp low frequency oscillation considering the multiple time delays so as to improve power system security. As shown in Fig.2, the general model of PSS consists of an amplifier, a reset element, a phase compensator and a limiter. In PSS feedback loop, the input signal is converted to a supplementary excitation signal.

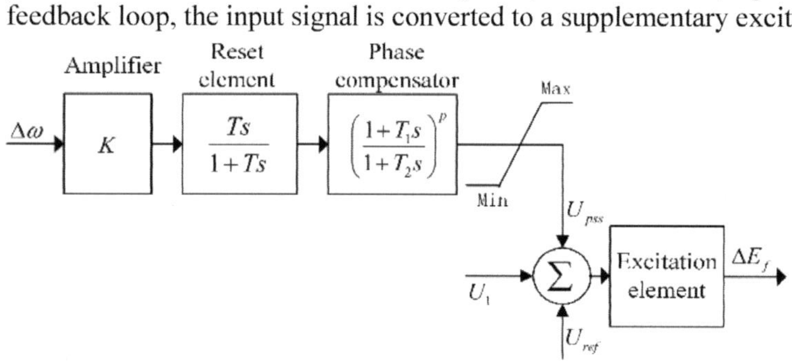

Figure 2. The Structure of Power System Stabilizer

In this study, PSS is introduced as supplementary control (both local and wide-area control) to damp the power system.
The local PSS could be expressed as

$$G_l(s) = K_l \left(\frac{Ts}{1+Ts}\right)\left(\frac{1+T_{1l}s}{1+T_{2l}s}\right)^p \quad (3)$$

and the wide-area PSS could be expressed as

$$G_w(s) = K_w \left(\frac{Ts}{1+Ts}\right)\left(\frac{1+T_{1w}s}{1+T_{2w}s}\right)^p \quad (4)$$

The goal in this study is to find the parameters of these two PSSs.

3.1 Pade approximation. Time delay, expressed as the exponential function $e^{-\tau_i s}$, could be also equivalently formulated as

$$\begin{cases} \dot{x}_\tau(t) = A_\tau x_\tau(t) + B_\tau y_w(t) \\ y_{w1}(t) = C_\tau x_\tau(t) + D_\tau y_w(t) \end{cases} \quad (5)$$

By using Pade approximation, the time delay module could be transferred into state space form, which is no longer difficult to deal with.
Consider the power system (1) with local control, wide area control and time delay module, the time-delayed power system with supplementary control is

$$\dot{x}_{all}(t) = A_{all} x_{all}(t) \quad (6)$$

where $x_{all}(t)=[x_p(t) \quad x_l(t) \quad x_w(t) \quad x_\tau(t)]$.

3.2 Parameter configuration with PSO. Both the local and wide-area supplementary controllers are to be designed. For the best performance of intelligent algorithm, some parameters of PSS given. Generally, the T_{1l}, T_{1w} of PSS are unknown and T_{2l}, T_{2w} are set to a specific value.

The design of PSS is therefore defined as an optimized problem with unequal constraints below

$$\begin{cases} \max \quad J \\ s.t. \quad K_{\min} < K_l, K_w < K_{\max} \\ \quad T_{1l\min} < T_{1l} < T_{1l\max} \\ \quad T_{1w\min} < T_{1w} < T_{1w\max} \\ \quad eig(A_{all}) < 0 \end{cases} \quad (7)$$

where $J = \min\{\xi_i, i \in electromechanical\ modes\}$ and $\xi = -\sigma_i \Big/ \sqrt{\sigma_i^2 + \omega_i^2}$.
Accordingly, the object function could be rewritten as $f = \min\ f_{fitness}$ where fitness degree function $f_{fitness}(x)$ is split as $f_{fitness}(x) = f(x) + g(x)$. The objective function $f(x)$ is denoted as $f(x) = -\max\ J$, while the penalty function $g(x)$ is defined as $g(x) = \sum_{i=1}^{n_p} N_p \cdot real(x_i)$.

3.3 Procedure of control design. The procedure of control design could be summarized as follows.

Input the data of power system and initialize the particle and speed of PSO and set the limit.
Calculate the augment state matrix of the time-delayed power system and its eigenvalue.
Check whether the constraints are satisfied.
Calculate the fitness value of each particle and rank them.
Continue iteration until the allowable maximum number of iteration.

4.Results

The two-area four-machine system with data mainly inspired from [1] and adopted for this study to test the proposed method. The modal analysis with results in Table I illustrates three electromechanical modes with two local modes (mode 1 and 2) and one inter area mode (mode 3).

Table 1. Electromechanical Modes of Power System

Mode	Eigenvalue	Damping ratio	Participant machines
1	-0.1030 ± 6.2610i	0.0164	G1,G2
2	-0.5192 ± 6.8519i	0.0755	G3,G4
3	0.1209 ± 3.6288i	-0.0333	G1-G4

As damping ratio of mode 2 and 3 are insufficient, 2 PSSs, namely PSS1 (as local PSS) and PSS2 (as remote PSS) are installed on the power system to damp the local mode inter-area mode. The PSSs here adopt angular velocity type PSS.

By calculating the controllability and observability of mode 1 and mode 3, we choose $\Delta\omega_2$ of G2 as the feedback signal to PSS1 on G1, and $\Delta\omega_4$ of G4 as the feedback signal to PSS2 on G2.
The goal is to optimize the T_{1l}, T_{1w}, K_l, K_w of the two PSSs. In this study, fifth order Pade approximation is used to replace the time delay module.
The given parameters of PSSs are: $T_{2l} = T_{2w} = 0.05s$, $T = 10s$, $\tau = 0.2s$ .In PSO, the number of particle $N = 40$, learning factor $c_1 = c_2 = 2$, maximum allowable iteration $M = 50$, dimension $D = 4$, penalty factor $N_p = 10$. The initial value of weight $w_0 = 0.5$, the allowable $F_0 = 0.5$.
According to the design procedure in Section III, the parameters of two PSSs could be designed as

$$\begin{cases} G_{pss1}(s) = 13.7712\dfrac{10s}{10s+1}\left(\dfrac{1+0.01s}{1+0.05s}\right)^2 \\ G_{pss2}(s) = 3.3523\dfrac{10s}{10s+1}\left(\dfrac{1+0.01s}{1+0.05s}\right)^2 \end{cases} \quad (8)$$

and the value of fitness function in iteration process is illustrated in Fig.3. As can be seen, the final value of fitness function is -0.0633, which indicates that the inter-area mode is damped effectively.

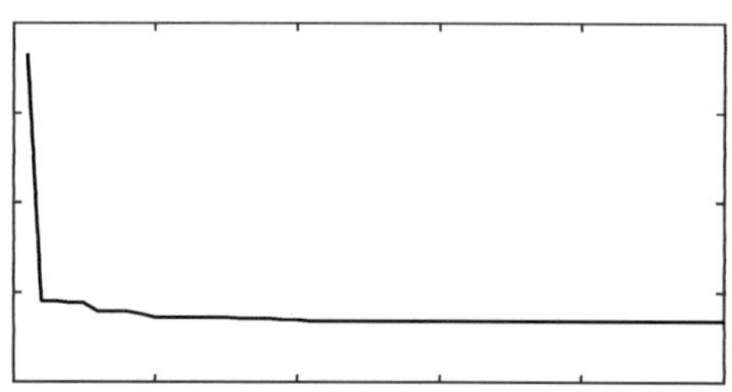

Figure 3. Fitness funtion value by using PSO

An eigenvalue analysis of the closed-loop system was carried out to examine the performance of the designed controllers where the time delay module is replaced by Pade approximation. Effect of these derived controllers is illustrated in Table II.

Table 2.Electromechanical Modes with Controller

Mode	Time delay	Eigenvalue	Damping ratio
1	With 0.1s delay	-0.4305 ± 6.7849i	0.0633
2		-1.0599 ± 6.9328i	0.1511
3		-0.1881 ± 3.6200i	0.0519
1	With 0.2s delay	-0.4176 ± 6.7772i	0.0615
2		-1.0430 ± 6.9092i	0.1493
3		-0.2659 ± 3.4857i	0.0761

The results in Table II reveals that the damping ratio of mode 3 improves to 0.0519 with 0.1s time delay and 0.0761 with 0.2s time delay. Besides, both mode 1 and mode 2 are damped in the presence of the controllers. This indicates that the proposed method is effectively enhance the power system stability and reduce the negative effect of time delay.

With different wide-area signal delays considered, the system response to a small disturbance and large disturbance at the reference voltage of excitation on G2 is shown in Fig. 4 and Fig.5. The power system is unstable without supplementary control. With the proposed PSSs, the power system could be damped effectively. Compare the curve of 0.2s time delay with 0.1s time delay, the former time delayed power system is much more stable than the latter.

In conclusion, the damping performance of the designed controllers is therefore considered effective to enhance the damping ratio and boost the small-signal stability of time-delay system against negative or low damping modes.

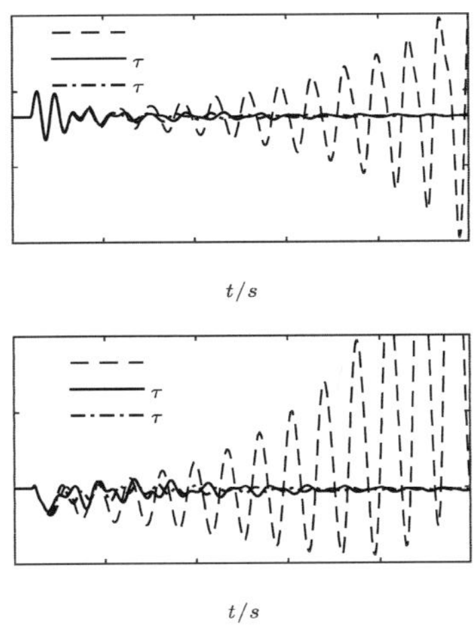

Figure 4. System response with different -wide-area signal delays (small disturbance)

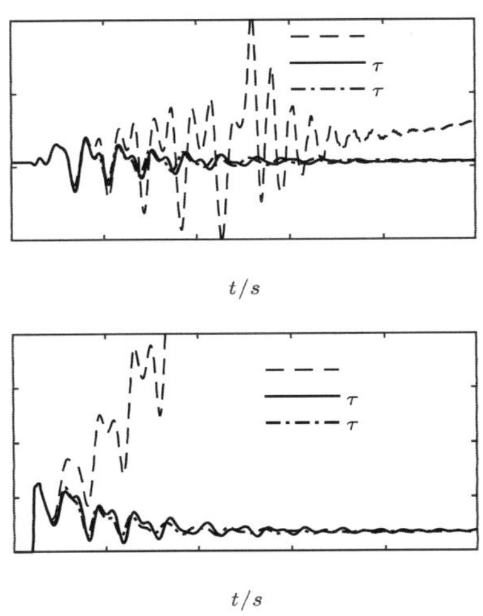

Figure 5. System response with different -wide-area signal delays (large disturbance)

5.Conclusion

This paper proposed a WAMS-based PSS design strategy with output feedback delay considered to ensure satisfactory dynamic performance of controllers in power systems. An integrated approach based on Pade approximation and PSO is employed for parameter configuration of PSS. By employing state space expression of fifth order Pade rational polynomials to approximate the value of time delay, the time delay link element could be transformed into a linear element and an augmented state space model of power system is formulated. The validity of this derived PSS is verified on a two-area four-machine power system. Through eigenvalue analysis and time domain simulation, the proposed approach could damp the inter-area oscillation effectively and enhance power system stability.

References

[1]KundurP, Balu N J and Lauby M G 1994 *Power system stability and control, vol.7.*New York: McGraw-hill.
[2] Zhu F, Zhao H, Liu Z and Kou H 2007 The influence of large power grid interconnected on power system dynamic stability. *Proceedings of the CSEE, vol.* 27.1-7.
[3]Jia H, Chen J and Yu X 2006 Impact of time delay on power system small signal stability. *Automation of Electric Power Systems, vol. 18.*no.3 pp389-391.
[4]Jia H., Xie X and YuX 2006 Power system small signal stability region with time delay considered. *Automation of Electric Power Systemsvol. 30* no.21 pp.1-5.
[5] JiangQ, Zhang P and Cao Y 2006 Wide-area FACTS damping control in consideration of feedback signals' time delays.*Proceedings of the CSEE vol. 44*no.7 pp.82-88.
[6]Mokhtari M, Aminifar F, Nazarpour D and Golshannavaz S 2013 Wide-area power oscillation damping with a fuzzy controller compensating the continuous communication delays. *IEEE Transactions on Power Systemsvol. 28*pp1997-2005.
[7] Azad S P, Iravani R and Tate J E 2013 Damping inter-area oscillations based on a model predictive control (MPC) HVDC supplementary controller. *IEEE Transactions on Power Systems vol. 28* pp3174-83.

The Inductance Parameter Study ofLoose Coupled Transformer

K RChen ,Y Han , L F Guo,N N Gao,Y Peiand P Li

[1]State Grid of China Technology College, China

Email: *630932982@qq.com

Abstract. We built the simulation model of loose coupled transformer in the foundation of the experimental model including the core of E-type. We analyzed the electro and magnetic relations of the loose coupled transformer, and by means of the energy method, the coupling coefficient is studied and inductances are calculated. Because of the abnormal working state of the WPT system in practical use, , we use ANSOFT to calculate both coupling coefficient and the inductances when longitudinal and lateral distance air gap is in different situation, in this way, we can study the coupling coefficient when the three factors changed, which can give a full reference for studying the coupling coefficient parameters of loose coupled transformer.

1.Introduction

As a cross subject in the field of electrical and electronic engineering, Wireless power transmission system is based on the principle of transformer in electrical engineering[1], adopting the power electronic technology such as power electronics, electromagnetic power transmission, implements the power by the original to the deputy while no electrical connection. As a brand-new electrician class discipline in resent years, WPT systems has received a lot of experimental research, its research results have been applied in all kinds of electronic products and electric power supply and charging systems.[2-4]In some special circumstances, such as mine, underwater, etc, WPT technology shows obvious advantage over traditional wired connection, especially in the avoid the switch switch, electrical, leakage of edm, and effectively prevent the security hidden danger, to curb the root cause of the accident.

We divide loose coupled transformer into three parts: including core type, resonance coupling and electromagnetic radiation. Among them, including core type core-type transmission system is more practical, relative to other two outstanding performance in the core of loose coupling transformer.[5-6]The same as the traditional transformer, there is a large permeability core in the loose coupling transformer, improved the effect of the permeability of the entire system effectively, which reduced due to the magnetic field and magnetic circuit magnetic resistance caused by the rising through the air, to a certain extent. However, due to the smaller permeability of the air, the coupling of the loose coupled transformer is still small, especially when the situation is bad, coupling coefficient will also produce great influence, the efficiency will also be affected by the corresponding.

To this end, the paper focus on the WPT system coupling coefficient of the loose coupled transformer. Considering the WPT systems working in the case of non-ideal state, through the loose coupled transformer model, using finite element software, the former vice edge analysis of the loose coupled transformer inductance parameters and coupling coefficient by air spacing, vertical lateral and horizontal lateral distance and the influence of the magnetic permeability.

2.Methodology

Loose coupling transformer basic electromagnetic relation is shown in figure 1.

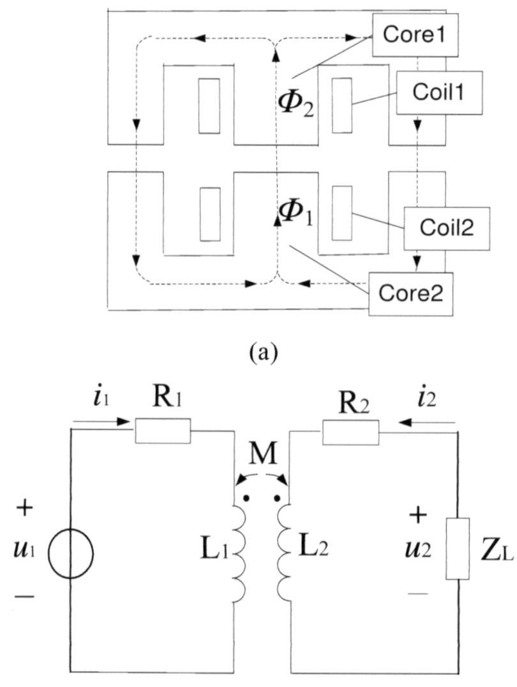

Figure.1 The Magnetic Relation and Electronic Relation of the System

Magnetic field Equations derived by maxwell's equations as follows:

$$\nabla \times \frac{1}{\mu} \nabla \times \boldsymbol{A} = \boldsymbol{J} \quad (1)$$

Current density is expressed by J, the vector magnetic potential is expressed by A, material permeability is expressed by μ. We can adopt current density J c by means of experimental measurement.

Unit interpolation as follows:

$$\boldsymbol{A} = \sum_{n=1}^{n_n} \boldsymbol{M}_n(x, y, z) A_n \quad (2)$$

A series of basis functions use the expression of n, The edge number is expressed by nn, The

base sequence of functions is expressed by {Mn, n=1, 2, …,nn}.
The weighted residual equation is obtained by using the Green's theorem:

$$\iiint_V \frac{1}{\mu}(\nabla\times \boldsymbol{M}_m)\cdot(\nabla\times \boldsymbol{M}_n)A_n\,\mathrm{d}V = \iiint_V \boldsymbol{M}_m\cdot \boldsymbol{J}\,\mathrm{d}V \quad (3)$$

Among them, the base sequence of functions is expressed by Mm, Weight function and odd function are equivalent. Under the condition that I is known, put weight function into the plug type, available in the vector magnetic potential of the node A, we can calculate the vector magnetic potential in the fields B and H.

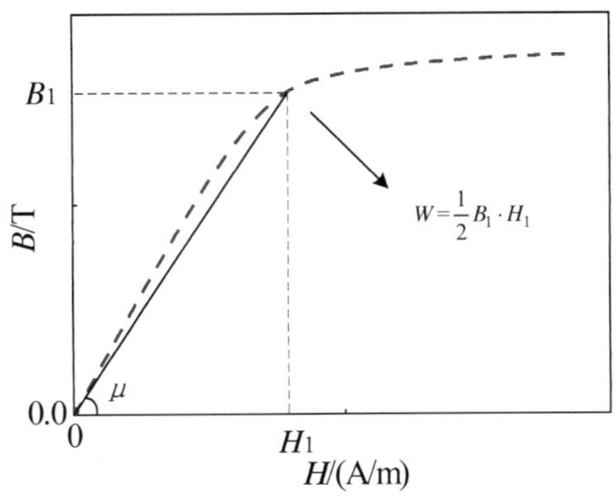

Figure.2 Energy Calculation Method

According to the energy method to adopt the magnetization curve by calculating the parameter. Figure 2 shows that B, H value can get from above can calculate magnetic energy value:

$$W = \frac{1}{2}\int_\Omega \boldsymbol{B}\cdot\boldsymbol{H}\mathrm{d}\Omega \quad (4)$$

At the same time, the transformer magnetic energy and inductance parameters and excitation current has the following relationship:

$$W = \frac{1}{2}L_{\mathrm{ab}}I_a I_b,(a,b=1,2) \quad (5)$$

Among them, on the condition that a=b, self inductance of the former and vice edge L11 and L22 can be calculated; on the condition that a≠b, mutual inductance of the former and vice edge M12 equals to L12.
Equation (4) energy result equals to equation (5), L1, L2, M12 can be obtained the same. This loose coupling transformer coupling coefficient can be obtained: $k=\frac{M_{12}}{\sqrt{L_{11}L_{22}}}$

3.The Study of Coupling Coefficient related Factors

Because of the loose coupled transformer on both sides of the coil winding on the two different core respectively, and an air gap exist between them. At the same time, the former and vice edge of lateral core may exist certain Angle, there may some errors spacing on two sides of the coil, which related to the efficiency, the inductance parameters and coupling coefficient. The paper researched on the former and vice edge coupling coefficient under different relative position.
At present, the commonly used core is U-U type, C-I type, E-E type, etc., we adopt the type

of E as an example. The former and vice edge under different relative position of the loose coupled transformer coupling coefficient for simulation analysis.

3.1 The Study of Air Gap Thickness influencing the Coupling Coefficient. As shown in figure 3, the loose coupled transformer model is built, analyzing coupling coefficient when the air gap changed.

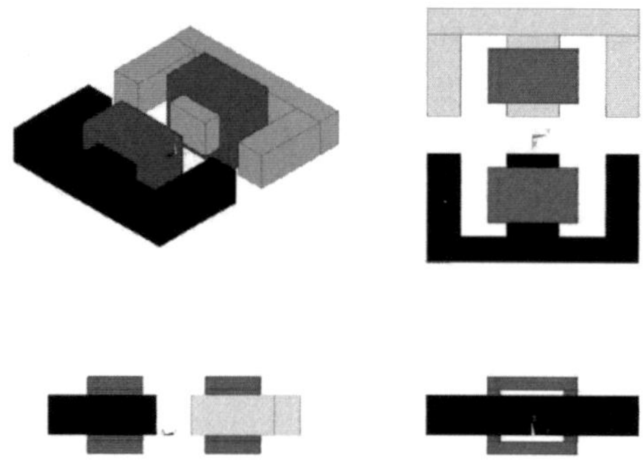

Figure.3 The 3D Model When the Air Gap is d

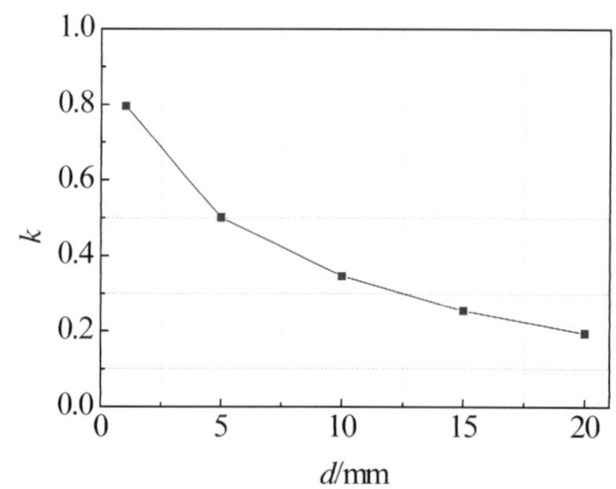

Figure.4 The relation between Coupling Coefficient and the Air Gap

The transformer core material is 30 q130 type silicon steel sheet, calculating finite element under 1 mm, 5 mm, 10 mm, 15 mm, 20 mm air-gaps. As is shown in figure 3, the coupling coefficient shows close to the air gap thickness, thickness of air gap increases in rapid attenuation of the inductance parameters and coupling coefficient greatly reduced.

3.2 Core Transverse Lateral Influence on Inductance Parameters. The figure 5 shows that loose coupled system model is built, studying when the air-gap is 10 mm, when the transverse lateral distance changed, coupling coefficient changes. The former and vice edge set core coaxial horizontal lateral distance is y = 0 mm.

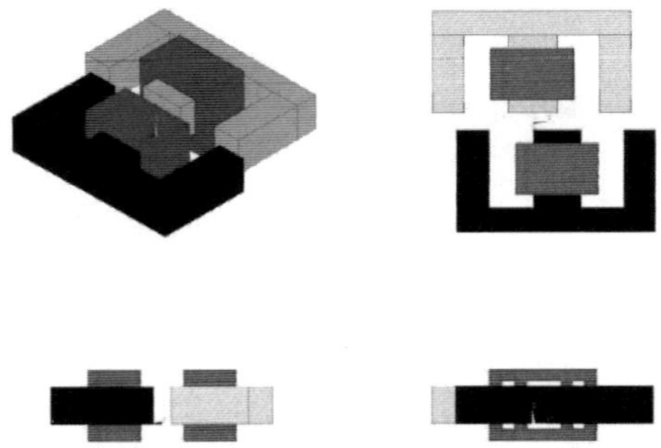

Figure.5 The 3D Model When the Transverse Lateral Distance is y

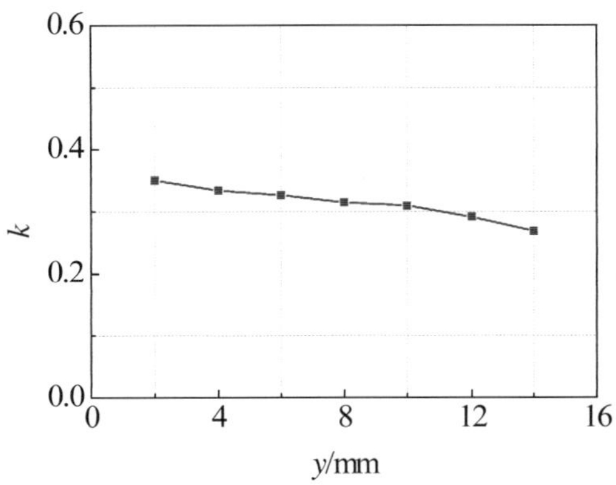

Figure.6 The relation between Coefficient and Transverse Lateral Distance

Calculating finite element under 2 mm, 4 mm and 6 mm, 8 mm, 10 mm, 12 mm, 14 mm lateral distance. The results shows that the loose coupled transformer air-gap thickness of certain circumstances, lateral distance show little influence on self-inductance, and mutual inductance reduced. Obviousely, the former and vice edge transverse lateral distance shows slight relation to the mutual inductance, coupling coefficient, lateral sway in the practical application of smaller cases, loose coupling transformer coupling coefficient of impact is small.

3.3 The Study of Longitudinal Lateral Influencing the Inductance Parameters. The figure 7 shows that loose coupled transformer model is built, studying when the thickness is 10 mm, when different longitudinal lateral distance coupling coefficient changes. The former and vice edge set core coaxial horizontal lateral distance is z = 0 mm, 2 mm, 4 mm and 6 mm, 8 mm, 10 mm five groups of longitudinal coupling coefficient under lateral distance value.

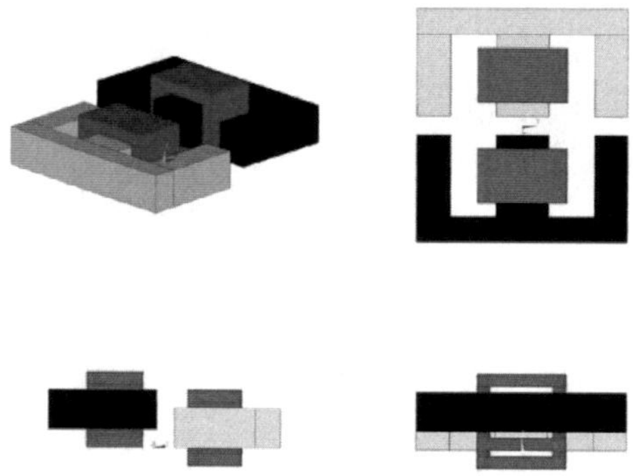

Figure.7 The 3D Model When the Longitudinal Lateral is z

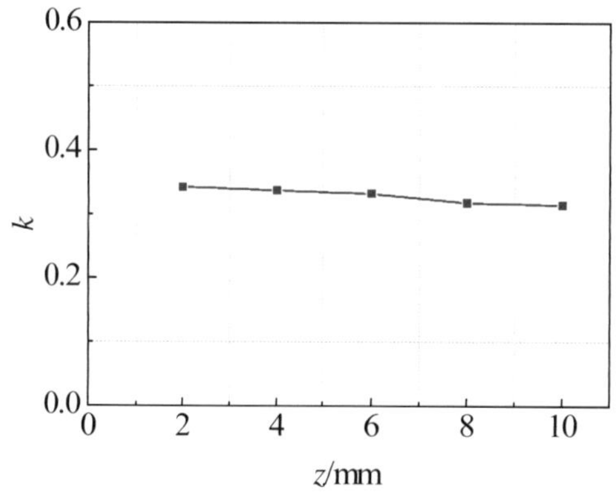

Figure.8 The relation between Coupling Coefficient and Longitudinal Lateral Distance

As shown in figure 6, when air-gap thickness of the loose coupled transformer is secure, coupling coefficient reduced when horizontal lateral distance increased, while it is slight, therefore, coupling coefficient shows some relation to the longitudinal lateral, but it's small.

4.Conclusion

This paper studies the impact of coupling coefficient in loose coupled transformer in Wireless power transmission system when the former and vice edge in different relative positions, the conclusions are as follows:

Using ANSOFT, the finite element method, to analyze the coupling coefficient of loose coupled transformer when the system work abnormal. Coupling coefficient shows great changed when the air gap changed. While it changed slowly when horizontal and vertical lateral distance increased, the coupling coefficient decrease slowly. The inductance shows little relate to the lateral change, in the case of small lateral situation, coupling coefficient error cannot ignore; because of the coupling coefficient shows great relation to the air gap , to reducing the air gap size or percentage of air gap in the magnetic field transmission circuit, such as increasing size of the core, improve its nominal capacity, etc, can effectively make up for the thickness of air gap caused by the coupling coefficient of attenuation.

References

[1] Wang C S, Stielau O H, Covic G A 2005 Design consideration for a contactless electric vehicle battery charger.*Industry Applications, IEEE Transactions on***52**1308-13.

[2] Sibué J R , Kwimang G and Ferrieux J P 2013 A global study of a contactless energy transfer system: analytical design, virtual prototyping prototyping, and experimental validation. *Power Electronics , IEEE Transaction on***28**4690-99.

[3] Bieler T, Perrottet M and Nguyen V 2002 Contactless power and information transmission. *Industry Electronics, IEEE Transactions on***38** 1266-72.

[4] Kacprzak D, Covic G A and Boys J T 2005 An improved magnetic design for inductively coupled power transfer system pickups. *7th International Power Engineering Conference*.1133-1106.

[5] Boys J T, Covic G A and GreenA W 2000 Stability and control of inductively coupled power transfer systems. *IEE-Elect. Power*37-43.

[6] Wang C W S, Stielau O.H and Covic G A 2000 Load models and their application in the design of loosely coupled inductive power transfer systems. *PowerSystem Technology*.1053-58.

A Systematic Multi-Time Scale Solution for Regional Power Grid Operation

W J Zhu [1], Z G Liu [2], T Cheng [3], B Q Hu [4], X Z Liu [5] and Y F Zhou[6]

[1 2 3 4]Hunan Electric Power Co. Ltd, Changsha, China

[5 6] Dept. of Electrical Engineering, Tsinghua University, Beijing, China

Email: *zhuwj2@hn.sgcc.com.cn

Abstract. Many aspects need to be taken into consideration in a regional grid while making schedule plans. In this paper, a systematic multi-time scale solution for regional power grid operation considering large scale renewable energy integration and Ultra High Voltage (UHV) power transmission is proposed. In the time scale aspect, we discuss the problem from month, week, day-ahead, within-day to day-behind, and the system also contains multiple generator types including thermal units, hydro-plants, wind turbines and pumped storage stations. The 9 subsystems of the scheduling system are described, and their functions and relationships are elaborated. The proposed system has been constructed in a provincial power grid in Central China, and the operation results further verified the effectiveness of the system.

1.Introduction

In the power dispatch department of a regional grid, there are many problems need to be considered while making schedule plans. Generally, there are mainly 4 concerns need to be taken into account:

The targets

The plan needs to focus on the schedule targets, in order to solve all the potential problems that may appear during operation. Usually, the objectives are different for problems with different time scales or considering different stations of the grid.

Time scales

We usually concern on different problems with the time scale of scheduling varies. Take wind power forecast as an example, the target and mathematical models can be very different considering different time scales, the accuracy of mid-long term forecast may be much worse than that of short-term forecast, and the result need to be corrected.

The objects

In a schedule plan, the objects are various type of sources, including thermal units, hydro-plants, wind turbines, all sorts of energy storage stations and other sources such as the integration of Ultra High Voltage (UHV). How to take advantages of their characteristics respectively and make a maximum coordination effect, has always been a research focus.

Grid status

In a schedule plan, different grid situations must be concerned. Reference [1] proposes the classical power system model consisting of three operating states. Different scheduling strategies need to be made in respond to these states.

The basic idea focusing on the questions above is decomposition. For example, control strategies with time-scale decomposition are widely used in many areas, such as power control system[2] and micro-grid operation[3]. In this paper, the entire schedule system in a regional grid is decomposed into 9 subsystems, solving problems focused on different time-scales and different grid operation status, trying to solve the problems introduced by the integration of large scale renewable energy sources and the UHV transmission.

This paper is organized as follows: Section II introduces the framework of the schedule system, and the detailed description of the 9 subsystems is given in Section III. Section IV concludes the work of the paper..

2. THE FRAMEWORK OF THE SCHEDULE SYSTEM

The framework of the proposed schedule system is shown in Fig. 1. It mainly consists of 9 subsystems, focusing on 9 sub-problems. The subsystems are listed below: 1) Economical storage strategy of thermal coal. 2) Analysis of peaking margin and strategy. 3) Intelligent scheduling of multi-type generators. 4) Wind power forecast based on multiple time scales. (Including middle-term, short-term and ultra-short-term forecast). 5) Automatic generation control (AGC), automatic voltage control (AVC), and their coordination control. 6) Synchronization and control of wind turbines. 7) Evaluation of the ability of primary and secondary frequency regulation. 8) Preventive control strategy and emergency control strategy. 9) Index system of evaluation on optimized operation.

These 9 subsystems form the framework of the schedule system. They solve the problems including forecast, control, dispatch and evaluation, through time scales from month, week, day-ahead, within-day to day-behind, considering various type of sources and concerning strategies under different grid stations. The system can also be used in regional grids with different backgrounds, such as grids with the integration of UHV.

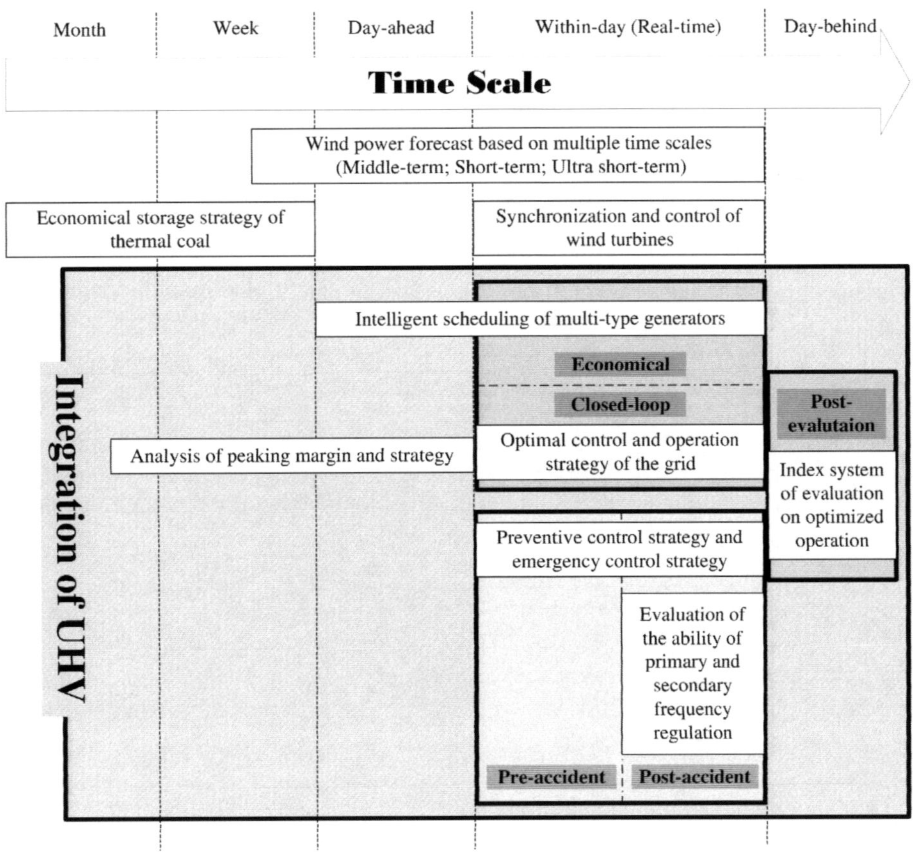

Figure 1.The framework of the schedule system

3.THE DETAILED DESCRIPTION OF THE SUBSYSTEMS

This part explains the detailed models and strategies of the 9 subsystems.

A. Economical Storage Strategy of Thermal Coal[4]

Storage strategy of thermal coal need to be made in a mid-long term schedule plan, to meet the need of thermal power generation.

The strategy consists of 3 parts: 1) Forecast of thermal coal, 3) Storage strategy of thermal coal, 2) Forewarning index of thermal coal.

1) Forecast of thermal coal

In this part, an indirect forecast method is proposed. First, a forecast of the whole consumption in the regional grid is calculated. Then a forecast of uncertain thermal generation is made, under different factors such as runoff, also taking into consideration the fluctuation of other clean energies.

2) Storage strategy of thermal coal

The storage strategy is decomposed into two parts: daily storage and safe storage. Daily storage ensures the daily thermal generation, which can be calculated through daily schedule. Safe storage avoids the supply gap in case of sudden increase of consumption or failure in coal supply.

3) Forewarning index of thermal coal

A forewarning index system is needed to dynamically evaluate the effectiveness of the storage strategy, and give forewarning before the stock reaches the margin. In terms of levels, the indices can be divided into coal supply index, coal consumption index, and coal storage index. In terms of spatial scales, the indices can be divided into plant storage index and regional storage index.

B. Analysis of Peaking Margin And Strategy[5]

In this subsystem, the operation scheme is examined to check if the peaking demand is met.

A decoupled peaking model is proposed in this subsystem to make a coordinated peaking plan for multi-type generators. The flowchart of the decoupled model is shown in Fig. 2.

In this model, a bi-level sub-model for thermal units is used to introduce deep peaking and give the start-stop peaking capacity of thermal units under different load peaks and valleys.

From the distribution map of peaking capacity, the peaking margin can be given and the operation strategy can be revised.

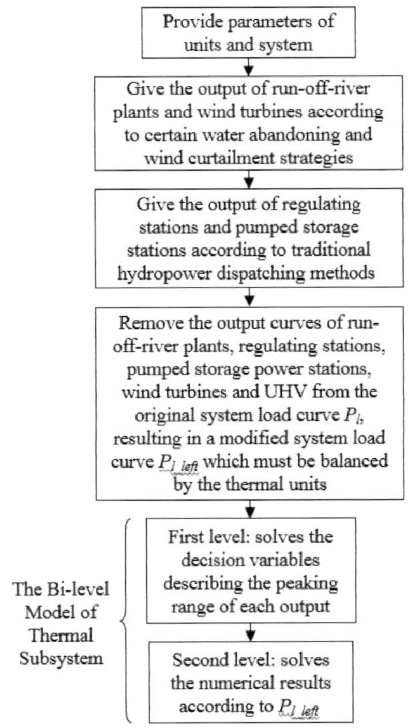

Figure 2.The flowchart of the decoupled peaking model

C. Intelligent Scheduling of Multi-type Generators[6]

This part is the major part of the whole system. In this part, multi time-scale multi power-type joint optimal scheduling approach is proposed. The basic scheme is shown in Fig.3. The schedules should consider the constraints of the whole system, including characteristics and functions of multi-type generators and systematical-level operating constraints, and then minimize the considered objective function, such as economic index, environmental index, and efficiency index[7].

Due to the complexity of this joint optimal scheduling problem and the requirement of practical operation, solving method should meet the need of rapidity, accuracy and practicability. Usually, improved intelligent algorithms (such as GA, PSO) can be used, especially in schedule plans with problems of unit commitment.

D. Wind Power Forecast Based on Multiple Time Scales[8]

Due to the characteristics of uncertainty, the forecast of wind power need to be focused. Wind power forecast has been researched much by many researchers. But in a regional grid, special concerns need to be taken. First, the forecast need to take into account the spatial correlation between different anemology stations. Second, the forecast should be decomposed into multiple time scales, to meet the need of different schedule plans with a series of time spans.

The model also uses the genetic algorithm and BP neural network, training the input data to output data, to satisfy the mapping relationship based on the historical data.

The framework of the subsystem is shown in Fig. 4.

IOP Conf. Series: Earth and Environmental Science **86** (2017) 012031 doi:10.1088/1755-1315/86/1/012031

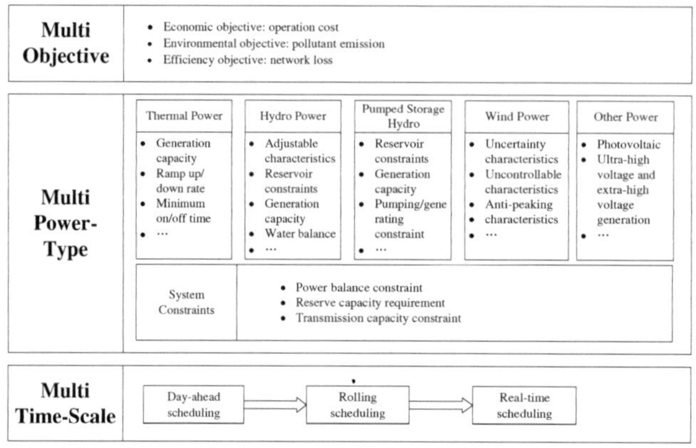

Figure 3.The scheme of multi time-scale multi power-type joint optimal scheduling

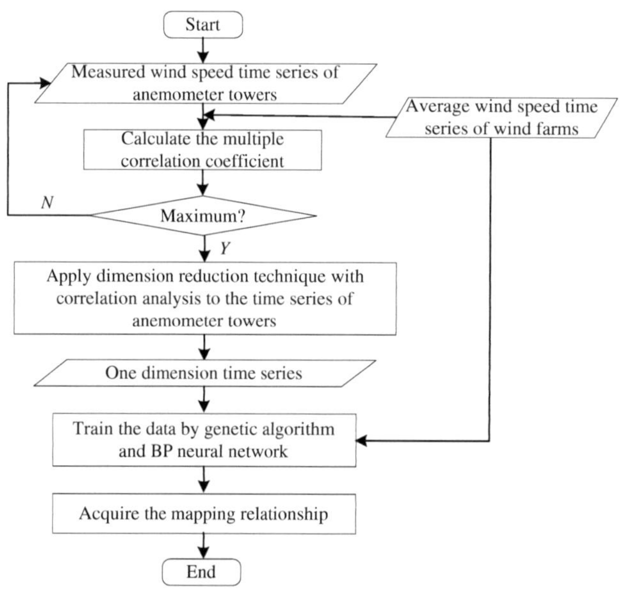

Figure 4.The flow chart of wind power forecast with consideration of the relationship of multiple temporal and spatial scales

E. AGC, AVC, And Their Coordination Control

In a running regional grid, Auto Generation Control (AGC) controls the active outputs of the generators, plays an important part in secondary frequency regulation and economic dispatch, while Auto Voltage Control (AVC) regulates the generator voltage. The coordination control model takes both the active power and voltage of the generators as control variables, to achieve the target of coordinative control.

The model takes economy, safety and quality as the targets. In practical terms, the objectives are transmission losses, reactive power balance, voltage deviation of hubs, and tie-line load bias.

The framework of the subsystem is shown in Fig. 5.

The coordinative control model realizes the closed-loop real-time schedule of the grid.

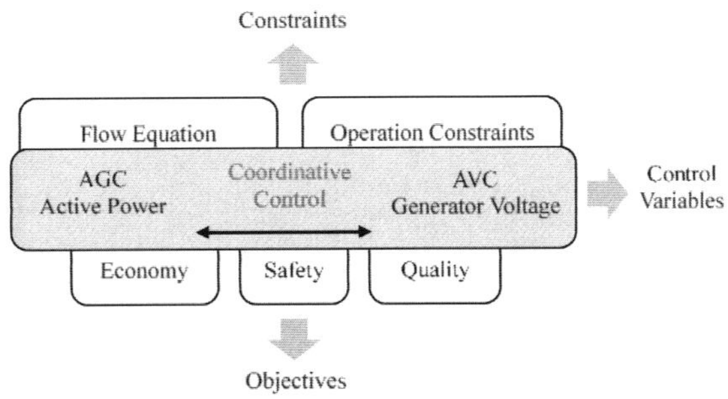

Figure5. The framework of the coordinative control model of AGC and AVC

F. Synchronization And Control of Wind Turbines

In the previous part, real-time control of regular generators is realized. But in a regional grid with high wind penetration rate, synchronization and control of wind turbines need to be considered specially. More efficient control strategy focused on the fluctuations and intermittency of wind power is needed.

In this subsystem, a probabilistic model is proposed to evaluate the influence of the characteristics of wind power on the operation of the grid. Based on the model, the control strategy is proposed. Some of the brief analysis is listed below.

1) Power balance

The fluctuation and uncertainty of wind power will have a great impact on power balance.

2) Voltage and reactive power control

Reactive power compensation equipment such as SVC, SVG, etc. can optimize the reactive power output to achieve the voltage of every bus within a reasonable range

3) Information and communication

The subsystem can report the operative status of wind turbines to the dispatch department, issuing the turbines to follow the scheduling target with consideration both various security constraints and economic benefits.

G. Evaluation of The Ability of Primary And Secondary Frequency Regulation[9]

Primary and secondary frequency regulation are important control means when the grid encounters a frequency deviation. The ability of frequency regulation depends on both the unit regulation performances and regulation strategy. An objective and comprehensive evaluation can make the grid more observable and controllable, and improve the grid's ability on dealing with sudden load change.

In this subsystem, an online assessment method for AGC regulation based on WAMS is proposed. Firstly, we put forwards the assessment index system which integrates the indices in time scale and state dimension, thus make a comprehensive appraisal for AGC units. Then responding to the high sampling rate and high precision character of WAMS data, the control process of AGC regulation is divided into three segments: fast segment, assessment segment and time-out segment. Different methods are applied in different segments, so the calculation amounts and data storage can be efficiently declined to meet the requirements of online evaluation. The flowchart of the assessment algorithm is shown in Fig. 6.

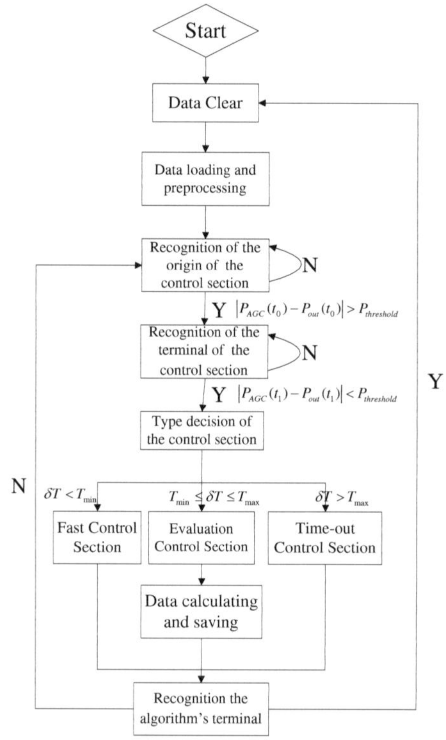

Figure 6 The flowchart of the decoupled peaking model

The evaluation system has been applied in provincial Power Grid and the results verify the feasibility and efficiency of proposed method.

H. Preventive and Emergency Control Strategy[10][11]

Guaranteeing safe and stable operation of the power system is an important goal for grid companies, and accomplishing this goal is of great importance to the development of economy together with people's lives. With the presentation of renewable energy generation and UHV technology, old weak links of the grid will be changed for the access of the UHV lines, and new weak links will also be shown. As a consequence, testing the network again and decide new controlling strategies will be a priority for grid companies. Based on weak links obtained from the PSASP data of a grid for the summer in 2015, the risk theory is used to study the cooperative control strategies of preventive control and emergency control in order to solve the overload problems, power angle problems and voltage problems. Some of the brief analysis is listed below.

1) Get the potential weak links

All the potential weak links have been checked using the PSASP planning data of Hunan grid for the summer in 2015, then the weak links of the provincial grid under the new condition have been obtained. By analyzing the results, overload problems, power-angle stability problems and voltage stability problems have been chosen to solve.

2) Solution to the overload problem

The risk equation describes the overload of UHV is set up to be the objective function. Time-domain simulation method and analysis method have been used to look for numerical relationships between the quantities of preventive control and emergency control. The restrictions of preventive control variables are decided according to the load shedding proportion regulations for different areas in Rule 599. The quantity of preventive control and emergency control are obtained when the objective function achieve its minimum value.

3) Solution to the power-angle stability problem

The risk equation describes power-angle stability problems in this grid is set up to be the objective function. Time-domain simulation method and polynomial fitting method have been used to look for numerical relationships between the quantities of preventive control and emergency control. The restrictions of preventive control variables are decided under the condition that the capacity which are decided to be shut down before and after a fault must be no less than the capacity of the whole power plant. The quantity of preventive control and emergency control are obtained when the objective function achieve its minimum value.

4) Solution to the voltage stability problem

The risk equation describes voltage stability problems in this grid is set up to be the objective function. Time-domain simulation method and polynomial fitting method have been used to look for numerical relationships between the quantities of preventive control and emergency control. The restrictions of preventive control variables are decided according to the load shedding proportion regulations for different areas in Rule 599. The quantity of preventive control and emergency control are obtained when the objective function achieve its minimum value.

I. Index System of Evaluation on Operation[12]

In order to assess the power grid's daily operation result, an assessment index system for power grid's daily operation is proposed and the assessment method is discussed in this subsystem. Firstly, based on general construction principles of an assessment index system, a three-level assessment index system is designed in consideration of the power grid's operation needs in aspects of security, economy, equity and environmental friendliness. This medium-sized index system can fully reflect the power grid's daily operation. As for index integration, Analytic Hierarchy Process (AHP) is used to determine the weight of each index, and Technique for Order Preference by Similarity to Ideal Solution (TOPSIS) is used to integrate the index in consideration of the real needs of the operators, and the practical steps of this method are also discussed. Finally, the whole index system is visualized using a radar map, to reflect the everyday situation of schedule and operation of the regional grid comprehensively. Case study results show that the proposed method can evaluate the power grid's daily operation. This assessment system can be used for comparison of the power grid's daily operation results, and provide guidance for optimal operation of power grids.

4.SUMMARY

In this paper, a schedule system for provincial power grid based on multi-time scale is proposed to deal with the problems the grid faced with the presence of large scale renewable energy resources and the integration of UHV. A systematic scheme framework with 9 different subsystems is presented in this paper. Detailed description of each subsystem is also given. The proposed schedule system has been constructed in a provincial power grid in Central China, and the operation results further verified the effectiveness of the system.

5.ACKNOWLEDGEMENTS

This work was supported by Science and Technology Projects of Hunan Electric Power Company (SGHN0000DKJS1300221).

REFERENCES

[1] Liacco D 1967 The adaptive reliability control system. *IEEE Transactions on Power Apparatus and Systems*. **5** 517-531.
[2] Wu W, Zhang B, Chen J and et al 2012 Multiple time-scale coordinated power control system to accommodate significant wind power penetration and its real application. *Power and Energy Society General Meeting, 2012 IEEE* 1-6.
[3] Bao Z, Zhou Q, Yang Z and et al A Multi Time-Scale and Multi Energy-Type Coordinated Microgrid Scheduling Solution. *Model and Methodology*.
[4] Yao J Y, Zheng L, Zhou J C, Wang J Z, Hu W M, Yang S Y 2014 Study on the Algorithm of Electricity Coal Inventory Based on the Balance of Power Supply and Demand. *ICEEAC 2014*.
[5] Liu X, Hu W, Hong C and et al Research of Modeling and Control Strategies of Multi-type Generators Peaking.
[6] Zhou Y F, Hu W, Hong C Q, Hu B Q and Cheng T 2014 Modelization and Optimization of Multi-Type Power Generators Joint Scheduling Based on Improved PSO. *APPEEC 2014*.
[7] Han Z H, Cheng T, Zhou Y F and Zhang P 2015 Multi-objective optimal scheduling for hydro-thermal-wind power system. *TENCON 2015*.

[8] Ye Y D, Lu Z X and Qiao Y 2016 Conditional probabilistic modeling of wind power forecast error. *2016 IEEE Power and Energy Society General Meeting*, Boston, MA, 2016.
[9] An Y K, Hu W, Song J Y, Ye Z and Cheng W Q 2014 A Multi-objective Assessment For Primary Frequency Of Generator. *ICEEAC 2014*.
[10] Wang Y G, Guan X F, Chen L, Wang J Z, Hu W M and Yang S Y 2014 Study on the Line Overload Harmonization and Optimization Control. *ICEEAC 2014*.
[11] Wang Y G, Zhou Z H and Chen L 2015 The Method to Obtain Quantity of Preventive Control and Emergency Control for Overload Problems of Tie Lines Based on Risk. *Shaanxi Electric Power*. **43** 8-14.
[12] Dai Y H, Chen L, Min Y and Cheng T 2015 Study on the Daily Operation Assessment System of Power Grid. *Power System Technology*. **39** 1611-16.

IOP Conf. Series: Earth and Environmental Science **86** (2017) 012032 doi:10.1088/1755-1315/86/1/012032

On Fusing Mechanism Involving the Fuses at Secondary Winding of PT Induced by VFTO

X H Ma [1], S Han [*2], Z Hu [2], J K Huang [1] and X B Liu [2]

[1]Nano/micro science and technology center, School of Mechanical Engineering, Jiangsu University, Zhenjiang 212013, PR. China

[1]Guizhou Electric Power Research Institute, Guizhou, China

[2]Department of Electrical Engineering, Guizhou University, Guizhou, China

Email: *shan@gzu.edu.cn

Abstract. The impact of very fast transient over-voltage(VFTO) on the primary and secondary equipment in gas insulated station(GIS) should not be ignored. In this paper, the research on fusing mechanism involving the fuses at secondary winding of PT induced by VFTO is carried out based on an actual case from the 500kV GIS substation belonged to Guizhou Tianshengqiao No.2 hydro power plant. As for the high voltage side, the transfer function about over-voltage can be formed in order to analyse the characteristics about high frequency over-voltage. As for the low voltage side, a three-phase equivalent model for secondary circuit of PT with the parallel arrangement at the A, B and C phases may be built for simulating the current distribution of the low voltage fuses in the situation of high frequency voltage and current. The results show that the maximum allowable current of fuse might be reduced greatly in the high-frequency circumstance due to skin effect and proximity effect, in comparison with nominal frequency condition.

1.Introduction

Gas-Insulated Switchgear (GIS) have been found a broad range of applications in power system over the last three decades because of their high reliability, easy maintenance, small space requirement, etc. There are also more and more GIS equipment in the power systems of China such as Guizhou Power Grid. Very Fast Transient Overvoltage (VFTO) may be generated in GIS due to voltage breakdowns (flashovers) in SF6 gas. They primarily occur during the operation of disconnector in GIS substation and can be hardly avoided. Due to the physical properties of breakdown in SF6 gas [1], the typical rise time of front VFTO is substantially smaller than the transient time of the associated electromagnetic wave through the components of GIS. This causes that each of the voltage breakdowns generates the travelling waves that could propagate back and forth in the GIS. They would be reflected and transmitted at the discontinued points of surge impedance, and eventually be superimposed to constitute the VFTO in the particular component of GIS. The induced VFTO would have impacts on primary equipment and secondary equipment as well [2].
For example, in the recent five years, there have been 7 abnormal events about the blown fuse at the secondary side of the potential transformer (PT) of 500kV # 2M bus in the 500kV GIS Switching Station of Tianshengqiao No.2 hydropower station. In fact, the protection devices at the 500kV bus usually acquire the voltage values from the PT. Therefore, it will lead to the malfunction of protection device and certainly jeopardize the power supply reliability of

IOP Conf. Series: Earth and Environmental Science **86** (2017) 012032 doi:10.1088/1755-1315/86/1/012032

power plant once if the secondary circuit goes abnormal. The abnormal events were likely caused by VFTO. Thus it is necessary to study on modeling, assessment, monitoring and control technology for VFTO in GIS substation.

As for the product design of GIS, it was started in the 1980s that some foreign manufacturers launched a lot of research on VFTO phenomenon resulted from switch operating in GIS involving the measurement system for VFTO, analogy experiment and digital simulation for analyzing the characteristics of VFTO, the effect of VFTO on insulation and the electromagnetic interference on secondary equipment, etc. Moreover, the suppression measures such as the switch with damping resistor for VFTO were proposed.

However, the studies on the destructive effects of VFTO in the recent years were mainly focused on the primary electrical equipment such as transformer [3] or other insulation-level evaluation issues. The main contribution from those works focused on the component model suitable for VFTO analysis including arrester model [4], arc model of switch [5-7], potential transformer model [8], the computing method for fast transient overvoltage [9] and so on. Recently, reference [10] conducted a test for transferring voltage between primary side and secondary side of the potential transformer in GIS utilizing a set of equipment capable of generating, controlling and measuring class B shock wave, which concludes that VFTO can pass through PT with the linear transferring characteristic.

Nevertheless, it seems that there are few papers have been published so far, addressing the secondary-side high-frequency overvoltage hazards of PT in GIS. Thus this paper carried out a study on the fusing mechanism involving the fuses at secondary winding of PT induced by VFTO based on a case of 500kV substation of Tianshengqiao hydropower plant in China.

2.The transferring characteristics of high frequency overvoltage in PT

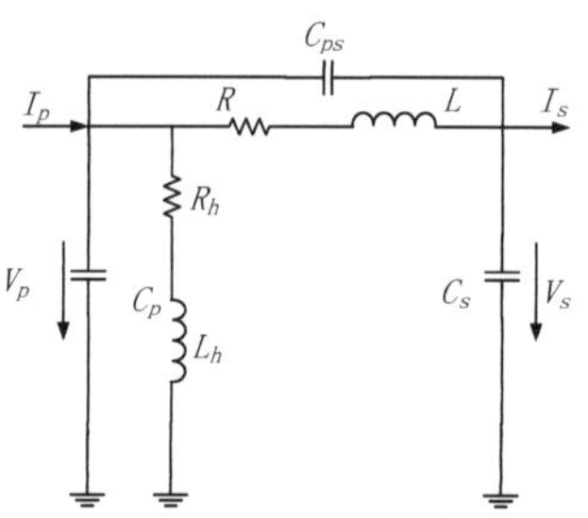

Fig.1 The equivalent circuit of PT

Referring to the principle that T-type equivalent circuit for single-phase transformer can be converted to Γ-type equivalent circuit, the equivalent model of potential transformer can be also converted into the equivalent circuit shown in Fig. 1 [11-12].

In Fig. 1, the resistance, inductance and capacitance of potential transformer are represented by R, L, and CPS, respectively. CP and CS stands for the capacitor at primary side and secondary side, respectively. Rh and Lh separately denote ferromagnetic loss resistance and excitation inductance. k stands for transformer ratio. Apparently, Vs=kV's, R=Rp+kRs, L=Lp+kLs, Cs=kC's. As a result of the equivalent circuit of PT displayed in Fig. 1, the transfer function of PT can be expressed as

$$\begin{aligned} H(s) &= \frac{V_s}{V_p} \\ &= \frac{1}{sC_s}\left[\frac{R+sL}{1+sRC_{ps}+s^2LC_{ps}}+\frac{1}{sC_s}\right] \\ &= \frac{1+sRC_{ps}+s^2LC_{ps}}{1+sR\left(C_s+C_{ps}\right)+s^2L\left(C_s+C_{ps}\right)} \end{aligned} \quad (1)$$

If a high-frequency signal is inputted in the primary side, especially at the megahertz level, s in the Equation (1) will be very large. At that time, the item with the highest exponent of s can be only considered. Consequently, Equation (1) might be simplified into (2) as follows.

$$H(s) = \frac{C_{ps}}{C_{ps} + C_p} \tag{2}$$

It can be seen from the transfer function that the signal transfer function of PT may be just related with the capacitor parameters at primary and secondary sides for the high-frequency signal. Therefore, if VFTO transmitted in the GIS encounters PT, those components with high-frequency physical characteristics would capacitively couple to the secondary wire cable through the distributed capacitance at primary and secondary sides, and then interfere with the secondary devices by the means of conduction [13].

3.The current distribution simulation in low-voltage fuse

When the high-frequency current caused by VFTO passes to the secondary side, the device on the secondary side mainly affected by its skin effect and proximity effect.

3.1 Skin effect and proximity effect. Skin effect is the tendency of an alternating electric current (AC) to become distributed within a conductor such that the current density is largest near the surface of the conductor, and decreases with greater depths in the conductor. The electric current flows mainly at the "skin" of the conductor, between the outer surface and a level called the skin depth. The skin effect causes the effective resistance of the conductor to increase at higher frequencies where the skin depth is smaller, thus reducing the effective cross-section of the conductor. The skin effect is due to opposing eddy currents induced by the changing magnetic field resulting from the alternating current. So each branch will establish an alternating magnetic field on other branches if a high-frequency alternating current flows through a set of parallel wires. Then the magnetic field would induce an electromotive force for preventing the electric current in each wire [14]. Among these wires, the wire in the middle has the largest induced electromotive force and also endures the greatest flux. As a consequence, the current flowing through the line may be minimum one.
Moreover, in a conductor carrying alternating current, if currents are flowing through one or more other nearby conductors, such as within a closely wound coil of wire, the distribution of current within the first conductor will be constrained to smaller regions. The resulting current crowding is termed as the proximity effect. This crowding gives an increase in the effective resistance of the circuit, which increases with frequency. A changing magnetic field will influence the distribution of an electric current flowing within an electrical conductor, by electromagnetic induction. When an alternating current (AC) flows through a conductor, it creates an associated alternating magnetic field around it. The alternating magnetic field induces eddy currents in adjacent conductors, altering the overall distribution of current flowing through them. The result is that the current is concentrated in the areas of the conductor furthest away from nearby conductors carrying current in the same direction. The proximity effect can significantly increase the AC resistance of adjacent conductors with frequency [15]. At higher frequencies, the AC resistance of a conductor can easily exceed ten times its DC resistance.
The high-frequency characteristics of VFTO can be conducted to the secondary side of PT, so that there is an uneven distribution of current. At that time, the inductance effect in this wire and the increased resistance caused by heat generated would be very small so as to be negligible [16-17]. However the thermal effect on the fuse is very sensitive. Hence high-frequency current play an inevitable role on the fusing. The part of fuse link would be overloaded and consequently blown out even if in the case with the current less than the rated value.

3.2 The distribution of current within the conductor based on the mechanisms of skin effect and proximity effect. It is supposed that a parallel arrangement of A, B, C three-phase line and return line could be depicted in Fig. 2. The parallel lines are equipped with the low voltage fuses with the same specifications and length.

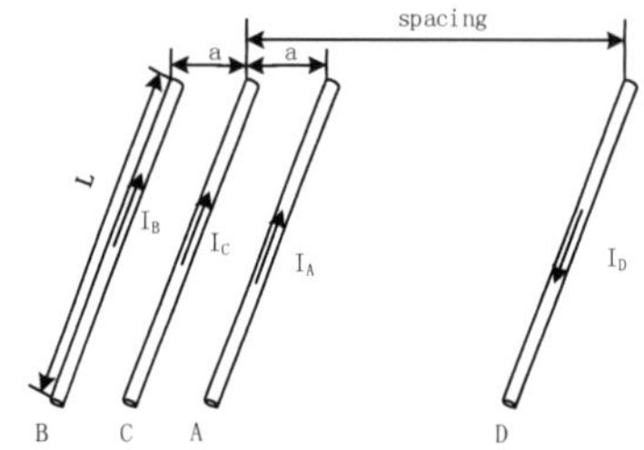

Fig.2 The scheme of line arrangement

When there is a high-frequency current in the line, the total flux ψDA for the loop AD according to an electric field theory could be calculated by the following Equation.

$$\psi_{DA}=\frac{\mu_0 L}{2\pi}\left(I_A \ln\frac{d_{AD}}{d_{AA}}+I_B \ln\frac{d_{BD}}{d_{BA}}+I_C \ln\frac{d_{CD}}{d_{CA}}-I_D \ln\frac{d_{DD}}{d_{DA}}\right) \quad (3)$$

where μ0 stands for the permeability of wire, L stands for the length of wire, the three-phase currents and the current of back line may be denoted by IA, IB, IC and ID, respectively. dAA and dDD represent the radius of A line and D line respectively, dAD, dBD, dBA , dCD , dCA and dDA represent the distance between A line and D line, B line and D line, B line and A line, C line and D line, C line and A line, D line and A line, respectively.
Similarly, the total flux ψDB for the loop BD can be listed as (4).

$$\psi_{DB}=\frac{\mu_0 L}{2\pi}\left(I_A \ln\frac{d_{AD}}{d_{AB}}+I_B \ln\frac{d_{BD}}{d_{BB}}+I_C \ln\frac{d_{CD}}{d_{CB}}-I_D \ln\frac{d_{DD}}{d_{DB}}\right) \quad (4)$$

Meanwhile ψDC for the loop BD can be also derived as (5).

$$\psi_{DC}=\frac{\mu_0 L}{2\pi}\left(I_A \ln\frac{d_{AD}}{d_{AC}}+I_B \ln\frac{d_{BD}}{d_{BC}}+I_C \ln\frac{d_{CD}}{d_{CC}}-I_D \ln\frac{d_{DD}}{d_{DC}}\right) \quad (5)$$

Based on the equivalent circuit depicted in Fig. 3, the induced voltage in the loop AD could be attained as (6).

$$V_A=-jw\psi_{DA} \quad (6)$$

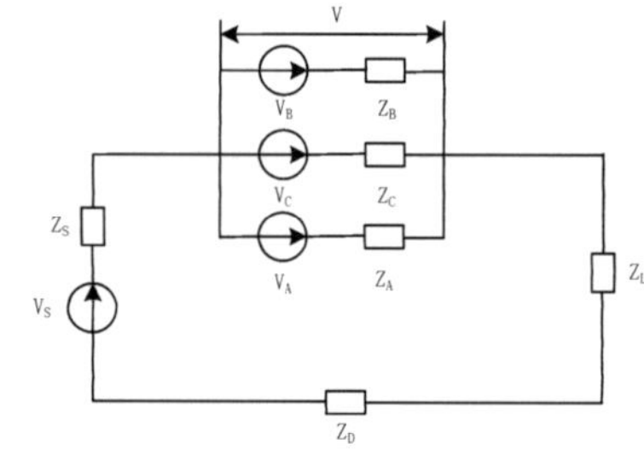

Fig.3 The equivalent circuit for the case in Fig.2

It can be clearly seen from Fig.3 that ID=IA+IB+IC. Therefore, Equation (6) can be rewritten in another form as (7)

$$V_A=\beta_{AA}I_A+\beta_{AB}I_B+\beta_{AC}I_C \quad (7)$$

where
$$\beta_{AA} = -\frac{jw\mu_0 L}{2\pi}\left(\ln\frac{d_{AD}}{d_{AA}}\frac{d_{AD}}{d_{DD}}\right);$$
$$\beta_{AB} = -\frac{jw\mu_0 L}{2\pi}\left(\ln\frac{d_{BD}}{d_{BA}}\frac{d_{AD}}{d_{DD}}\right);$$
$$\beta_{AC} = -\frac{jw\mu_0 L}{2\pi}\left(\ln\frac{d_{CD}}{d_{CA}}\frac{d_{AD}}{d_{DD}}\right)。$$

Then VB and VC might be also gotten in the same way.

It may be observed from Fig.3 that

$$V = V_A - I_A Z_A = V_B - I_B Z_B = V_C - I_C Z_C \quad (8)$$

in which ZA，ZB and ZC separately represent the total impedance of each branch.

$$V = (I_A + I_B + I_C)(Z_L + Z_D + Z_S) - V_S \quad (9)$$

Substituting (7) into (8), we obtain (10) as follows

$$V = (\beta_{AA} - Z_A)I_A + \beta_{AB}I_B + \beta_{AC}I_C$$
$$= -V_S + (I_A + I_B + I_C)Z \quad (10)$$

where $Z = Z_L + Z_D + Z_S$.

The corresponding equations for the other two loops are presented as follows.

$$V = \beta_{BA}I_A + (\beta_{BB} - Z_B)I_B + \beta_{BC}I_C$$
$$= -V_S + (I_A + I_B + I_C)Z \quad (11)$$

$$V = \beta_{CA}I_A + \beta_{CB}I_B + (\beta_{CC} - Z_C)I_C$$
$$= -V_S + (I_A + I_B + I_C)Z \quad (12)$$

Then these equations can be expressed in the form of a matrix as (13).

$$\begin{bmatrix} (Z+Z_A-\beta_{AA}) & (Z-\beta_{AB}) & (Z-\beta_{AC}) \\ (Z-\beta_{BA}) & (Z+Z_B-\beta_{BB}) & (Z-\beta_{BC}) \\ (Z-\beta_{CA}) & (Z-\beta_{CB}) & (Z+Z_C-\beta_{CC}) \end{bmatrix} \times \begin{bmatrix} I_A \\ I_B \\ I_C \end{bmatrix} = \begin{bmatrix} V_S \\ V_S \\ V_S \end{bmatrix} \quad (13)$$

According to the different distance from the three-phase line to return line, the current through the line can be transformed into the values with per unit. The matrix in (13) can be transformed into a function involving the distance between center line and back line and the rated current of fuse, which may be fitted into a curve, as shown in Fig. 4.

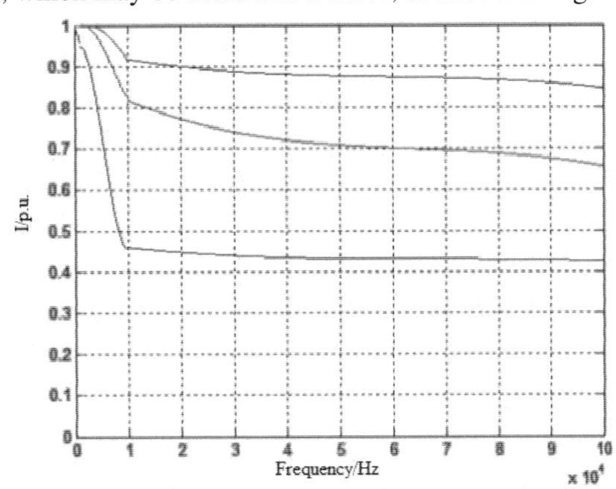

Fig.4 The distribution characteristics of current for fuse link varied by frequency at the different distance between center line and back line

3.3 The analysis about fusing mechanism on the high frequency condition. When low-voltage fuses work at the low frequency condition, the distribution of current in the three-phase lines could be affected by their impedances. Nevertheless, when the operating frequency rises up to

100kHz, the current distribution is totally different. As shown in Fig. 2, the current flowed through phase C line will be only 0.43 times rated current if the spacing distance between phase C line and the return line is only 30mm. But the phase A line will undertake the greatest impact since it is closest to the return line. It also means that its current could reach 1.5 times rated current. In other words, the reduced current in phase C line flows to the other two phases so that there is an excessive current in phase A line. Consequently, the heat productivity of fuse is greater than its heat release. Meanwhile its temperature rapidly rises so as to be blown out.

Therefore, in the case of high frequency, the maximum allowable current of fuse would be obviously decreased due to skin effect and proximity effect so that the reasonable arrangement of the distance between low-voltage fuses is critical in this working condition. Thus it should be noted that the distance between the low voltage fuses at the secondary side of PT should be increased as much as possible. Since most of low voltage fuses are currently designed without considering the VFTO working condition and are usually verified in the nominal frequency or the setting frequency less than MHz [18], the unpredictable consequences will happen once if they are used in the high-frequency circuits. Hence it should be suggested that the fuse for high-frequency working condition may be customized according to the additionally standard.

4.Case study

An actual case study has been carried on 500kV GIS substation belonged to Guizhou Tianshengqiao No.2 hydropower plant. There have been six events about blown fuse at secondary winding side of No.52YH PT at the No.2M 500kV bus in the five-year period from 2007 to 2011.The detailed data about these abnormal events are listed in Table 1.

Table 1: The operation mode of 500kV GIS substation in Tianshengqiao No.2 hydro power plant

No.	Fuse	Operation Descrip-tion	Operation Order	The status of 500kV GIS before the fuse is fused
1	52YH Phase B PF2	Open Switch 50132	No.5013 breaker runs from SB to OOS	No.5012 and 5013 breakers and No.50122,50121,50132 and 50131 switches open, others close.
2	52YH Phase B PF2	Close Switch 50132	No.5013 breaker runs from OOS to SB	No.5011,5012 and 5013breakers and No.50132, 50111 and 50112 switches open, others close.
3	52YH Phase B PF3	Close Breaker 5013	No.5013 breaker runs from SB to IS	No.5013,5023, 5033 and 5043 breakers open, others close.
4	52YH Phase B PF2	Open Switch 50122	No.5012 breaker runs from SB to OOS	No.5012 breakers and No. 50121switch open, others close.
5	52YH Phase B PF2	Open Switch 50132	No.5013 breaker runs from SB to OOS	No.5012, 5013breakers and No.50122,50121 and 50131 switches open, others close.
6	52YH Phase B PF2	Open Switch 50222	No.5022 breaker runs from SB to OOS	No.5021, 5022 breakers、No.50221switch open, others close.

Note1:OOS denotes ‘out of service’, SB denotes ‘standby’, IS denotes ‘in service’.

PSCAD/EMTDC has be taken as the simulation platform. The simulation time step is selected as 45μs and the calculation step is selected as 50ns. The model data comes from the specific

IOP Conf. Series: Earth and Environmental Science **86** (2017) 012032 doi:10.1088/1755-1315/86/1/012032

parameters of major electrical components provided by the hydropower plant and the operational and planning database of China Southern Power Grid. A simulation model is built as Fig. 5 where the model of PT is described in Fig. 6.

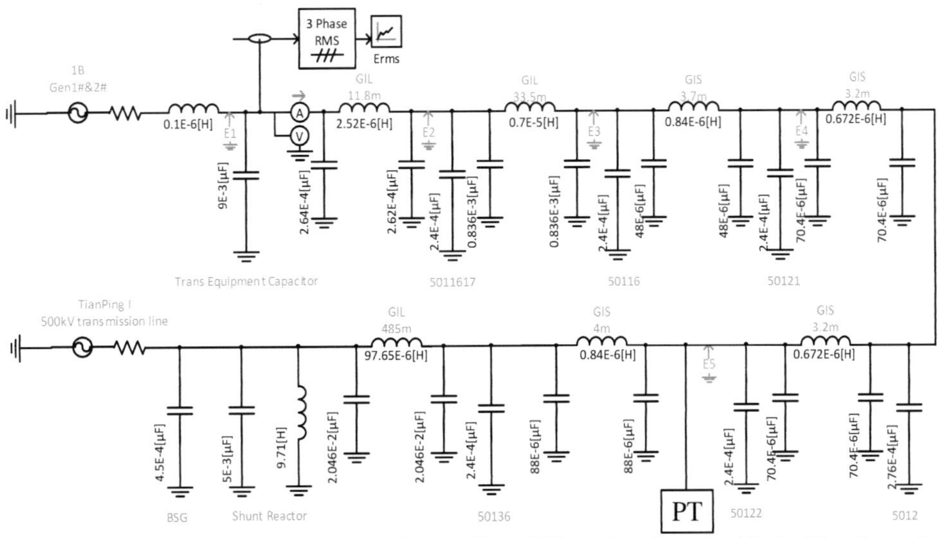

Fig.5 The VFTO analysis model for 500kV GIS substation in No.2 Tianshengqiao hydropower plant

PSCAD/EMTDC has be taken as the simulation platform. The simulation time step is selected as 45μs and the calculation step is selected as 50ns. The model data comes from the specific parameters of major electrical components provided by the hydropower plant and the operational and planning database of China Southern Power Grid. A simulation model is built as Fig. 5 where the model of PT is described in Fig. 6.

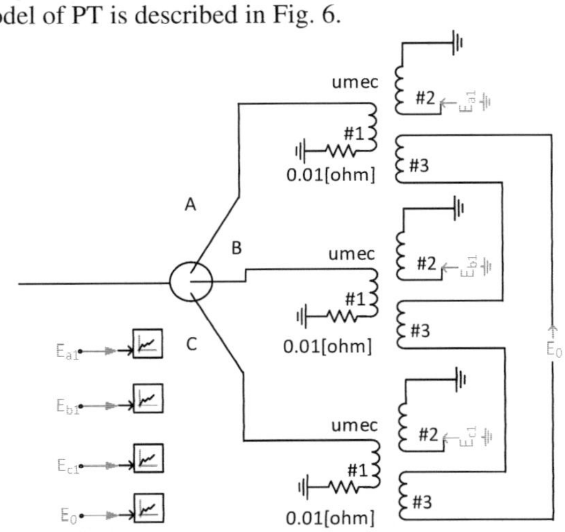

(a) The model of PT with conventional opening delta connection

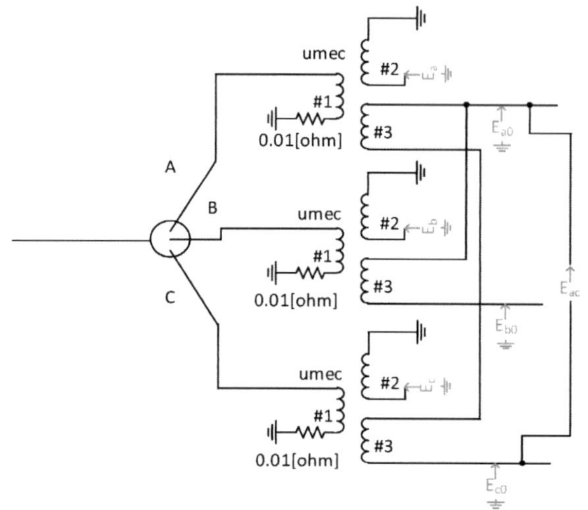

(b)The model of PT for Tianshengqiao No.2 hydro power plant

Fig.6 The model of PT

The first event in Table 1 is taken as an example for the VFTO analysis model. In this case, No.5012, No.5013 No.50122, No.50121, No.50132, No.50131 disconnectors are opened, and other breakers as well as switches are closed. There is a typical serious transient voltage case obtained in the numerous VFTO simulations, displayed as follows.

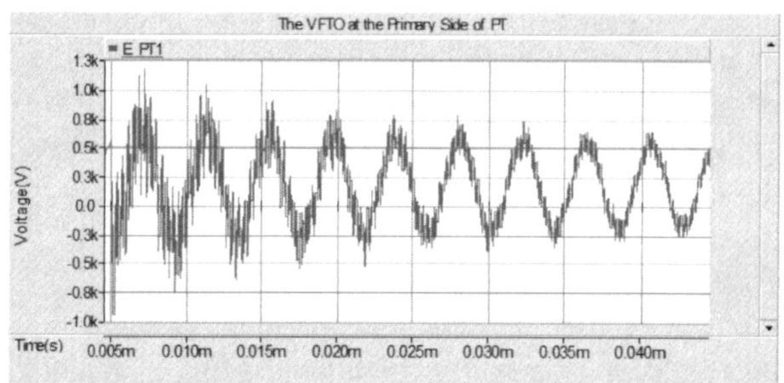

Fig.7 The VFTO curves for Phase A at primary winding of 52YH

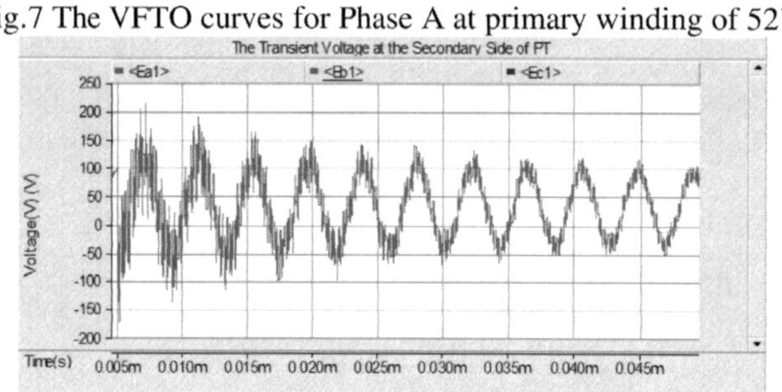

Fig.8 The VFTO curves for Phase A at secondary winding of 52YH

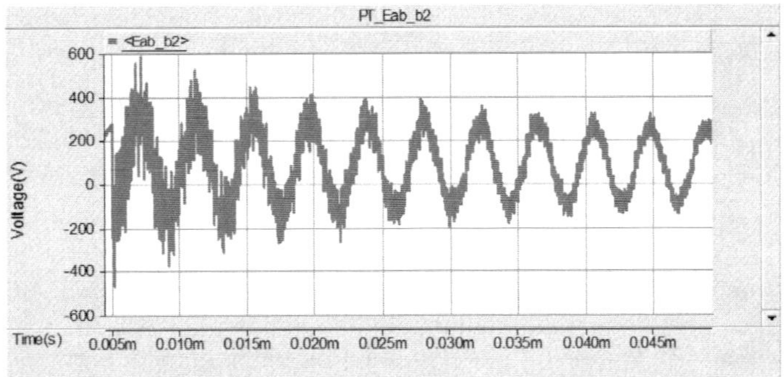

Fig.9 The overvoltage between B-phase of the primary winding and B-phase of the third winding with opening delta connection in 52YH

As seen from Figure 7, the peak value of phase voltage at the primary side of No. 52YH PT may rise up to 1168 kV ~ 1172kV, i.e., about 2.60 p.u. ~ 2.61p.u.. Furthermore, the peak value of the phase voltage at its secondary side displayed in Figure 8 can attain 215V, which might belong to the operational over-voltage. According to the reports from field survey, there is a 2mm air gap between B phase of the first winding of No. 52YH PT and the connected part for phase A and phase B involving the third winding which is also the opening triangle winding. Meanwhile, the related wires are arranged in parallel. As a result, it can be inferred that the gap between the fuses at the ends is also 2mm. The breakdown voltage resulted from surge overvoltage at the primary side can be ranged from 670V to 760V for root-mean-square value with 960V to 1060V for peak value. Considering the possible existence of impurities in the gap between the two fuses, the equivalent distance of air gap may be appropriately reduced to 1.5mm. In this way, the peak value of the discharge breakdown voltage could drop to 720V ~ 795V. Figure 9 shows that the peak overvoltage between B phase of the first winding and B phase of the third winding with opening delta connection of No. 52YH PT may move up to about 585V that would be close to the breakdown voltage of air gap. If VFTO becomes more severe, it is possible for the fuses at both ends to breakdown. Consequently, it will be able to result in the malfunction of protection devices, which could make a serious threat to reliable operation in the power plant.

5.Electronic submission of the full paper

In this paper, the research on fusing mechanism involving the fuses at secondary winding of PT induced by VFTO has been carried out based on an actual case from the 500kV GIS substation belonged to Tianshengqiao No.2 hydropower plant in China. There are several findings and conclusions as follows.

1) As for the high voltage side of PT, the transfer function about over-voltage has been formed in order to analyze the characteristics about high frequency over-voltage.

2) As for the low voltage side of PT, a three-phase equivalent model for secondary circuit of PT with the parallel arrangement at the A, B and C phases may be built for simulating the current distribution of the low voltage fuses in the situation of high frequency voltage and current. The results show that the maximum allowable current of fuse might be reduced greatly in the high-frequency circumstance due to skin effect and proximity effect, in comparison with nominal frequency condition.

3) The results from this case study indicate that the impact of VFTO on the primary and secondary equipment in gas insulated station should not be ignored. In the next step, a case study on 500kV Liupanshui substation would be done for proving the conclusions again.

6.Acknowledgements

This work was supported in part by Science& Technology Program of Guizhou Power Grid Co., Guizhou Science & Technology Collaboration Foundation (No. 20157635) and the Training Program for Outstanding Young Scientists and Educators of Guizhou (No.2012151).

References

[1] Gu D X, Xiu M H, Dai M and et al 2007 Study on VFTO of 1000kV GIS substation. *High Voltage Engineering, volume 11* 27-32.
[2] Chen W J, Yan X L, Wang S W and et al 2011 Recent progress in investigations on very fast transient overvoltage in gas insulated switchgear. *Proceedings of the CSEE, volume 31* 1-11.
[3] Zhang X L, Liang G S, Sun H F and et al 2005 Computation of very fast transient overvoltage in transformer windings. *High Voltage Engineering, volume 08*, 4-6.
[4] Lin X, Wang J, Xu J Y and et al 2012 Influence of metal oxide arrester on very fast transient overvoltage in GIS. *High Voltage Engineering, volume 10*, 2655-61.
[5] Liu D D, Wu W H, He R W and et al 2015 Calculation of very fast transient overvoltage with the arc resistant model considering inductance characteristic. High *Voltage Apparatus, volume 01*, 105-109.
[6] Liu J M, Du W, Li Y G and et al 2013 Analysis on the influence of the arc model and its parameter of disconnector on very fast transient overvoltage. *Journal of North China Electric Power University, volume 03*,17-24.
[7] Liu D D, He R W, Huang K and et al 2015 Research of gas insulated substation switching arc model in the simulation of very fast transient overvoltage. *Journal of Electrical Engineering, volume 06*, 88-93.
[8] Chen W J, Wang H, Wang Z J and et al 2013 Withstand characteristics analysis of inductive voltage transformer in UHV gas insulated switchgear under very fast transient overvoltage. *High Voltage Engineering, volume 06*, 1402-10.
[9] Wang N, Lin X, Xu J Y and et al 2012 Simulation and characteristics analysis of very fast transient overvoltage in GIS insulated substation with ultra-high voltage. *High Voltage Engineering, volume 12*, 3310-15.
[10] Huang H, Dai J, Liu Y and et al 2014 Experimental study on transfer overvoltage of voltage transformer in GIS. *Electrical Measurement & Instrumentation, volume 12*, 1-6.
[11] Ying X C, Wang J H, Zhang B H 2006 Analysis for the High-Frequency Transmission Characteristics of Electromagnetic Voltage Transformer in Protection Equipment. *Journal of Xi'an Jiao Tong University, volume 12*, 1467-69.
[12] Zhang Z Y, Yan J and Li W F 2009 A high frequency mathematical model for potential transformer based on the scattering parameter. *Power System and Clean Energy, volume 02*, 9-12.
[13] Zhang W D, Chen P L, Chen W J and et al 2013 Measurement and simulation of disturbance voltage generated by VFTO in UHV GIS substation on the secondary cables. *Proceedings of the CSEE, volume 16*, 187-196.
[14] Wang J F, Hu X K, Wei N S and et al 2014 Analysis and discussion on the skin effect of zine oxide nonlinear resister. *Insulators and Surge Arresters, volume 03*, 88-95.
[15] Zhao Z H, Yuan J S and MA W M. 2007 The influence of proximity effect on the steel impedance and its equivalent circuit model. *Tsinghua University(Sci& Tech), volume 04*, 490-493.
[16] Howe A F 1982 Derating of semiconductor fuselinks for use in high-frequency applications. *IEE Proceedings B - Electric Power Applications, volume 129* 111-116.
[17] Wright A and Newbery PG 2004 Electric Fuses. *The Institution of Engineering and Technology*, 27-29.

IOP Conf. Series: Earth and Environmental Science **86** (2017) 012033 doi:10.1088/1755-1315/86/1/012033

A Novel Three-Phase Three-Wire Dynamic Voltage Restorer

W P Zhou [1], Z Y Wang [2] and Z G Wu [2]
[1] Naval University of Engineering, Wuhan, China.
[2] Zhejiang Hangshen Electrical Group Ltd., Hangzhou, Zhejiang, China.
Email: *cnwhwp@126.com

Abstract. A novel optimal control method of three-phase three-wire Dynamic Voltage Restorer (DVR) and its simplified topological structure are proposed. Based on virtual-floating-ground analysis method, a time-domain mathematical model and the control equation of DVR's converter are established; the optimal feasible solution with minimum objective function value of the control equation is deduced. Compensation voltage detection is based on digital phase lock, and the switching time is calculated by line voltages, the control performance under the proposed control strategy is analyzed, which greatly improve the utilization rate of DC side voltage and the DVR's compensation capability. The simulation and experiment results of three-phase three-wire DVR under the proposed algorithm are given; the results show its validity and feasibility.

1.Introduction

With large numbers of industrial sensitive load in use, power users become more and more sensitive to power quality problem. The power quality disturbance problems such as the grid voltage sag, harmonic and flicker, and even instant loss of electric are inevitable, and voltage sag has become the most prominent and widespread problem which influence the safe operation of the power load, at the meantime, power quality problems will shorten the life of equipment and even lead to disastrous economic losses. Dynamic Voltage Restorer (DVR) is considered as the most economic and effective user power device to solve the problem of Voltage sag at present [1-16], the DVR by controlling the power electronics inverter produces a set of waveform controlled ac voltage, which can ensure the stability of the voltage of the sensitive load when there is a disturbance source voltage.

Dynamic voltage restorer in general can be divided phase voltage compensation type DVR and line voltage compensation type DVR. Phase voltage compensation type DVR is easy to control, can split phase compensation, can compensate zero sequence voltage, but also has disadvantages of more power switching devices demanded, big volume and high cost. Line voltage compensation type DVR has the advantages of simple structure, less power devices, relatively small volume, but has the disadvantages of complex control, can compensate zero sequence voltage. Because the non-ground neutral system is widely used in medium voltage distribution network[1], and three-phase three-wire system power supply is widely used in many occasions, so the line voltage compensation type DVR have broad application prospects.

In this paper, the line voltage compensation type three-phase three-wire DVR topology is analyzed and simplified, by analyzing the DVR time-domain circuit equations, the voltage control equation is obtained under the optimized index of the optimal solution. The optimal solution is based on the DVR output line voltage to calculate switch time, reducing the line voltage can be converted into phase voltage of the link, and can improve the utilization rate of DC side voltage and the compensation ability of DVR. Compensation voltage detection is based on digital phase locking method, the proposed control strategy and implementation scheme was analyzed, and the simulation and experimental results are given, the results show that the method is feasible.

IOP Conf. Series: Earth and Environmental Science **86** (2017) 012033 doi:10.1088/1755-1315/86/1/012033

2. SIMPLIFIED MAIN CIRCUIT TOPOLOGY STRUCTURE

Main power circuit of three-phase three-wire line voltage compensation type DVR is shown in figure 1, the rectifier module is a rectifier, and inverter module adopts three-phase Inverter Bridge. Set the imaginary point O as grid voltage reference point, $u_{a'o}$, $u_{b'o}$, $u_{c'o}$ for grid side voltage, u_{ao}, u_{bo}, u_{co} for the load, u_{dc} for DVR dc side voltage value, assuming that injection transformer turn ratio of 1:1.

$$\begin{aligned} u_{ao} &= u_{a'o} + u_{aa'} = u_{a'o} - u_{A'o'} \\ u_{bo} &= u_{b'o} + u_{bb'} = u_{b'o} - u_{B'o'} \\ u_{co} &= u_{c'o} + u_{cc'} = u_{c'o} - u_{C'o'} \end{aligned} \quad (1)$$

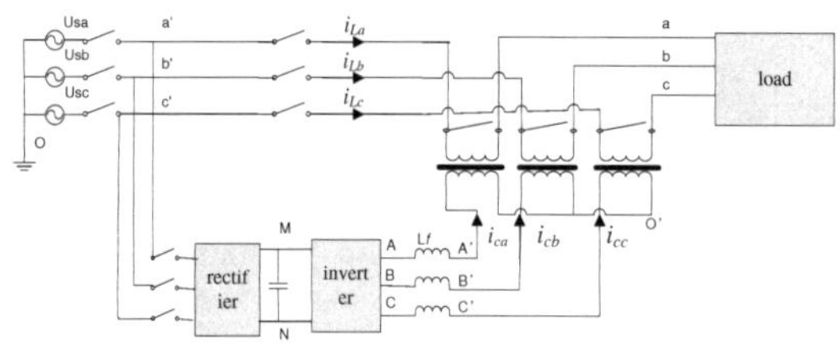

Figure1.Main power circuit of three-phase three-wire DVR

Set the DVR filtering inductances as L_f, and considering inductance resistance r, circuit equations can be written as:

$$\begin{aligned} u_{AO'} - u_{A'O'} &= L_f \frac{d\,i_{ca}}{d\,t} + r\,i_{ca} \\ u_{BO'} - u_{B'O'} &= L_f \frac{d\,i_{cb}}{d\,t} + r i_{cb} \\ u_{CO'} - u_{C'O'} &= L_f \frac{d\,i_{cc}}{d\,t} + r\,i_{cc} \end{aligned} \quad (2)$$

Actually for three-phase three-wire power system, only two equations are independent in equation (2), only two of the compensation voltages are independent, thus to ensure that conform to the requirements of the load line voltage, it can choose a kind of control method makes the $u_{cc'} = 0$, so the topology of the figure 1 can be simplified as shown in figure 2.

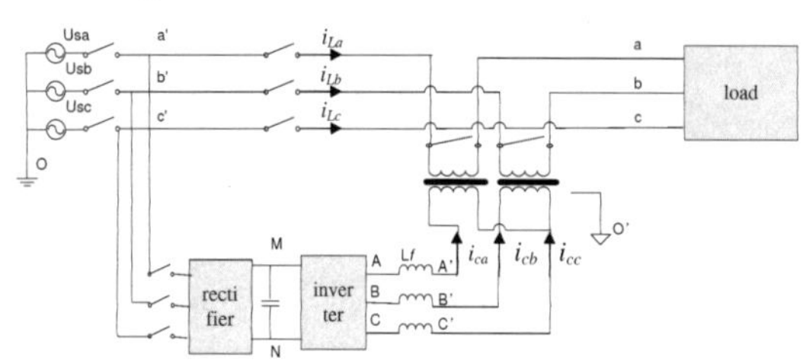

Figure2.Simplified topological structure of three-phase three-wire DVR

3.DVR OPTIMIZATION CONTROL STRATEGY

3.1. Offset voltage detection and the control equation. Voltage disturbance can be detected based on coordinate transformation method [5]. But in this paper, by using the orthogonality of trigonometric function principle, the detection method based on digital PLL is adopts. The fundamental wave phase of line voltage can be detected, and the ideal value of load end line voltage can be written as follows:

$$u_{ac}^{*} = 380\sqrt{2}\sin(\omega t + \varphi)$$
$$u_{bc}^{*} = 380\sqrt{2}\sin(\omega t + \varphi - \pi/3) \quad (3)$$

In which, ω for fundamental wave angular frequency, and φ for the fundamental phase of $u_{a'c'}$. $u_{A'C'}^{*}, u_{B'C'}^{*}$ (That is $u_{a'a}^{*}, u_{b'b}^{*}$) are compensation voltage, there are

$$u_{A'C'}^{*} = u_{a'a}^{*} = u_{a'c} - u_{ac}^{*}$$
$$u_{B'C'}^{*} = u_{b'b}^{*} = u_{b'c} - u_{bc}^{*} \quad (4)$$

In figure 2, to control the DVR is equivalent to control the output voltage of the three phase inverter bridge. Set O' point for virtual floating location, and $u_{AO'} + u_{BO'} + u_{CO'} = 0$ can be met at any moment, and the ideal value of $u_{AO'}, u_{BO'}, u_{CO'}$ is:

$$u_{AO'}^{*} = u_{A'O'}^{*} + L_f \frac{d\,i_{ca}}{dt} + r\,i_{ca}$$
$$u_{BO'}^{*} = u_{B'O'}^{*} + L_f \frac{d\,i_{cb}}{dt} + r i_{cb} \quad (5)$$
$$u_{CO'}^{*} = u_{C'O'}^{*} + L_f \frac{d\,i_{cc}}{dt} + r\,i_{cc}$$

Ignore the power frequency voltage drop on the filtering inductance and resistance, combine with (4)

$$u_{AC}^{*} = u_{A'C'}^{*} = u_{a'a}^{*} = u_{a'c} - u_{ac}^{*} \quad (6)$$
$$u_{BC}^{*} = u_{B'C'}^{*} = u_{b'b}^{*} = u_{b'c} - u_{bc}^{*}$$

Define the switch state of the switch tube of the inverter bridge in the upper arm is connected with A, B, C for the switch state, when the switch state (j_a, j_b, j_c), $j_a, j_b, j_c \in \{0,1\}$, u_{dc} for dc side voltage value.

$$u_{AO'} = u_{dc}(2j_a - j_b - j_c)/3$$
$$u_{BO'} = u_{dc}(-j_a + 2j_b - j_c)/3 \quad (7)$$
$$u_{CO'} = u_{dc}(-j_a - j_b + 2j_c)/3$$

Switch state (j_a, j_b, j_c) has eight kinds of switch state, according to the corresponding binary $(j_a\ j_b\ j_c)$ defines them as numerical order k0 ~ k7, the voltage vector of each switch state can also be controlled by (7). DVR time-domain circuit of voltage control equation is obtained:

$$\frac{1}{T_S}[\mathbf{u} \cdot \Delta\mathbf{t}] = \begin{bmatrix} u_{AO'}^{*} \\ u_{BO'}^{*} \\ u_{CO'}^{*} \end{bmatrix} \quad (8)$$

In which $\mathbf{u} = [\mathbf{u}(0), \mathbf{u}(1), \ldots, \mathbf{u}(6), \mathbf{u}(7)]$
$\Delta\mathbf{t} = [\Delta t(0), \Delta t(1), \ldots, \Delta t(6), \Delta t(7)]^T$

3.2.the optimization control based on the equation optimal solution. According to the built DVR time-domain circuit equations of voltage control, to discuss the solution of the equation under $u_{AO'}^{*} > -u_{BO'}^{*} \geq -u_{CO'}^{*} > 0$. Due to the absolute value of $u_{AO'}^{*}$ is the largest, and positive, and thus switch state of k4(1,0,0), k5(1,0,1),k6(1,1,0) can be selected, which can make u_A increasing, that is:

$$\begin{bmatrix} 2/3 & 1/3 & 1/3 \\ -1/3 & -2/3 & 1/3 \\ -1/3 & 1/3 & -2/3 \end{bmatrix} \begin{bmatrix} \Delta t(4) \\ \Delta t(5) \\ \Delta t(6) \end{bmatrix} = \frac{T_s}{u_{dc}} \begin{bmatrix} u^*_{AO'} \\ u^*_{BO'} \\ u^*_{CO'} \end{bmatrix} \quad (9)$$

The above equations is strange, if add the conditions optimization index of $J = \sum_{i=1}^{6} \Delta t(i) = \min$, it leads to

$$\begin{cases} \Delta t(4) + \Delta t(5) + \Delta t(6) = T_s u^*_{AB} / u_{dc} + \Delta t(6) \geq T_s u^*_{AB} / u_{dc} \\ \Delta t(4) + \Delta t(5) + \Delta t(6) = T_s u^*_{AC} / u_{dc} + \Delta t(5) \geq T_s u^*_{AC} / u_{dc} \\ \Delta t(5) = T_s u^*_{CB} / u_{dc} + \Delta t(6) \geq T_s u^*_{CB} / u_{dc} \end{cases} \quad (10)$$

The optimal solution under $\sum_{i=1}^{6} \Delta t(i) = \min$ is:

$$\begin{aligned} \Delta t(4) &= \frac{T_s}{u_{dc}}(u^*_{AO'} - u^*_{CO'}) = \frac{T_s}{u_{dc}} u^*_{AC} \\ \Delta t(5) &= \frac{T_s}{u_{dc}}(u^*_{CO'} - u^*_{BO'}) = \frac{T_s}{u_{dc}} u^*_{CB} \\ \Delta t(0) &= Ts - \Delta t(4) - \Delta t(5) \end{aligned} \quad (11)$$

The optimal solution of equation (8) can be list as table 1.

Tab.1 Optimal solution of control equation

case	u^*_{AB}	u^*_{BC}	u^*_{CA}	optimal solution
I	+	+	-	k4× t(*A*,*B*)+k6× t(*B*,*C*)
II	-	+	-	k2× t(*B*,*A*)+k6× t(*A*,*C*)
III	-	+	+	k2× t(*B*,*C*)+k3× t(*C*,*A*)
IV	-	-	+	k1× t(*C*,*B*)+k3× t(*B*,*A*)
V	+	-	+	k1× t(*C*,*A*)+k5× t(*A*,*B*)
VI	+	-	-	k4× t(*A*,*C*)+k5× t(*C*,*B*)

In which, $t(m,n) = \frac{T_s}{u_{dc}}(u^*_{mO'} - u^*_{nO'})$, $m,n \in \{A,B,C\}$

The proposed approach is based on the DVR output line voltage command to calculate switch time, reducing the line voltage can be converted into a first phase voltage; This article is based on the optimization index for effective vector K1 ~ K6 "minimum level", to meet the requirements of the output voltage, the purpose of making effective vector can have the ability to make a larger output voltage. The method is based on the optimal solution of control equation is obtained, for the rest of the Ts in added by k0, to ensure the DC side voltage must improve the utilization rate of DC side voltage and to improve the compensation capacity of DVR (9, 18).

4.THE SIMULATION AND EXPERIMENTAL VALIDATION

In order to verify the proposed method, the simulation and experimental validation studies. Figure 3 and figure 4 is three-phase three-wire simulation results of dynamic voltage restorer, figure 3 is a three phase three wire line voltage is falling slowly compensation results (Uac), voltage get a good compensation. Figure 4 is three-phase DVR compensate three-phase voltage instantaneous fall by more than 40%, after DVR compensating load voltage becomes normal.

IOP Conf. Series: Earth and Environmental Science **86** (2017) 012033 doi:10.1088/1755-1315/86/1/012033

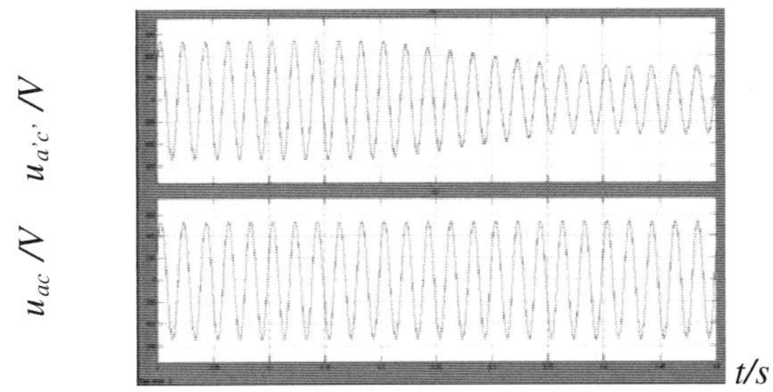

Figure 3.Simulation waveforms of voltage up:source voltage down: load voltage

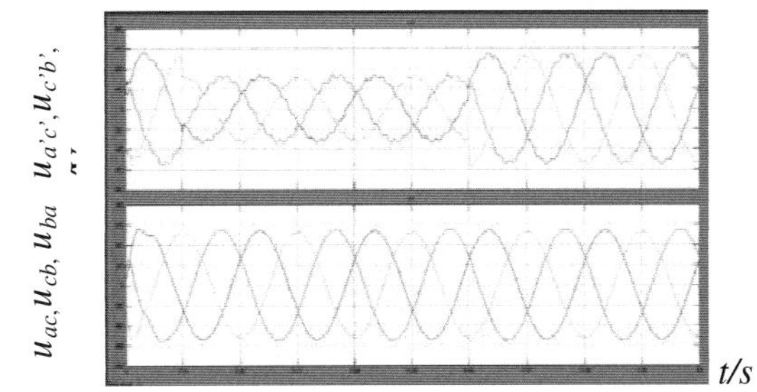

Figure 4.Simulation waveforms of voltage up: source voltage down: load voltage

Experimental platform of controlled power tube for SKM75GB128D IGBT tube, maximum switching frequency of 10 kHz; Detection and control core is TMS320F28335 DSP unit. Figure 5 is the experimental results of three-phase voltage falls slowly compensation; Figure 6 is three-phase voltage sags and severe distortion of the experimental results, the voltage amplitude of short-term fell nearly 50%, and serious distortion, THD value reached 30%, after DVR compensating load voltage to maintain in the given value, THD value also dropped to nearly 3%, it can be found that the experimental results and simulation results are consistent, DVR can compensate voltage harmonic distortion and voltage sag at the same time.

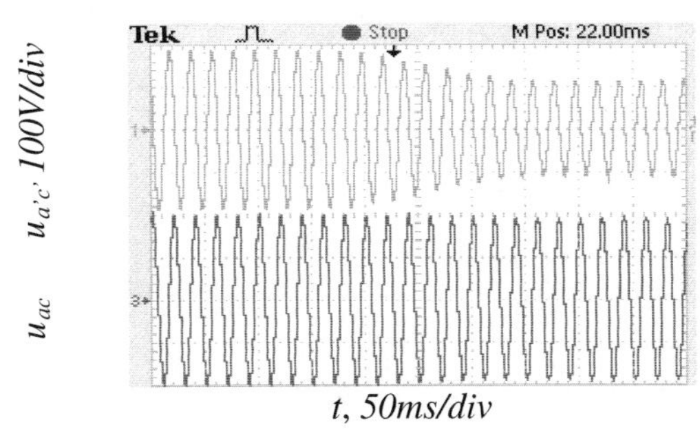

Figure 5 Experiment waveforms of voltage up:source voltage down: load voltage

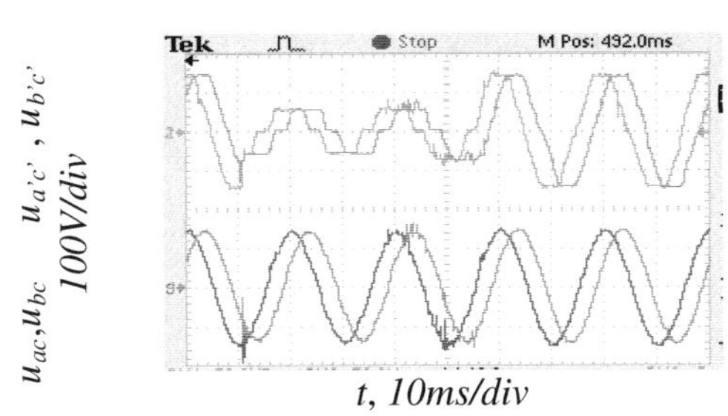

Figure 6.Experiment waveforms of voltage up:source voltage down: load voltage

5.CONCLUSION

In this paper, the line voltage compensation type three-phase three-wire DVR topology is analyzed and simplified, the time domain analysis equation of DVR is established, and the optimal solution of the voltage control equation is obtained under the optimized index, which calculate switch time based on the line voltage, and can improve the utilization rate of dc side voltage and the compensation ability of DVR. The optimal performance can be obtained by using the proposed performance optimization objective function. And the optimal solutions can be got by simple judgments and computations, which can significantly decrease computational complexity. The simulation and experimental results are given; the results show its validity and feasibility.

References

[1] Liu H, Han M X, You Y and Xiao X G 2003 Analysis Of Compensation Capability Of Line Voltage Compensation Based Dynamic Voltage Recovery (DVR). *Automation of Electric Power Systems.* **27** 54-57.

[2] Huang B R, Xia L, Wu Z G and etc 2011 Double Feed forward Plus Feedback Control Strategy Based On Line Voltage Compensation For Dynamic Voltage Restorer. *Electric Power Automation Equipment* **31** 61-64.

[3] Wang Z Y, Wu Z G and Zhou W P 2009 A Deadbeat Control Based Dynamic Voltage Restorer With Line Voltage Detection. Power System Technology.**33** 106-112.

[4] Guo W Y,Xiao L Y,Guo J D and et al 2009 An Optimal Control Strategy for Dynamic Voltage Restorer With an Optimal Filter. *Proceedings of the CSEE*, **29** 48-54.

[5] Zhou X S, He J, Ma Y J and et al 2008 Main Circuit Topology and Control Algorithm of Dynamic Voltage Restore. *High Voltage Engineering*, **34** 753-758.

[6] Wang Z Y,Wu Z G, Zhou W P and et al 2009 A Novel Reference Wave Generating Algorithm Based on Characteristic Vector Extracted for DVR. *Transactions of China Electrotechnical Society.* **24** 168-174.

[7] Wang Z Y,Wu Z G and Zhou W P 2009 Study on the dynamic voltage restorers with capacitor direct-coupled. *Power System Protection and Control*. **37** 82-87.

[8] AWAD H and BLAABJ ERG F 2004 Transient performance improvement of static series compensator by double vector control. *Proceedings of 19th IEEE Applied Power Electronics Conference and Exposition Vol 1 Anaheim, CA , USA. Piscataway* 607-613.

[9] Li T,Wang B,Guo C P and Huang B J 2008 Simulation Research on Three-phase Four-leg DVR Based on Sliding Mode Variable Structure Control. *Electric Power Science and Engineering*, **24** 5-8.

[10] LIU N, ZHOU Li D,YAO G and et al 2012 Dynamic voltage restorer based on robust H∞ controller. *Electric Power Automation Equipment*, **32** 98-102.

[11] CHEN G D,ZHANG L and CAI X 2012 Compensation strategy based on positive and negative sequence extraction for dynamic voltage restorer. *Electric Power Automation Equipment*, **32** 87-91.

[12] XIAO C,TANG Z D and JIN X 2010 A Novel Dynamic Voltage Restorer and Simulation. *Journal of Jiamusi University (NaturalScience Edition)*, **28** 891-893.

[13] ZHU J X, WANG T X 2010 A novel dynam ic voltage restorer w ith flywheel energy storage system. *Journal of Mechanical & Electrical Engineering*, **27** 93-96.

[14] GE C H, CHENG H Z, WANG X H and et al 2009 DVR controlal gorithm based on minimal energy. *Electric Power Automation Equipment*, **29** 70-74.

[15] WANG S C, YU K S and TANG G F 2008 Performance Analysis and Improvement of Digital Vector Control Algorithm for Dynamic Voltage Restorer. *Proceedings of the CSEE*, **28** 64-71.

[16] HOU S Y, LIU Z C, JI L M and et al 2009 Cascaded inverter dynamic voltage restorer based on minimum active power injection strategy. *Electric Power Automation Equipment*, **29** 33-36.

[17] Zhou W P, Wu Z G and Xia L 2004 Harmonic and reactive current detection in APF based on high-accuracy phase and frequency detection. *Proceedings of the CSEE*, **24** 91-96.

[18] Zhou W P, Wu Z G, Tang J S and et al 2006 Novel algorithm of SVPWM and the study on the essential relationship between SVPWM and SPWM. *Proceedings of the CSEE*, **26** 133-137.

Design of energy saving monitoring system on isothermal oilless lubricated air compressor

S N Ma [1], Q Liu [1], B L Xu [2], Y Liu[2] and F Hu [2]

[1]Jiangxi Engineering Research Center of Process and Equipment for New Energy, East China University of Technology, Nanchang, China.

[2]East China University of Technology, Nanchang, China.

Email: *Corresponding author:masnong@sohu.com

Abstract. This design introduces a kind of STM32F051C8T6 circuit monitoring system which is based on the ARM core. According to the operating principle of air compressor, the reduction of temperature and current is converted to save electricity and directly displayed, save electricity information stored at the same time, to achieve real-time monitoring air compressor energy saving effect in the process of operation.

1.Introduction

The isothermal oilless lubricated air compressor is internationally accepted contract energy management mode was adopted to realize user zero investment, zero risk, high yield of the implementation of the method, need through the energy display monitor system to implement, development and design a set of isothermal oilless lubricated air compressor according to monitoring and control system for energy conservation.

By technical improvement on each air compressor of the station, the energy saving monitoring system takes the decreasing temperature and electricity during running time transform to saving electric energy and displays immediately.

For instance, the 1# air compressor's exhaust temperature decreases from 150 degree to 125 degree and its current decrease from 23A to 20.7A through the improvement, but the air flow and pressure are not change. It requires the energy saving monitoring system is able to display the temperature and the current differences, and figure out the energy-saving effect according to the volumetric expansion coefficient of compressed air(every degree increase in temperature, the compressed air expanded 1/273.15M^3, or vice versa).

2.System requirements

Based on the description of the energy saving monitor, this system should be able to achieve the following requirements:

IOP Conf. Series: Earth and Environmental Science **86** (2017) 012034 doi:10.1088/1755-1315/86/1/012034

(1) Collecting the operating current and outlet temperature of the air compressor, displaying in real time and updating per second. The accuracy of current and temperature should reach 0.1A and 0.1 degree respectively.

(2) Calculating and displaying the saving electric energy according to the temperature and the current differences per minute.

(3) Displaying the cumulative running time.

(4) Achieving storage function when the system losing power.

(5) Before the improvement, the operating current and temperature of the air compressor should input by administrators, and they should be observed during the system runtime.

(6) The system must be equipped with good fault warning.

Considering the high voltage motor with 6000V will cause great electromagnetic interference, the operating current and outlet temperature monitor of the air compressor must adopt industrial level transmitter which has strong anti-disturbance capability, and the data communication method have to uses the 485 buses and 4~20MA data communication with strong anti-disturbance either. This system apply the industrial grade perforated guide rail mounting type current transmitter to its current acquisition module, which has transmission distance and immunity to interference. And the temperature acquisition module makes by the industrial grade thread mounting type light small integrated temperature transmitter, this kind of transmitter is able to use in the difficult conditions and proof of wet, explosion, water, earthquakes, harmful gas and so on.

3.System solutions

Figure 1 is the topology structure graph of this system:

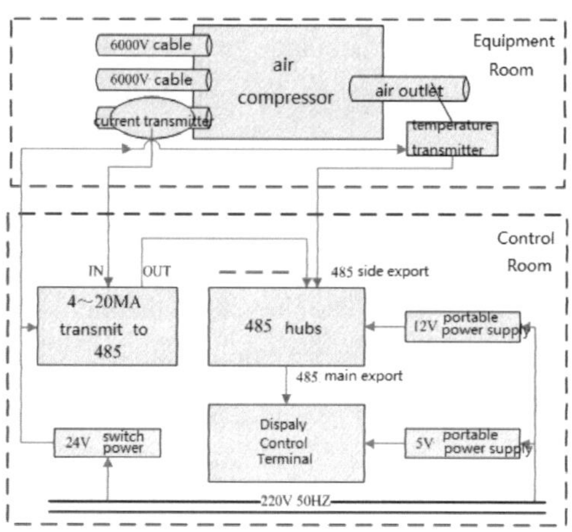

Fig. 1 Topology Structure Graph

(1) Design of monitoring system

Display control terminal is mainly to complete data acquisition of each node from 485 bus, and calculate the saving energy of per minute, as well as display the result with a display panel. Users are able to check each air compressor by shifting the terminal control panel and by the LED display board, and also allowed to set the reference value of operating current and outlet temperature before improvement. The control terminal circuit also need to achieve storage function when the system losing power, such as the cumulative saving energy, cumulative running time, operating current, outlet temperature and the reference value before improvement.

The display control terminal is equipped in the control room and avoided from strong interference, therefore it is based on the ARM Cortex-M0 processor. This kind of processor is available the most minimum volume, minimum energy consumption and the most energy-saving on the market. With the very low energy consumption, developers could obtained the performance of the 32 bit processor at the price of 8 bit processors.

Due to the power down save function, the EEPROM memory is adopted the AT24CXX series of ATMEL, which could save data up to one hundred years and one million times repeatedly erased. Accordingly, this memory can realize a complete, accurate and long-term storage. At the same time, the memory is built with IIC data communication and simplified PCB routing, in order to reduce the design cost of the circuit under the premise of function completely realized.

Figure 2 is the circuit graph of the monitoring system:

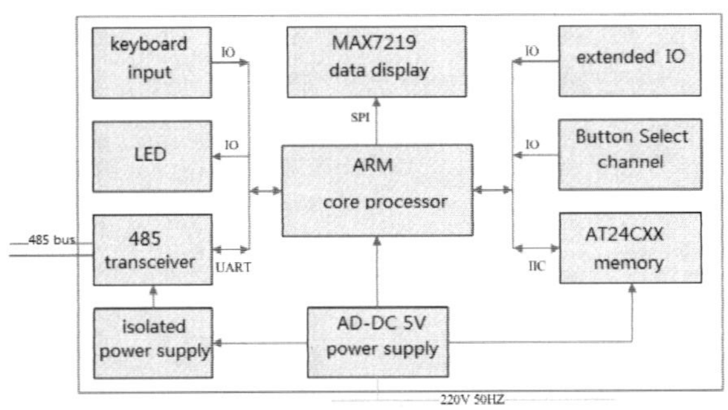

Fig. 2 Monitoring System Circuit Graph

According to the shell box structure of terminal, the monitoring system circuit is required to fall into 2 parts, the ARM core circuit include the STM32F051C8T6 core circuit, the 485 bus transceiver circuit, the AT24CXX data storage circuit and the power isolation circuit. The other part is display circuit consist of MAX7219 drive digital tube, the button selection circuit and the LED indicator circuit of air compressor.

Display circuit as shown in figure 3:

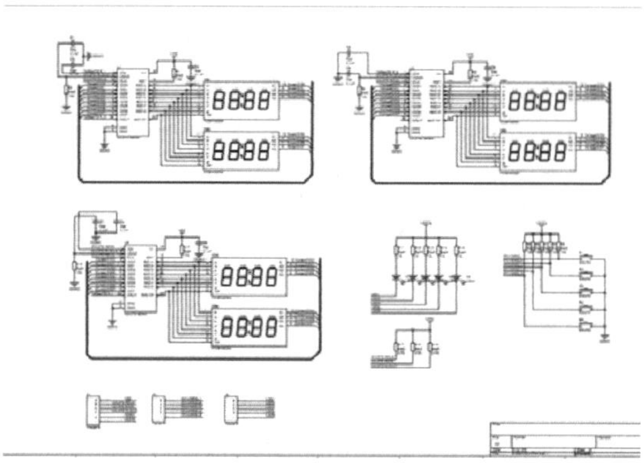

Fig. 3 Display Circuit

(2) Software Design

The software design is mainly to analyze and calculate from data acquisition of each node. The following tree distribution is abstracted from the topology structure graph(as shown in figure 4):

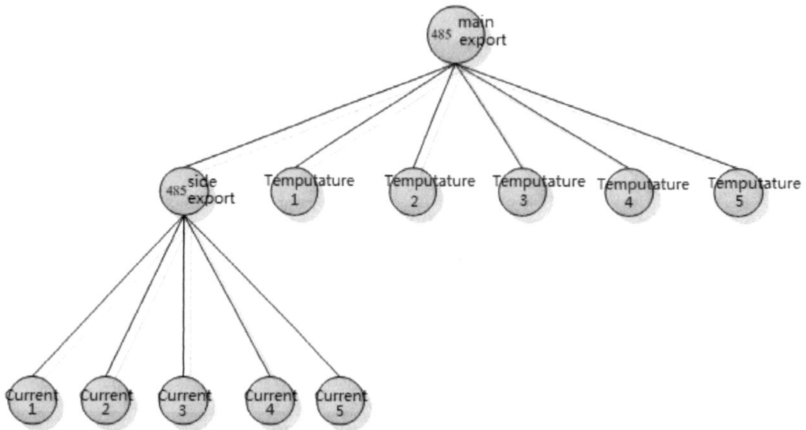

Fig. 4 Tree Distribution

It can be seen that we can complete data acquisition as long as traverse every node of the tree distribution.

The software distribute a data structure which include the cumulative saving energy, cumulative running time, operating current, outlet temperature and the reference value before improvement. The core function of software system is to maintain these data structures. With powered on, the software system analysis if the air compressor had reference values, and make calculations on the basis of reference values. The program traverses every node of the tree distribution per second, and assigns to corresponding data structure, meanwhile displays the present air compressor's current and outlet temperature. When the time accumulates to one minute, the program will figure out the saving energy in this period and display the total result on the panel. Certainly, the program will not work when the air compressor is shutdown equates to the acquired current is 0A. At the end of this one minute, the system will store the cumulative saving energy, cumulative running time, operating current, outlet temperature into the EEPROM and provide to the next computing cycle.

Together the workflow of software program and the tree diagram, Figure 5 is the design of software process:

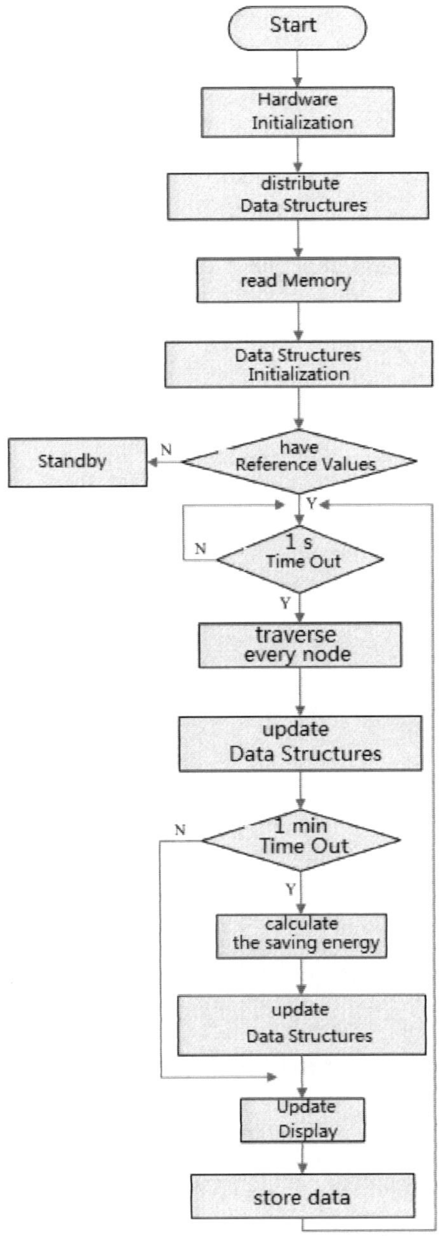

Fig. 5 Design of Software Process

The saving energy of air compressor which display on the panel mainly comes from two aspects, for one thing is the energy consumption caused by current decrease, for another is caused by temperature decrease. The calculation algorithm is as follows equation (1).

$$P = P_T + P_C \qquad (1)$$

In the above equation, P_T represent the energy consumption caused by temperature decrease, P_C represent the energy consumption caused by current decrease.

The change of compressed air temperature must lead to the change of volume, therefore the temperature decrease will cause the increase of gas production. In accordance with 40m³ per minute gas production, the actual added gas production(ΔV) caused by temperature decrease per minute is equation (2).

$$\Delta V = \Delta T \times \frac{1}{273.15} \times 40M^3 \quad (2)$$

In the above equation, ΔT represent the temperature difference of the air outlet between before and after the improvement, 1/273.15 represent the volumetric expansion coefficient of compressed air, 40m³ represent the exhaust volume per minute.

Considering of the 250KW motor power, the consumed power(ΔP) per cubic meters is equation (3).

$$\Delta P = \frac{250KW}{60MIN} \times \frac{1}{40M^3} \quad (3)$$

Therefore the saving energy per min is equation (4).

$$P_T = \Delta V \times \Delta P = (\Delta T \times \frac{1}{273.15} \times 40M^3) \times (\frac{250KW}{60MIN} \times \frac{1}{40M^3}) \quad (4)$$

According to the collected single-phase current and the consumed power of air compressor, the saving energy caused by current decrease is equation (5).

$$P_c = \Delta C \times 6000V \times \frac{1}{1000} \times \sqrt{3} \times PF \times \frac{1}{60MIN} \quad (5)$$

In the above equation, ΔC represent the operating current difference between before and after the improvement, PF is power factor(obtain from the power factor table), $\sqrt{3}$ represent the coefficient of three phase electric, then divided by 60 is the saving energy per minute. And then transform to the unit of electrical energy divided by 1000.

4.Conclusion

The innovation points of this system: (1) Considered the high voltage of air compressor station, the collectors of current and temperature both use the acquisition transmitters with strong anti-interference ability; (2) The data communication method adopt the 4~20MA and 485 buses with strong anti-disturbance, which are widely used in industry and adapt to rough conditions. (3)Due to the calculation of consumed power involved floating point operation, the control terminal adopts ARM processor which has excellent capacity of rapid floating-point.

5.Acknowledgements
I would like to acknowledge The National Natural Science Foundation of China (Grant No. 61161010) and The Open Project Program of Jiangxi Engineering Research Center of Process and Equipment for New Energy（Grant No. JXNE2015-11）for helping and supporting.

References

[1] Zhou M, Wang Z F and Liu W B 2012 The research of fault diagnosis model for reciprocating compressor. *Compressor Technology*, **04** 13-15.
[2] Gan F C, Liu B F and Lv F X. 2009 The design of energy-saving analysis and control system for air compressor. *Automation of industry and mining,* **3** 72-74.
[3] Yang M 2009 Air compressor efficiency in a Vietnamese enterprise. *Energy Policy*, **37** 2327-37.
[4] Joseph Y C 2009 ARM Crotex - M3 authoritative guide . Song Yan. translation. *Beijing Aeronautics and Astronautics University Press*
[5] Dana S, Sagahyroon A., Elrayes A, AI-Ali A R, Al-Aydi R 2008 Development of a monitoring and control platform for PLC-based applications. *Computer Standards&Interfaces*,**30** 157-166.
[6] REN S L, LU H, WANG Y Z and FU H Y 2007 Development of PLC-based Tension Control System. *Chinese Journal of Aeronautics*, **20** 266-271.
[7] Michele A, Alessandro B and Cristian B 2009.Advanced control systems for single compressor chiller units. *International journal of refrigeration*, **32** 1068-76.
[8] Helge R, Matthias K, Alexander L and Jorg O 2007 Making human-machine interfaces more brain-adequate . *International Congress Series*, **1301** 15-21.
[9] ARM Ltd 2006 Crotex-M3 Technical Reference Manual Rev r0p0.
[10] STMicroelectronics Ltd 2009 RM0008 Reference manual: STM32F101XX, STM32F101XX, STM32F102XX, STM32F103XX, STM32F105XX, STM32F107XX advanced ARM-based 32- bit MCUs Rev 9.

IOP Conf. Series: Earth and Environmental Science **86** (2017) 012035 doi:10.1088/1755-1315/86/1/012035

Design of MPPT Controller Monitoring Software Based on QT Framework

X Z Meng[1] and P G Lu [2]

[1]Qingdao University of Science & Technology, Qingdao,China

[2] Qingdao University of Science & Technology, Qingdao,China

Email: *942105626@qq.com

Abstract. The MPPT controller was a hardware device for tracking the maximum power point of solar photovoltaic array. Multiple controllers could be working as networking mode by specific communicating protocol. In this article, based on C++ GUI programming with Qt frame, we designed one sort of desktop application for monitoring and analyzing operational parameter of MPPT controller. The type of communicating protocol for building network was Modbus protocol which using Remote Terminal Unit mode and The desktop application of host computer was connected with all the controllers in the network through RS485 communication or ZigBee wireless communication. Using this application, user could monitor the parameter of controller wherever they were by internet.

1.Introduction

The output characteristics of photovoltaic array are a typical nonlinear model [1], and are easily disturbed by various conditions such as temperature, light intensity and variable Load. The MPPT controller can use properly algorithm to real-timely adjust the working point of photovoltaic array(byadjusting output voltage), to make sure the working point is near at maximum power point. Therefore, it's necessary tomonitor the real-time operational parameter of MPPTcontroller.

Qt developing frame is designed by C++ language which is supplied by third party. Using Qt developing frame to design Graphical User Interface is convenient and safe.Qt developing frame is suppling cross-platform [2] development.Now, the cross-platform feature of Qt supplies x86、SPARC、ARM and so on. It is also compatible with various kinds of operating system such as Windows, Mac OS X, Linux,Android, iOS and VxWorks, Embedded Linux. In this paper,we based on the Qt development library, developed a desktop application named Solar MPPT Controller Monitoring Software under the Windows system. The application contained serial read and write module, could parse and organized Modbus instruction, and the network function of the maximum power controller also could be implemented according to the master-slave communication mode in the Modbus protocol.

2.Monitoring System Structure

Every solar photovoltaic array would configure one MPPT controller, when all controller wereallocated a unique address, they could work in a same network. The host computer monitoring software was connected with all the controllers in the network through RS485 communication or ZigBee wireless communication. Due to that the MPPT controller has been embedded the Modbus communication protocol, Modbus specification pointed it out clearly that each slave device had a unique address. In this way, host computer monitoring software could use this address to distinguish each device.

In this system, the MPPT controller was used as the server and the desktop application played as client so it could access the server by polling mode. The working parameters of each controller will be processed, displayed and stored separately in desktop application. Besides ,we design one kind of Network management model, This model is used to build and Managethe MPPT controller network, Desktop applications need to dynamically add or exit a controller in the network, and also need to determine whether the communication ofcontrollers are in the regular state in order to change the network topology properly. In totally, the host computer and the MPPT controller constitute the entire monitoring systemby Modbus protocol. This type of building MPPT controller system is flexible and stable,user can rebuild the system easily according theirs specific purpose. Figure 1 shows thewhole system structure including MPPT controller, solar photovoltaic array and monitoring software.

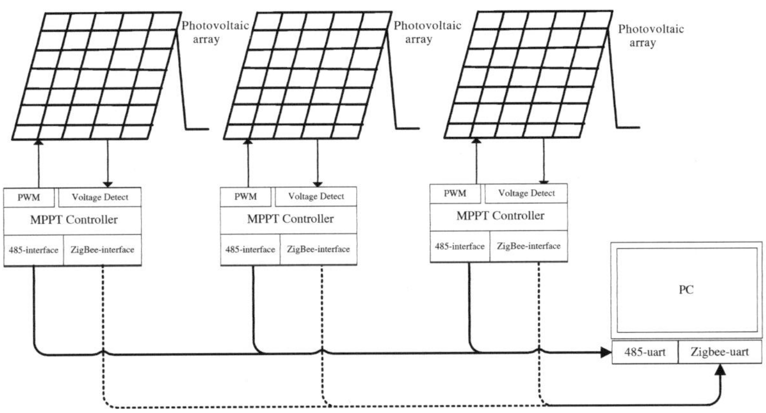

Figure 1. Monitoring System Structure

3.Monitoring Software Structure

3.1 Communication Module. MPPT controller uses the Modbus protocol which bases on serial port or ZigBee technology, and it will save allparameters of itself to the registers which are possessing continuous register address. Accordingly, the monitor alsousing serial port or ZigBee technology to adapt to controllers communication port and suppling the function of resolving data instructions which are stipulating by Modbus protocol, The monitor plays master part for requesting data from slave part i.e. MPPT controller. Figure 1 shows this model.

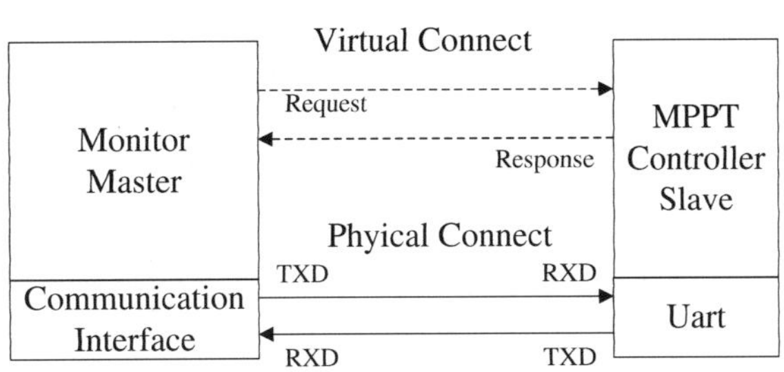

Figure 2. Communication Module

*3.2 Network Management.*The way to manger network is based on a specific kind of registry [3]. When multiple MPPT controller are joining to the same monitor software , the monitor software can recognize every single MPPT controller by slaver's address, and it works by the way of periodically polling .This need the monitor software to maintain a registry which contains the status of each salver, The status will illustrates whether slaver joins the polling cycle or not.

Background program will refresh the registry according to the action of interface, only if one certain slaver is confirmed by user to join to the network, monitor will send data request instructions to the slaver. Besides, when monitor has been sent the data request instructions to the slaver but never received response instructions from slaver during the Tolerance time, then background program will mark the status of this slaver as false to indicate that current slaver is in bad communication situation. Figure 3illustrates the condition of 5 slavers join in the network and 2 slavers whose address is 1 and 3 are in bad communication situation.

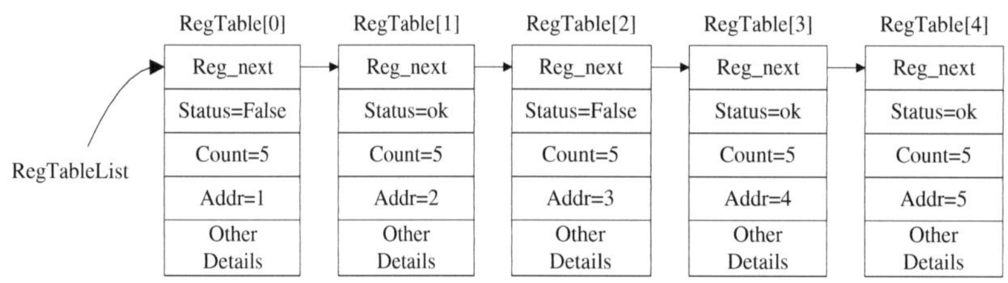

Figure 3 Network Status Registries

As show by the figure above, the whole registry is organized based on linked list structure, Registers List represent the address of first table, and every single table contains a special pointer which point to the next table, status of communication condition, total numbers of slaver, address of slaver and some other details. GUI application possess main thread named GUI thread, because of that GUI thread is the only thread to be allowed to run the operation about GUI. For some operations which need to respond swiftly or may be very time consuming, we must move this operation to an individual threat we call that is multithread programming.

To implement the multithread programming [4] based on Qt frame is very convenient. We just need to instantiate the QThread class and overload its run () function of the instantiated objects. In addition, the QSerialPort class is a thread-safe class, and all the member function must be thread-safe and re-entrant, so move all the operation such as serial port to single thread is absolutely safe.

*3.3 Database Storage.*For the purpose of obtaining long-time running parameter of MPPT controller, it's necessary to store long-time running parameter of MPPT controller by using data base technique. We choose MySQL [5] database supplied by Oracle Company. Qt library contains QtSql module which supply platform-independent interface for accessing to database. This module is constituted by two parts, one part is Model-View-Controller, and the other one is user interface class.

When users configure the Qt library, there are two choices for getting database driver, one choice is using the database driver which is included by Qt library, the other choice is to compile database driver by the form of plug-in. For now, Qt5.5 has been supplied the database driver to MySQL, so using Qt library to develop database is more convenient. Figure 4 shows the window of database configuration and the window of loading data from database.

IOP Conf. Series: Earth and Environmental Science **86** (2017) 012035 doi:10.1088/1755-1315/86/1/012035

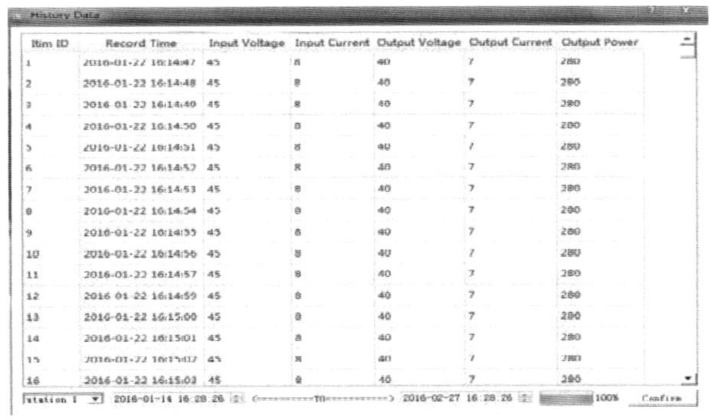

History Data

Itim ID	Record Time	Input Voltage	Input Current	Output Voltage	Output Current	Output Power
1	2016-01-22 16:14:47	45	8	40	7	280
2	2016-01-22 16:14:48	45	8	40	7	280
3	2016-01-22 16:14:49	45	8	40	7	280
4	2016-01-22 16:14:50	45	8	40	7	280
5	2016-01-22 16:14:51	45	8	40	7	280
6	2016-01-22 16:14:52	45	8	40	7	280
7	2016-01-22 16:14:53	45	8	40	7	280
8	2016-01-22 16:14:54	45	8	40	7	280
9	2016-01-22 16:14:55	45	8	40	7	280
10	2016-01-22 16:14:56	45	8	40	7	280
11	2016-01-22 16:14:57	45	8	40	7	280
12	2016-01-22 16:14:59	45	8	40	7	280
13	2016-01-22 16:15:00	45	8	40	7	280
14	2016-01-22 16:15:01	45	8	40	7	280
15	2016-01-22 16:15:02	45	8	40	7	280
16	2016-01-22 16:15:03	45	8	40	7	280

station 1 2016-01-14 16:28:26 <---------TO---------> 2016-02-27 16:28:26 100% Confirm

Figure 4 Loading data for database

The MySQL database can be used as a web server and allowing remote client accesses to the database of server by internet. For to implement this function, user should make a simple configuration just using the standard SQL language as bellows:

MySQL> GRANT ALL PRIVILEGES ON *.* TO 'user'@'%' IDENTIFIED BY '123' WITH GRANT OPTION;

*3.4 Workflow of Software.*The workflow of monitor software contains serial data dispatch, Modbus protocol instructions analysis, data storage and so on. In Qt frame, all interface action is transmitted by event mechanism in the whole application system. The QApplication class is used to control the event mechanism and when a specific signal of event is emitting, event mechanism can transfer the signal to the corresponding slot, then the slot will call specific function unit to handle the event immediately. This process is shown in Fig. 5.

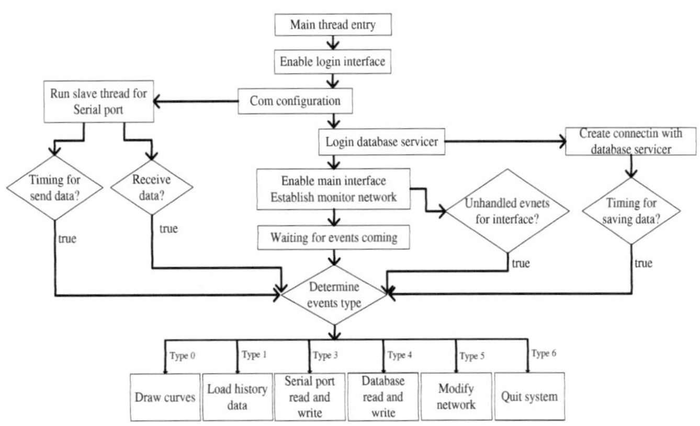

Figure 5 Workflow of Monitoring software.

When users log into the monitor software, the main windowwill display. The main window contain Data display unit, Alarm parameter setting unit, and debug information. Someothers function are embodied in the menu bar. When users choose specific slave devices to join into the working network and click the Start button, the software will be running the event loop and process the ready event according to priority. The main window is shown in figure 6.

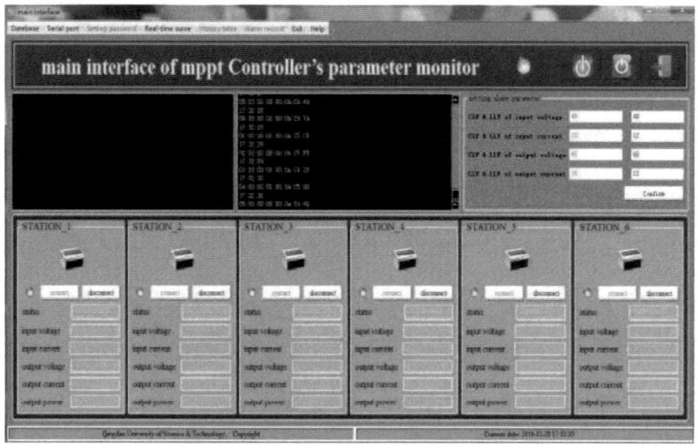

Figure 6Main-window

4.Conclusions

This article used Qt frame to design cross-platform softwareto monitor the MPPT controller's operational parameter. Through analysis and comparison of test results, this software can build a stable network between multiple MPPT controllers, and also can indicate the status of controller timely, additionally, this application of monitoring software is based on multiple threads programming technology, and this makes the software more stable and responsive.

References

[1]Zhao Z M, , Liu J Z, Sun X Y 2005 Solar photovoltaic and applications.*Beijing Science Press*,201-255.

[2]Gordon H 2007 C++/C the Visual C++ LanguageFor *.Net. Wiley India Pvt Ltd*, 168-205.

[3]Jeff C, John P M 2009 C++ All-In-One DeskReference for Dummies. *Wiley publishing*.

[4]Stephen R. D 2008 C++ for Dummies. *Wiley publishing*, 544-561.

[5]Thomas M. C, Carolyn E B 2009 Database systems: a practical approach to design, implementation, and management. *Addison-Wesley*.

Development and Application of integrated monitoring platform for the Doppler Weather SA-BAND Radar

Q Zhang [1], J Sun [1], C C Zhao [2] and H Y Chen [3]

[1]Meteorological Observatory of Shandong Province, Jinan, China

[2] Meteorological Bureau of Weifang, Shandong Province, Weifang, China

[3]Meteorological Bureau of Guan'nan, Jiangsu Province, Lianyungang , China

Email: *nuist_zq@126.com, 512913621@qq.com

Abstract. The doppler weather SA-band radar is an important part of modern meteorological observation methods, monitoring the running status of radar and the data transmission is important.This paper introduced the composition of radar system and classification of radar data,analysed the characteristics and laws of the radar when is normal or abnormal. Using Macromedia Dreamweaver and PHP, developed the integrated monitoring platform for the doppler weather SA-band radar which could monitor the real-time radar system running status and important performance indicators such as radar power,status parameters and others on Web page,and when the status is abnormal it will trigger the audio alarm.

1.Introduction

The doppler weather SA-band radar (abbreviation radar) is one of the most effective means for monitoring and early warning of sudden disaster weather, it plays a important role in monitoring storms, typhoons, hail, tornado and other severe weather system, and provides favorable support forecasters to analysis of changes in weather forecast and artificial influence on weather operations. Radar has achieved remarkable economic and social benefits, is one of an important symbol of the meteorological modernization of our country [1].

The doppler weather SA-band radar is a complex structure which will generate reflect radar products and the working state of the various parameters which are very important to determine radar operating status and do maintenance for the system and improve the availability of radar system and reduce the failure incidence[2].

Therefore, it is important to monitor the operation status of the system and the upload of the radar data and the operating parameters of the system.

2. BRIEF INTRODUCTION OF RADAR SYSTEM

2.1 System components. The radar system is composed of three parts:
RDA(Radar data acquisition):RDA consists of four parts: transmitter, antenna, receiver and signal processor.Its main function is to produce and transmit radio frequency pulse, and receive the scattering energy of these pulses, and form the basic data.
RPG(Radar products generator):radar product generation system, which is the command center of the whole radar system which receives the basic data from the RDA, carries on the processing to it and generates various kinds of products, and passes the products to users.
PUP(Principal user processor): the main user terminal system, which obtains, stores and displays products [1].

2.2 Data classification
2.2.1 Radar status information. Radar status information are stored in/opt/rda/log folder.
These log files include:
Calibration.log,FC.log,Alarm.log,Operation.log,Status.log,PL.log,Rad.log,which FC.log and Rad.log generate a new file each hour with the format of YYYYMMDDHH_ file name. Log, other documents, such as Calibration.log generate a new file every day with the format of YYYYMMDD_filename. Log, which YYYYMMDDHHmm respectively year, month, day, hour, minute.

2.2.2 Radar base data. Stored in RPG Computer Archive2 folder format Archive YYYYMMDDHH.mmA, such as Archive2013091011.12A,is Sept.10, 2013 11:12 based data,(world time).

2.2.3 Radar Product data. 39 kinds of products such as radar reflectivity factor(R),the average radial velocity(V) stored in the PUP Computer folder named after the product and the day time(Beijing time)[1].

3.Monitoring platform development

The doppler weather SA-band radar integrated monitoring platform could monitor the operation status of the radar system and data generation and uploaded, determine the radar operating status and the performance of the system with radar state information on the web.

3.1 Development tools. We choose Macromedia Dreamweaver(DW) to produce the web page, using PHP to prepare the core program and Apache to build Web services .

3.2 Platform construction
3.2.1 Design web page framework. Designed the framework and the layout of the interface for the new generation of weather radar monitoring platform. Using DW to build the basic framework of the doppler weather SA-band radar monitoring platform.
3.2.2 Build server environment. The core code of the monitoring platform is written by PHP, so it needs to build the PHP operating environment. Installed Apache (web server software) and embedded PHP language environment, system configuration and network port settings

including PHP working environment, set the home page for the http://127.0.0.1/index.php.

3.2.3 Core programs. The core programs of the monitoring platform were written by PHP, including system operating status displaying program, the alarm sound triggering program, the status information to reading and displaying program, numerical comparison program, the alarm information displaying program and others.

3.3 Operation monitoring principle

3.3.1 Information acquisition. The monitoring platform read RDA information in the status.log, calibration.log, alarm.log, IQ62.log and other documents. To obtain information from status.log for monitoring RDA operation status, accessing power radar, clutter suppression and calibration data from calibration.log and IQ62.log, getting alarm information from alarm.log.

As shown in Figure 1, take the RDA "OPERATIONAL status "as an example, first positioning "OPERATIONAL status" in the text, using PHP we can read the the state numerical value and the recorded time of " OPERATIONAL status ".We can read out the recorded time of "STATE OPERATING" state values and "PEAK POWER TRANSMITTER" numerical value in the same way[2].

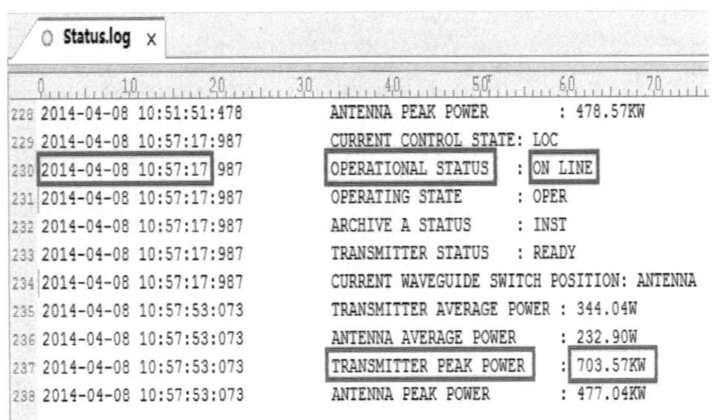

Status.log

```
228 2014-04-08 10:51:51:478   ANTENNA PEAK POWER        : 478.57KW
229 2014-04-08 10:57:17:987   CURRENT CONTROL STATE: LOC
230 2014-04-08 10:57:17:987   OPERATIONAL STATUS    : ON LINE
231 2014-04-08 10:57:17:987   OPERATING STATE       : OPER
232 2014-04-08 10:57:17:987   ARCHIVE A STATUS      : INST
233 2014-04-08 10:57:17:987   TRANSMITTER STATUS    : READY
234 2014-04-08 10:57:17:987   CURRENT WAVEGUIDE SWITCH POSITION: ANTENNA
235 2014-04-08 10:57:53:073   TRANSMITTER AVERAGE POWER : 344.04W
236 2014-04-08 10:57:53:073   ANTENNA AVERAGE POWER     : 232.90W
237 2014-04-08 10:57:53:073   TRANSMITTER PEAK POWER    : 703.57KW
238 2014-04-08 10:57:53:073   ANTENNA PEAK POWER        : 477.04KW
```

Figure 1 location, read the value of the parameter

3.3.2 Monitoring principle of RDA. All the radar systerm operating status information of each scan will be recorded in status.log,the recording time accurate to the millisecond, including the operation status, scanning mode, transmitter power etc.[1]. In the state information, according to the RDA "OPERATIONAL status " and " operationg state" and " transmitter peak power" to judge the running status of the RDA. RDA"OPERATIONAL status "has five status: shutdown(SHTDN), not operation(INOP), must maintain(MNT man), need to maintain(MNT REQ) and online(on line), "operating state" has six status : startup (stup), stand by (STBY), restart(RSTUP）, operate(OPER) and playback (PLYBK) and off-line operate(OLOPER).

When the radar system is in normal operation state, the "transmitter peak power" should be greater than 750kW and less than 900kW, "OPERATIONAL status " is "on line", the "operationg state" is "OPER". When the radar is in abnormal operation and performance in general for two cases: first,the radar system is running, but the transmitter power abnormal either below 750kW or higher than 900kW, second,the radar antenna switch stop running,

RDA " OPERATIONAL status "and "operating state" are no longer " on line "and" OPER". Using PHP to read the value of the last recorded transmitter peak power in status.log, if the peak power is not between the 750-900KW, that means RDA is abnormal operation, if peak power is between the 750-900kW,then by judging state of the last recorded value of "OPERATIONAL status" and "operationg state", if the states are "on line" and "OPER" respectively, we identified radar system is in normal operation, otherwise it is in abnormal operation.

3.3.3. Monitoring principle of RPG. When PRG is in normal operation, radar will produce the base data stored in the RPG host Archive2 folder, one base data for one scan time. At present, the wide application of scan mode is VCP21, an individual sweep time is about 6 minutes. We could get the running state of RPG by judging the latest base data generation time. If the difference between the current time and the latest data recorded time is about 6 minutes, we judged RPG is normal, otherwise abnormal.

The monitoring principle of PUP is as the same as RPG and do not introduce.

4.FUNCTIONS

4.1 Monitoring platform display. The overall framework of the monitoring platform is build by Macromedia Dreamweaver. Figure 2 for the monitoring platform interface.

Fig. 2 interface of monitoring platform

Platform interface is divided into the name area and functional area ,the functional area included of the link area, the monitoring area and alarm information displaying area.

4.2 Monitoring platform functions

4.2.1 URL link function. The monitoring platform links the common business platforms such as the China Meteorological Administration Integrated Meteorological System Operation Monitoring Platform (ASOM) and others.

4.2.2 Status diagnosis function. According to the radar status information,the base data and the products, the monitoring platform judges the radar system operation status. If any monitoring status is abnormal, the monitoring platform will trigger the audio alarm, as shown in Figure 3 below. Take RDA as example, when the operation of the RDA is normal, shows

IOP Conf. Series: Earth and Environmental Science **86** (2017) 012036 doi:10.1088/1755-1315/86/1/012036

RDA ●, if abnormal shows RDA ● and audio alarm starts.

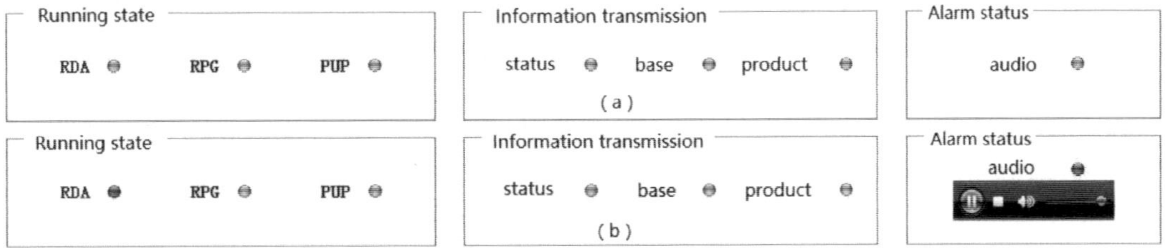

Figure 3 state monitoring of radar systems: (a) normal (b) abnormal

4.2.3 Parameter monitoring function. The monitoring platform makes real-time monitoring of the system parameters of the radar operation, including system parameters and radar transmitter power, noise temperature, syscal value, and show numerical values. To continuous wave (CW), velocity and spectrum width (phase and velocity) and other parameters, the monitoring platform not only shows the expected value and the measured value, and laso compared and made judgments. The difference of the measured value and the expected value set a threshold of 0.5 and 1.0, if the difference absolute value less than 0.5, shows●, between 0.5 and 1.0, judge as warning state, shows●, greater than 1.0 for the abnormal state, shows ●, else shows "null" and ●. The following figure 4, take monitoring parameter "CW" for example.

CW ●	CW ●	CW ●	CW ●
MEASURED=41.50	MEASURED=41.50	MEASURED=41.50	MEASURED=null
EXPECTED=41.79	EXPECTED=42.10	EXPECTED=42.60	EXPECTED=null

Fig. 4 monitoring parameter "CW" of radar system

4.2.4 Alarm information display function. When radar system has alarm information, in the alarm information displaying area will show them,else shows "no alarm".

5.Conclusion

The doppler weather SA-band radar integrated monitoring platform is in the form of web page display, for the first-line technical staff of radar judgment on radar system operating state and have a guiding role for maintenance. The monitoring platform can monitor various operating parameters of the radar system in real time, and the response time of the audio alarm is less than 60 seconds when the operation state is abnormal.

The monitoring platform is a stand-alone version of the platform currently, could only monitor one radar, if you want to control the other radar only need to change the corresponding parameters in the platform. And affected by the factors such as network circuit and transmission software instability, the monitoring platform will have a short time delay occasionally.

IOP Conf. Series: Earth and Environmental Science **86** (2017) 012036 doi:10.1088/1755-1315/86/1/012036

References

[1]Yu X D, Yao X P and Xiong T N 2006 The principle and applications of Doppler weather radar. *China Meteorological Press*, 5-10

[2]Wang Z W, Zhong T, Wang Z W and Lin Z N 2008 Calibration Method of Echo Intensity for China Doppler weather SA-band radar-S. *Modern radar*, **30** 30-33

[3]Huang X, Zhang P Y and Xiong Y 2005 Transmitter Phase Noise of Doppler Weather Radar Measurement. *Modern radar*, **07** 62-66.

Maximizing Hydropower Generation in Flood Control Operation using Preference Based Multi-objective Evolutionary Algorithm

Y T Qi [1,*],J G Luo[2], Y X Wei [1],Q G Miao [1]

[1]School of Computer Science and Technology, Xidian University, Xi'an, China.

[2]Institute of Water Resources and Hydro-Electric Engineering, Xi'an University of Technology, Xi'an, China

Email: ytqi@xidian.edu.cn

Abstract. Reservoir flood control operation (RFCO) is a challenging optimization problem with multiple objectives which are conflicting with each other. In this work, a tri-objective optimization model which maximizes hydropower generation during flood control is investigated and a multi-objective optimizer based on decomposition technique and the decision maker's preference information is developed. Experimental studies on typical floods at Ankang reservoir indicate the effectiveness of the proposed algorithm. It successfully provides preferred scheduling plans with more hydro- power generations than those considering the safety objectives only.

1.Introduction

Reservoir food control operation (RFCO) problem is a nonlinear and non-convex optimization problem. It involves multiple long-term and short-term objectives which are conflicting with each other [1]. During the process of food control, the safety of both upstream and downstream are of course the two major issues. Research works have been done to guarantee the flood control safety by minimizing the peak values of discharge water volume and the upstream water level simultaneously [2]. However, few efforts have been devoted to maximizing the hydropower generation in flood control operation. This work considers the objective of hydropower generation in flood control and investigates a tri-objective optimization model for RFCO problem.

As a consequence of therapid progress in the last few years in multi-objective optimization techniques [3], more and more research efforts have been devoted to optimizing the conflicting optimization objectives in RFCO problems simultaneously, instead of converting it into single objective optimization problem as before. Yu et al. developed a multi-objective fuzzy decision-making model for RFCO problem [4]. Kim et al. proposed a multi- objective optimizer for RFCO problem using genetic algorithm [5].Nagesh-Kumar et al. presented a particle swarm optimization algorithm to optimize the reservoir operation policies [6].Afshara et al. suggested a multi-objective optimizer for RFCO problem using ant colony optimization algorithm [7]. Li et al. developed a shuffled frog leaping algorithm for multi-objective RFCO problem [8]. Qin et al. proposed a bi-objective model for RFCO problem and developed a cultured differential evolution algorithm to solve it [2]. Qi et al. developed an immune inspired algorithm for multi-objective RFCO problem [9]. Luo et al. suggested a hybrid multi-objective algorithm for RFCO problem by combing the particle swarm optimization and the estimation of distribution algorithm together [10].

Despite the success of multi-objective optimization algorithms for RFCO problem, almost all of them were developed with the aim of obtaining a set of trade-off solutions that approximates the entire Pareto front (PF) which is the set of all the best trade-off solutions. Few efforts have been devoted to incorporate the decision maker's (DM) preference into multi-objective optimizers for RFCO problems, although the preference based multi- objective optimization is not brand new in the community of multi-criteria decision making [11]. According to our previous investigations [9][12], multi-objective RFCO problems have irregularly shaped

PFs which pose a big challenge to multi-objective optimization algorithms [13]. Moreover, in multi-objective RFCO problems, the difficulties of obtaining solutions at different PF regions could be various. In this case, multi-objective optimizers can hardly obtain a set of non-dominated solutions with good enough coverage on the PF. Instead, algorithms can converge to different PF regions on different flood instances without control. Such situation becomes even worse when solving a tri-objective optimization model for RFCO problem which considers hydropower generation in flood control.

Since it is challenging and unnecessary to approximate the entire PF in multi-objective RFCO problems, a preference based optimizer which obtains trade-off solutions at the preferred PF region could be a promising substitute. In our previous work, we considered the final upstream water level (FUWL) preference in bi-objective RFCO problem and developed a preference based selection mechanism for immune inspired multi-objective optimizer [14]. On the other hand, we developed preference based multi-objective optimization algorithm using decomposit- ion technique and applied it to solve bi-objective RFCO problem. In this algorithm, the DM's preference information is expressed by using the light beam search preference model [17] in the objective space. However, in RFCO problem, the FUWL preference is implicit, it is difficult to articulate in the objective spaceby using existing preference representation techniques. This paper extends our previous work and contributes in three aspects. 1) The objective of hydropower generation is considered in flood control and a tri-objective optimization model for RFCO problem is developed.2) A preference modeling method which converts the FUWL preference into a preferred region in the objective space is designed. 3) According to the preferred region which is dynamically determined by the proposed preference modeling method, a preference based multi-objective evolutionary algorithm is developed for RFCO problem.

The remainder of this paper is organized as follows. Section 2 describes the tri-objective optimization model for RFCO problem. Section 3 presents the preference modeling method for presenting the FUWL preference in RFCO problem. Section 4 describes the flowchart of the proposed algorithm. Section 5 verifies the effectiveness of the proposed algorithm. Section 6 conclusions this paper.

2.Multi-objective Model for RFCO Problem

Taking thehydropower generation and the FUWL preference into consideration, this work investigates the following tri-objective optimization model for RFCO problem.

$$
\begin{aligned}
&\text{Minimize} \quad F(Q) = (f_1(Q), f_2(Q), f_3(Q)) \\
&\quad f_1(Q) = \max(Z_t) \qquad t = 1, 2, \ldots, T \\
&\quad f_2(Q) = \max(Q_t) \qquad t = 1, 2, \ldots, T \\
&\quad f_3(Q) = 1/E \\
&\text{Subject to:} \\
&\quad Z_{\min} \le Z_t \le Z_{\max} \\
&\quad 0 \le Q_t \le Q_{\max} \\
&\quad V_t = V_{t-1} + I_t - Q_t \\
&\text{Preference} \\
&\quad Z_T \to Z_{FL}
\end{aligned}
\tag{1}
$$

where $Q = (Q_1, Q_2, \ldots, Q_T)$ is the water release volumes at T scheduling periods. Each Q_t (t=1, 2,...,T) has a non- negative value no larger than $Q_{\max}$. Z_t is the upstream water level of the t-th scheduling period,it has a value between $Z_{\min}$ and $Z_{\max}$. V_t and I_t are respectively the reservoir storages and the reservoir's inflow volume of the t-th scheduling period. Z_{FL} isthe final target upstream water level at the end of the scheduling. E is the hydropower generation, it can be calculated as following.

$$E = \sum_{t=1}^{T} (N_t \cdot \Delta t) \quad (2)$$
$$N_t = K \cdot Q_t \cdot H_t$$

where Δt is the time interval of the scheduling period. K is the efficiency coefficient. Q_t and H_t are respectively the water release volume and the water head of the t-th scheduling period. The average output N_t has a lower bound $N_{\min}$ and an upper bound $N_{\max}$.

In the optimization model in equation (1), $f_1(Q)$ means the highest upstream water level. It should be minimized to guarantee the safety of the upstream side. $f_2(Q)$ is the largest water release volume. We minimized it to protect the downstream side. $f_3(Q)$ is the reciprocal of hydropower generation. We convert the maximization of hydropower generation during flood into minimization of its reciprocal.

3. Presentation of DM's Preference

The proposed preference modeling method converts the FUWL preference into a preferred region in the objective space. At each iteration of the multi-objective optimization algorithm for RFCO problem, the individuals whose final upstream water levels lie in between $Z_{FL} - Z_{PT}$ and $Z_{FL} + Z_{PT}$ are identified from the evolving population to form the preferred solution set S^P. According to S^P, the preferred region P^R can be determined by a middle point $M = (M_1, M_2, M_3)$ and a veto threshold vector $V = (V_1, V_2, V_3)$ as following.

$$P^R = \left\{ Q \middle| \left| f_1(Q) - M_1 \right| \le V_1 \ and \ \left| f_2(Q) - M_2 \right| \le V_2 \right\} \quad (3)$$

$$M = \left\{ F(Q) \middle| \min \left\{ \left| FL(Q) - Z_{FL} \right| \right\}, Q \in S^P \right\} \quad (4)$$

$$V = \left\{ (V_1, V_2, V_3) \middle| V_1 = 1.2\hat{V}_1, V_2 = 1.2\hat{V}_2, V_3 = \infty \right\} \quad (5)$$

$$\hat{V}_i = \max \left\{ \left| f_i(Q) - M_i \right| \right\}, Q \in S^P, i = 1, 2 \quad (6)$$

As shown inequation (3), the preferred region is defined based on the first two objectives, because the objective value of hydropower generation is the larger the better. In equation (4), $FL(Q)$ denotes the final upstream water level of the individual Q. The middle point is defined as the objective vector of the individual whose final upstream water level is the closest to Z_{FL}. The veto threshold vector extends the calculated result $\hat{V}$ by 20%. The motivation behind this extension is to provide better coverage of the preferred region.
Based on the definition of the preferred region, a measure of preference level, denoted asPL(Q), is defined on each individual Q to rank the preference degree of individuals.

$$\mathrm{PL}(Q) = \sqrt{\sum_{i=1}^{2} \left(\frac{f_i(Q) - M_i}{V_i} \right)^2} \quad (7)$$

4.The Proposed Algorithm

Due to its simplicity and efficiency, the multi-objective evolutionary algorithm based on decomposition (MOEA/D) has achieved a great success and attracted a lot of attention [17]. MOEA/D decomposes the target multi-objective optimization problem (MOP) intoa number of scalar optimization sub-problems by employing an aggregation approach with a set of evenly scattered weight vectors. And then, these decomposed sub- problems are solved simultaneously in a collaborative manner by using an evolutionary algorithm.
Given the target MOP with$m$ objectives $f_1(\mathbf{x}), \ldots, f_m(\mathbf{x})$ and a weight vector $\boldsymbol{\lambda} = (\lambda_1, \ldots, \lambda_m)$, ($\sum_{i=1}^{m} \lambda_i = 1$, $\lambda_i \ge 0$,

$i = 1,...,m$) the weighted Tchebycheff aggregation approach decomposes the target MOP into the following scalar optimization problem.

$$\min_{\mathbf{x}\in\Omega} g^{tc}(\mathbf{x} \mid \boldsymbol{\lambda}, \mathbf{z}^*) = \min_{\mathbf{x}\in\Omega} \max_{1\le i\le m}\{\lambda_i \times \mid f_i(\mathbf{x}) - z_i^* \mid\} \quad (8)$$

In which, $\mathbf{z}^* = (z_1^*,...,z_m^*)$ ($z_i^* < \min_{x\in\Omega}\{f_i(x)\}$,i= 1, … ,m) is the ideal point. It has been proved that each optimal solution to equation (8) is a Pareto optimal solution to the target MOP. For each Pareto optimal solution $\mathbf{x}^*$,there must be a scalar optimization problem with weight vector $\boldsymbol{\lambda}^*$ which takes $\mathbf{x}^*$ as the optimal solution.

In our previous work, we have investigated the geometric relationship between sub-problems' weight vectors and their optimal solutions. We have proved that, if the vector $\boldsymbol{\lambda}'=(1/\lambda_1,...,1/\lambda_m)$ starting from $\mathbf{z}^*$ has an intersection with the PF of the target MOP, then the intersection point is the optimal solution of the scalar optimization sub- problem with weight vector $\boldsymbol{\lambda}=(\lambda_1,...,\lambda_m)$. We have also defined the WS-transformation which is a map from $\boldsymbol{\lambda}$ to $\boldsymbol{\lambda}'$ as following [13].

$$\lambda' = \mathrm{WS}(\lambda) = \left(\frac{1/\lambda_1}{\sum_{i=1}^{m} 1/\lambda_i}, ..., \frac{1/\lambda_m}{\sum_{i=1}^{m} 1/\lambda_i} \right) \quad (9)$$

It should be noted that the WS-transformation is self- inverse (i.e. $\lambda = \mathrm{WS}(\lambda') = \mathrm{WS}(\mathrm{WS}(\lambda))$). Based on this property, if we want to obtain some non-dominated solutions located within preferred PF region, we can generate some scalar optimization sub- problems with specific weight vectors which take those preferred solutions as optimal solutions. That is the basic idea of the preference based MOEA/D (p-MOEA/D) we developed in our previous work [15]. p-MOEA/D guides the search of MOEA/D towards the preferred PF region by removing scalar sub-problems from unwanted PF areas and add new ones into the preferred region.

This work inherits the preference based weight adjustment strategy in p-MOEA/D, design a new preference modeling method for representing the FUWL preference in tri- objective RFCO problem, and develop a preference and decomposition based multi-objective optimizer for RFCO problem (PD-RFCO). At each iteration, PD-RFCO maintains the following. 1) An evolving population with N individuals $pop = \{ind^1,...,ind^N\}$, each ind^i records the current solution $\mathbf{x}^i$ to the$i$-th scalar sub-problem and its objective values FV^i . 2) A set of N weight vectors $\Lambda = \{\boldsymbol{\lambda}^1,...,\boldsymbol{\lambda}^N\}$ and a neighborhood list for each sub-problem $\mathbf{B} = \{B(1),...,B(N)\}$. 3) An estimated ideal point $\mathbf{z}^*$. 4) An external population EP.The pseudo code of the proposed algorithm is as follows.

***Algorithm* 1**: PD-RFCO
Input: Population size N, external population size N^E, neighborhood size T,initial middle point M^0 and veto threshold vector V^0, maximal number of adjusted scalar sub-problemsN^{WA}, iteration interval of weight adjustment I^{WA}, and the maximal function evaluation number $\mathrm{FE}_{\max}$.
Output: $\{\mathbf{x}^1,...,\mathbf{x}^N\}$ and $\{FV^1,...,FV^N\}$
1. $Initialization(\Lambda, \mathbf{B}, pop, \mathbf{z}^*, EP, M, V)$;
2. *Gen*=0;
3. ***While*** function evaluations <$\mathrm{FE}_{\max}$***do***
4. $\mathrm{Evolving}(pop, \mathbf{z}^*)$;
5. $\mathrm{UpdateEP}(EP, pop, N^E)$;
6. $\mathrm{UpdatePreferredRegion}(EP, M, V, S^P)$;

7. ***If*** $|EP| > N$ and $|S^P| > 2$ and$Gen \bmod I^{WA} = 0$ ***do***
8. RemoveSubproblems$\left(N^{WA}, M, V, pop, \Lambda, \mathbf{B}\right)$;
9. AddNewSubproblems$\left(N^{WA}, M, V, pop, \Lambda, \mathbf{B}\right)$;
10. ***End If***
11. *Gen++;*
12. ***End While***
13. ***Return*** $\{\mathbf{x}^1, ..., \mathbf{x}^N\}$ and $\{FV^1, ..., FV^N\}$

In algorithm1, the initialization step (line 1) and the evolving step (line 4) are exactly the same as those in MOEA/D [17]. In this work, the simulated binary crossover and the polynomial mutation are employed to generate new individuals. The ideal point is also updated in the evolving step.

In the UpdateEP step (line 5), non-dominated solutions are first identified from the union set of current *EP* and *pop*, giving rise to a temporary set of *NS*. If the size of *NS* is no larger than N^E, then copy *NS* to *EP*. Otherwise, calculate the preferencelevel of each individual in *NS* using to equation (7), and update *EP* by the individuals with the first N^Esmallest preference levels in *NS*.

In the UpdatePreferrdRegion step (line 6), the preferred individuals with preference levels smaller than 1 are first identified from EP to form the population SP. If SP contains more than two individuals, then update the middle point M and the veto threshold vector V according to equations (4) and (5-6) respectively.

In the RemoveSubproblems step (line 8), N^{WA} scalar sub-problems are removed from the evolving population. It chooses the scalar sub-problems that will be removed according to their preference neighbor distances. The ones locate outside the preferred PF region will be selected in priority. However, when preferred solutions will have to be removed, the ones with smaller crowding distances will be selected in the first place.

In the AddNewSubproblems step (line 9), N^{WA} individuals in the external population will be recalled into the evolving population and new weight vectors will be generated and added to the sub-problem set. On the contrary to the RemoveSubproblems step, the preferred solutions with smaller crowding distances will be selected in priority. When there are not enough preferred solutions, the ones with smaller preference neighbor distances will be selected in the first place.

5.Experimental Studies

In this section, typical floods at Ankang reservoir on the Hanjiang river in Shanxi Provence of China are investigated. The Ankang reservoir has amaximum water capacity of 2.585 billion cubic meters, a normal water level of 330 meters, a flood control limit level of 325 meters, a dead water level of 300 meters and a designed discharge capacity of 37474 cubic meters per second. The annual hydropower generation of Ankang station is about 2.857 billion KWH.

The parameters of the proposed PD-RFCO are set as follows. The evolving population size *N* is set to 100 and the external population size N^E is set to 250. The neighborhood size *T* is set to 10. The initial middle point M^0=(325,10000,-) and the veto threshold vectorV^0=(4, 1000,-). The maximal number of adjusted scalar sub- problems N^{WA} is set to 20. The iteration interval of weight adjustment I^{WA} is set to 500. The maximal function evaluation number FE_{max} is set to 10^6 times. The final target upstream water level at the end of the scheduling Z_{FL} is set to 325. Thepreference threshold Z_{PT} is set to 2.

IOP Conf. Series: Earth and Environmental Science **86** (2017) 012037 doi:10.1088/1755-1315/86/1/012037

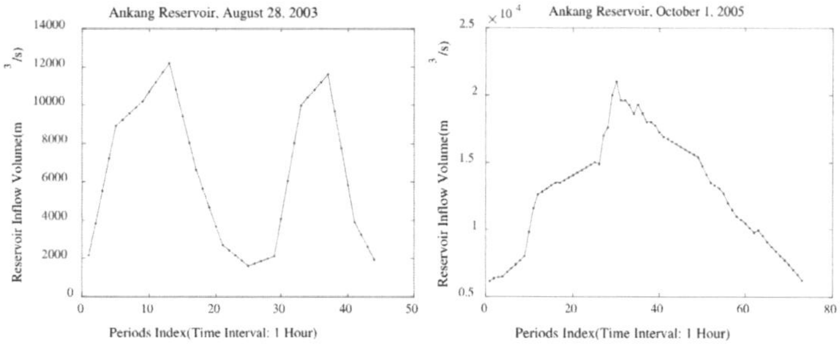

Figure 1.Forecasted inflow of investigated floods.

Figure 1 is the forecasted water inflow volumes of the investigated floods. The flood on August 28, 2003 (left) lasts for 44 hours, it has two flood peaks and relatively low maximum inflow volumes per second. The flood on October 1, 2005 (right) lasts for 73 hours, it hasone higher flood peak and larger total water volumes.

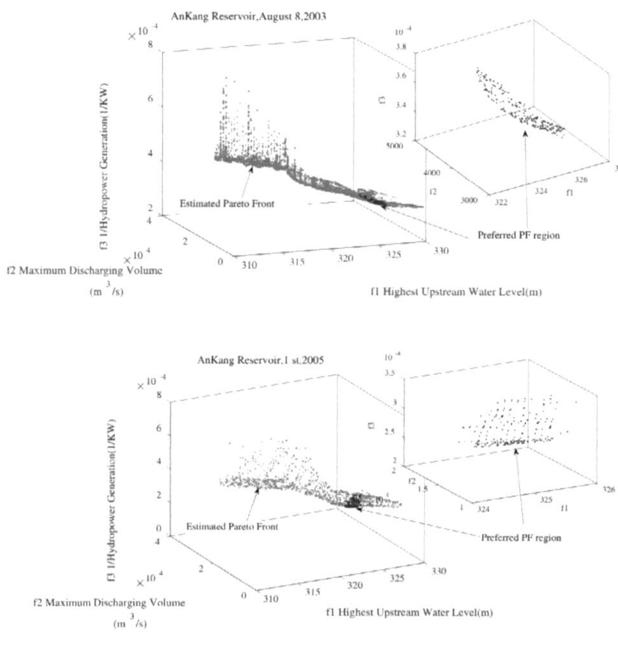

Figure 2.Obtained non-dominated solutions.

Figure 2 illustrates the Pareto optimal solutions obtained by the proposed PD-RFCO. The estimated Pareto fronts in this figure are obtained by running MOEA/D-AWA [13] with 6,000,000 function evaluations over 30 runs. It can be seen that the estimated PFs are complex in shape for both of the two flood instances. PD-RFCO successfully converge to a preferred region on the estimated PF. When looking at the distribution of the obtained non-dominated solutions within the preferred region, it has a much simpler shape comparing with the entire PF. Since the PF shape has a significant influence on the difficulty of a MOP, it is very challenging for a multi-objective optimizer to approximate the entire PF of the tri-objective RFCO problem. If we focus the search effort of the algorithm on a sub-region which is less complex in shape, the problem becomes easier. In addition, it is unnecessary to obtain non-dominated solutions that cover the entire PF, because only the ones within the preferred region are likely to be put into practice. Therefore, computing

IOP Conf. Series: Earth and Environmental Science **86** (2017) 012037 doi:10.1088/1755-1315/86/1/012037

efforts are more efficiently used the proposed PD-RFCO.

Figure 3 is the discharge volumes of the non-dominated solutions obtained by PD-RFCO. It can be seen that the scheduling schemes have maximum discharging volumes of less than 6000 m^3/s and 15000 m^3/s respectively for the two investigated floods, which are much lower than the peak inflow volumes of the two floods, say 12200 m^3/s and 21000 m^3/s respectively. Therefore, the proposed PD-RFCO successfully provides scheduling schemes that reduce the flood peak significantly.

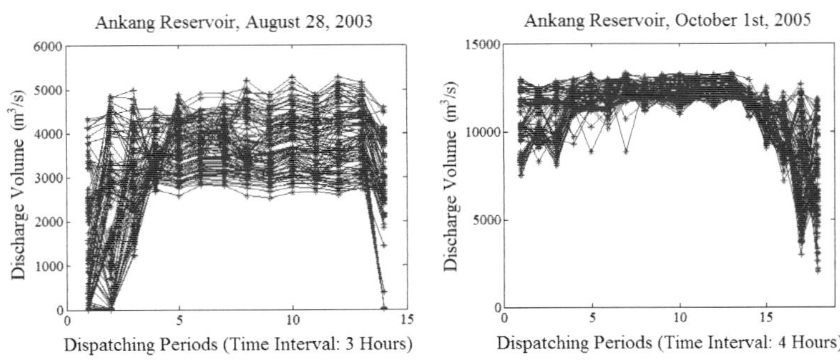

Figure 3.Discharge volumes of obtained solutions.

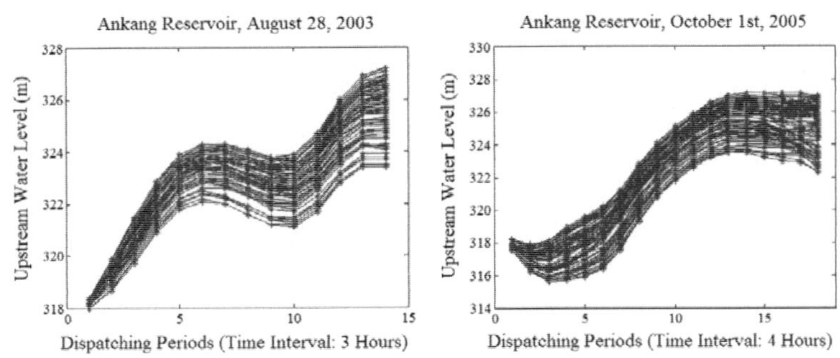

Figure 4.Upstream water levels of obtained solutions.

Figure 4shows the upstream water levels of the non- dominated solutions obtained by PD-RFCO. As shown in this figure, the final upstream water levels of the curves locate evenly within 2 meters above and below the flood control limit level of 325 meters as expected. These results indicate that PD-RFCO successfully obtains a set of preferred non-dominated solutions which evenly scattered over the specified preferred region.

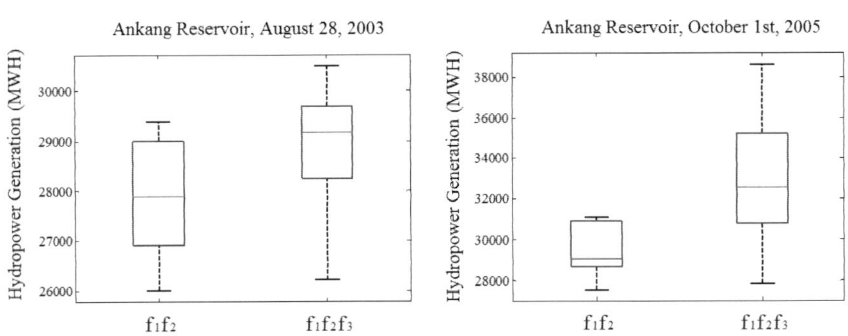

Figure 5.Comparisons on hydropower generation.

Experimental studies have been conducted to make a comparison between the tri-objective optimization model for RFCO in equation (1) and the existing bi-objective optimization model which does not consider hydropower generation and minimizes $f_1(Q)$ and $f_2(Q)$ in equation (1) only. Figure 5compares the hydropower generation of the non-dominated solutions obtained by PD-RFCO for solving the two optimization models respectively. In the box plots in this figure, the bottom and top of the box are respectively the first and third quartiles. The band inside the box is the second quartile, say the median. The ends of the whiskers represent the minimum and maximum of all the hydropower generation data. It can be seen that the boxes for the tri-objective model (right) are higher than those for the bi-objective model (left). This indicates the fact that by considering the maximization of hydropower generation in flood control, the optimizer can provide scheduling schemes with more hydropower generations.

6. Conclusions

In this work, a tri-objective optimization model which maximizes hydropower generation in flood control is developed for reservoir flood control operation (RFCO) problem. Considering the final upstream water level (FUWL) preference in RFCO problem, a new preference modeling method is designed by convertingthe FUWL preference into a preferred region in the objective space. The preference model is then incorporated into the algorithmic framework of the decomposition based multi- objective optimization algorithm (MOEA/D), giving rise to a preference and decomposition based multi-objective optimizer for RFCO problem (PD-RFCO).

Experimental studies have been done on two typical floods at Ankang reservoir to verify the effectiveness of the proposed PD-RFCO. Experimental results indicate that PD-RFCOobtains a set of non-dominated solutions which evenly scattered over the specified preferred region. And the scheduling schemes provided by PD-RFCO reduce the flood peak significantly. Comparison studies on multi-objective optimization models which consider and not consider the objective of hydropower generation have also been conducted. The results indicate that by considering the maximization of hydropower generation in flood control, the optimizer can provide scheduling schemes with more hydropower generations.

7.Acknowledgments

This work was supported by the National Natural Science Foundation of China under Grant Nos. 61303119 and 61202040, the Science and Technology Program of Shaanxi Province under Grant Nos. 2014K09-07 and 2015KJXX-30.

References

[1] MalekmohammadiB, ZahraieB and KerachianR 2011Ranking solutions of multi-objective reservoir operation optimization models using multi-criteria decision analysis. *Expert Systems with Applicat- ions, Vol. 38*, 7851-63.

[2] QinH., ZhouJ, LuY, Li Yand ZhangY 2010 Multi-objective cultured differential evolution for generating optimal trade-offs in reservoir flood control operation.*Water Resources Management, Vol. 24*, 2611- 32.

[3] ZhouA, QuB, LiH, ZhaoS, SuganthanPN and ZhangQ 2011 Multiobjective evolutionary algorithms: A survey of the state of the art.*Swarm and Evolutionary Computation, Vol. 1, No.1*32-49.

[4] YuY, WangB, WangG and LiW 2004 Multi-person multiobjective fuzzy decision-making model for reservoir food control operation. *Water Resources Management, Vol.18, No.2*111-124.

[5] Kim Tand HeoJ 2006 Application of multi-objective genetic algorithms to multi-reservoir system optimizat- ion in the han river basin. *KSCE Journal of Civil Engineering, Vol.10, No.5* 371-380.

[6] Nagesh-Kumar Dand Janga-ReddyM 2007 Multi- purpose reservoir operation using particle swarm optimizatio.*Journal of Water Resources Planning and Management, Vol.133, No.3*192-201.

[7] AfsharaA, SharifiabF and JalaliabMR 2009 Non- dominated archiving multi-colony ant algorithm for multi-objective optimization: Application to multi- purpose reservoir operation. *Engineering Optimizat- ion, Vol. 41, No.4*313-325.

[8] LiYH, ZhouJ Z, ZhangY C, Qin Hand LiuL 2010 Novel multi-objective shuffled frog leaping algorithm with application to reservoir food control operation. *Journal of Water Resources Planning and Management, Vol.136, No.2*217-226.
[9] QiY, LiuF, LiuM, GongM and JiaoL 2012 Multi-objective immune algorithm with baldwinian learning.*Applied Soft Computing, Vol.12, No.8*2654-74.
[10] LuoJ, QiY, XieJ andZhangX 2015 A hybrid multi-objective PSO-EDA algorithm for reservoir flood control operation. *Applied Soft Computing, Vol. 34*526–538.
[11] RachmawatiL and SrinivasanD 2006 Preference incorporation in multi-objective evolutionary algorithms: A survey.*IEEE Congress on Evolutionary Computation*962-968.
[12] QiY, BaoL, SunY, LuoJ and MiaoQ 2014 Amemetic multi-objective immune algorithm for reservoir flood control operation.*Water Resources Management. Published online at 18 April.*
[13] QiY, MaX, LiuF, JiaoL, SunJ and WuJ 2014 MOEA/D with adaptive weight adjustment. *Evolutionary Computation, Vol.22, No.2*231–264.
[14] LuoJ, ChenC and XieJ 2015 Multi-objective immune algorithm with preference-based selection for reservoir food control operation. *Water Resources Manage- ment, Vol.29, No.5*1447-66.
[15] MaX, LiuF, Qi Y and et al 2015 MOEA/D with biased weight adjustment inspired by user preference and its application on multi-objective reservoir flood control problem. *Soft Computing, Published online at18 July.*
[16] DebK. andKumarA 2007 Light beam search based multi-objective optimization using evolutionary algorithms.*IEEE Congress on Evolutionary Computation* 2125-2132.
[17] ZhangQ and LiH 2007 MOEA/D: A multiobjective evolutionary algorithm based on decomposition. *IEEE Transactions on Evolutionary Computation, Vol.11, No.6* 712-731.

Allocation of Load-Loss Cost Caused by Voltage Sag

X Gao

State Grid Energy Research Institute, Beijing 102209, P.R.China

Email: gaoxiao@sgeri.sgcc.com.cn

Abstract. This paper focuses on the allocation of load-loss cost caused by voltage sag in the environment of electricity market. To compensate the loss of loads due to voltage sags, the load-loss cost is allocated to both sources and power consumers. On the basis of Load Drop Cost (LDC), a quantitative evaluation index of load-loss cost caused by voltage sag is identified. The load-loss cost to be allocated to power consumers themselves is calculated according to load classification. Based on the theory of power component the quantitative relation between sources and loads is established, thereby a quantitative calculation method for load-loss cost allocated to each source is deduced and the quantitative compensation from individual source to load is proposed. A simple five-bus system illustrates the main features of the proposed method.

1.Introduction

According to IEEE Standard 1159-1995 [1], a voltage sag is defined as an rms variation with a magnitude between 10% and 90% of nominal and a duration between 0.5 cycles and one minute. In distribution networks, voltage sag is an increasingly important issue to electricity consumers at all levels of voltage. It causes malfunction of equipments and economic losses of consumers. A serious voltage sag at the terminals of customer may lead to machine/process downtime, scrap cost, clean up costs, product quality and repair costs. For example, voltage sags can cause the most sensitive equipments like ASD's, PLC's to trip, thus affecting industrial production process leading to revenue loss. The possible action for customers to decrease the effects of voltage sag is to purchase auxiliary equipments like UPS, adapt the producing process, and so on. The total cost of all these actions is called load-loss cost, which contribute to make voltage sags costly to the customers. It is obviously unreasonable that customers undertake the total load-loss cost themselves under electricity markets. So who are the objects responsible for the load-loss cost and how to quantitatively allocate this cost to the objects are urgent problems in electricity markets.

Although no indices are part of any standard document, a set of voltage sag indices are widely used, especially by utilities. The most commonly referred index is the System Average RMS variation Frequency Index (SARFI) [2], which provides a count or rate of voltage sags, swells, and interruptions for a system. The sag energy method of characterization gives total "lost energy" on a bus, such as Sag Energy Index (SEI), which is the sum of the voltage sag energies for all qualified events at a given bus during a given period [3]. For three-phase systems the sag energy is added for the three phases. Some other methods for the evaluation of voltage sag are proposed in reference [4]-[9]. However, no direct index is available for the customers' loss due to voltage sag yet.

In this paper, the quantitative evaluation of voltage sag is illustrated at first. Then the objects responsible for load-loss cost and the allocation principle are analyzed and proposed. What is more, the load-loss cost allocated to each load is determined by the quantitative evaluation index of voltage sag. At last, each source compensate for load-loss cost to load based on the complex-power relation between sources and loads.

2.Economic Evaluation of Voltage Sag

Voltage sags may have significant economic effects on some facilities. Based on the voltage sag indices the average LDC [10], an evaluation index load-loss cost is identified to quantify the economic effects from voltage sag. In power systems, load customers are usually divided into four types: resident, commercialist, small industries and large industries. LDC indicates the average cost of one type of customer at the same bus when the expected demand affected by sags is 1kW. LDC is associated with actions that customer take on their own behalf to reduce the effects of voltage sags. Thus the load-loss cost of one type of customer can be expressed as:

$$\sigma = LDC \cdot P \qquad (1)$$

Where, σ is the load-loss cost, P is the expected demand affected by sags of the customers of the same type. Because the LDC is the average cost of one type of customer, the load-loss cost calculated in this way is imprecise. The total customers' load-loss cost of one bus is the sum of the cost of four types of customers. Of course, the total cost of a power system is the sum of the cost over all buses in this system.

3.Load-loss Cost Allocation Rules

In our opinion, a load-loss cost should satisfy the following two rules. First, the payers responsible for the load-loss cost are defined; second, how much each entity has to pay is identified.
In liberalized power markets, an unbundled power supply utility is the owner of transmission and distribution grids and is responsible for the power quality service. The power is provided by the generators, then transmitted by the transmission network and distributed by the distributed system, at last delivered to the loads. In this process, voltage sag can be caused by fault conditions within the plant or power system and lasts until the fault is cleared by protective device. So each entity can cause voltage sags. The utility owning the grids is also one source of voltage sags. Obviously, all the probable sources causing voltage sag should be responsible for load-loss cost. Therefore, the payers responsible for the load-loss cost are the independent generators, the utility owning distribution and transmission grids and the loads. The first two entities serve loads together.
How much each entity should pay for the load-loss cost is a complicated problem concerning circuit and economics theory. Three relevant rates are defined. One, denoted by λ_u denotes the rate at which utility owing grids is charged for the load-loss cost. The second, denoted by λ_s denotes the rate at which generators are charged for the load-loss cost. The third, denoted by λ_l denotes the rate at which loads are charged for the load-loss cost. It is difficult to assign financial responsibility for voltage sag to utility, generators and loads exactly since all of them maybe the source of voltage sag. Thus the load-loss cost could be allocated to entities by their contribution to voltage sags. The rates can be calculated as follows.

$$\lambda_u = \frac{t_u}{t_u + t_s + t_l}$$

IOP Conf. Series: Earth and Environmental Science **86** (2017) 012038 doi:10.1088/1755-1315/86/1/012038

$$\lambda_s = \frac{t_s}{t_u + t_s + t_l}$$

$$\lambda_l = \frac{t_l}{t_u + t_s + t_l}$$

Where, t_u is the total number of sags caused by utility, t_s by all generators, and t_l by all loads. The entity that causes more voltage sags would be charged more for load-loss cost by this method. It is reasonable and predictable. But these equations only give a simple way to determine the three rates. In the actual network operation, more factors might effect the determination of them. The factors include the amount of equipments sensitive to voltage sags, the sensitivity of loads, the reliability of grids, and so on.

Another effective method to obtain the rates is authority survey. The questionnaires are designed to seek the respondent's opinion and judgement on issues about these rates. According to statistic and analysis on the response, a set of reasonable rate can be determined. This method avoids the errors in the calculation of rates. In our regard, the unbundled power supply utility shoud be charged for most of the load-loss cost, then loads for less, and sources for the least. λ_u value is set to 0.5, λ_s is 0.15, and λ_l is 0.35. The reasons are as follows. One, the utility is responsible for the voltage support service in liberalised power markets round the world. Two, those facilities improving voltage quality are attached to the responsibility of the transmission and distribution utilities. Three, usually voltage sags are mostly caused by faults in electric transmission and distribution grids, such as short-circuits, switchings, lightings. Four, loads are anther major source causing voltage sags, such as large motor starting and load switching. Five, power sources can cause voltage sags in case of internal short-circuit.

Consequently, the allocating rules about load-loss cost are proposed. The load-loss cost of individual load is divided into three parts. The load is charged for λ_l of the load-loss cost by itself, the utility owing grids is charged for λ_u, and all generators are charged for λ_s together. The load-loss cost allocated to generators or load is less than that to utility. And this case coincides with the physical fact that the overall contribution of transmission and distribution grids to voltage sags is much more than generators or loads. The cost each entity responsible for is also reasonable in economical benefit.

4.Allocation of Load-loss Cost to Customers

In the power grid, the loads at individual buses are different in composition. So the load vulnerability and LDC at buses are not the same, and this lead to the difference of load-loss cost among each bus. The bigger the LDC at a bus, the bigger the load-loss cost of customer at this bus. The above analysis focuses on a particular type of customer. Likewise, the four types of customers can obtain their own load-loss cost. If residential, commercial, small industrial and large industrial customers at a bus obtain the load-loss cost σ_R, σ_C, σ_{SI} and σ_{LI}, the load-loss costs that these customers should be charged for by themselves are $\lambda_l\sigma_R$, $\lambda_l\sigma_C$, $\lambda_l\sigma_{SI}$ and $\lambda_l\sigma_{LI}$ respectively. Among the same type of customers, each should be charged for λ_l of its load-loss cost by itself.

5.Allocation of Load-loss Cost to Sources

5.1 Allocation Method. The sources for a distribution grid commonly include independent generators and upgrade systems in the system. If a generator is the only source for the loads in the distribution grid with one power injecting bus, this generator should be charged for λ_s of the total load-loss cost of the distribution grid alone. When two or more sources connected to the distribution system, λ_s of the load-loss cost should be allocated to each source.

In this case, how to accurately allocate the total load-loss cost of all customers at one bus to each individual source is the key problem in this cost allocation. Therefore, this paper begins the analysis with a simple power system. For the simplicity of analysis, the steady-state model is used in the discussion. The load is represented by constant MVA. And annual average loads are used for cost allocation. All the customers connected to one load bus are considered as one equivalent load. The demand of an equivalent load is equal to the sum of total demand of all loads connected to this bus. According to the theory of complex power components incurred by individual generators [11, 12], the allocation procedure is as follows. Firstly, the power that one source provides for one equivalent load is calculated. Secondly, the relation between source and load connected to a bus is quantified. At last, λ_s of the load-loss cost of all customers at each bus is allocated to individual sources according to the above quantitative relation.

5.2 Power Contribution of A Source to A Load. A 3-bus distribution system in Figure 1 consisting of two sources and one load is used in this section. An independent generator is connected to the end of the distribution network, which connects to another source S via two lines. The impedance of each line and the power from each source to the network are also shown in Figure 1. The equivalent load at bus 2 is $P_L + jQ_L$ (MVA). Bus 3 is selected as slack bus, its voltage is $U_3\angle 0$ kV.

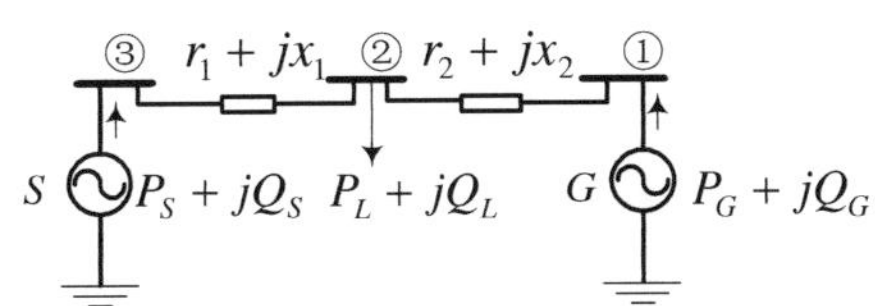

Figure 1.3-bus distribution system

We denote the power flowing from source S to the distribution network by $P_S + jQ_S$ (MVA), the voltages at buses 1 and 2 by $\boldsymbol{U}_1 = U_1\angle\theta_1$ and $\boldsymbol{U}_2 = U_2\angle\theta_2$, respectively. The total current injecting into the load at bus 2 is

$$\boldsymbol{I}_L = ((P_L + jQ_L)/\boldsymbol{U}_2)^* = a_L + jb_L$$

This current is supplied by the source S and the independent generator together. Therefore the equivalent impedance of the load is

$$\boldsymbol{Z}_L = \boldsymbol{U}_2 / \boldsymbol{I}_L = r_L + jx_L$$

In the calculation of the power relation of one source to a load, we suppose that only the considered source work. So the power contribution of this single source to load is determined. Assuming that bus 3 disconnects to the source S, only the independent generator G will supply power to the load in this distribution network. In this case, the response current of $\boldsymbol{Z}_L$ is

$$\boldsymbol{I}_G = ((P_G + jQ_G)/\boldsymbol{U}_1)^* = a_G + jb_G$$

Based on the complex power components incurred by individual sources, the active power delivered to the load and withdrawn from the generator is

$$P_{GL} = r_L(\boldsymbol{I}_G \cdot \boldsymbol{I}_L) = r_L(a_G a_L + b_G b_L) \qquad (2)$$

The linear admittance model of network is simple and unambiguous. Although this relation is based on linear network, it illustrates the power contribution of the source to the load directly.
Likewise, the active power delivered to the load and withdrawn from source S is

$$P_{SL} = r_L(\boldsymbol{I}_S \cdot \boldsymbol{I}_L) = r_L(a_S a_L + b_S b_L) \qquad (3)$$

Where $\boldsymbol{I}_S$ is the response current of $\boldsymbol{Z}_L$ when the distribution network connects to the source S and disconnects to generator G. And it is expressed as follows:

$$\boldsymbol{I}_S = ((P_S + jQ_S)/\boldsymbol{U}_3)^* = a_S + jb_S$$

5.3 Compensation to Loads by Each Source. Supposing that there are *n* sources in an *m*-bus system and source G_i (i=1,2,…,n) (including generator, transmission grid and adjacent distribution grid) supplies active power P_{GLij} to the equivalent load L_j (j=1,2,…,m). Given that only the source G_i supplies power to loads and other sources open, the circuit then contains only source G_i and all equivalent impedances of loads, the current through an equivalent impedance from source G_i is determined by KCL and KVL. The active power to the same load from G_i is also obtained according to equations (2) and (3). At a load bus, λ of the sum of load-loss cost of all customers at load bus L_j is allocated to individual sources in proportion to P_{GLij}. We denote by η_{ij} the ratio of load-loss cost charged by the source G_i, then we have

$$\eta_{ij} = P_{GLij}/(\sum_{h=1}^{n} P_{GLhj}) \qquad (4)$$

Provided that the sum of load-loss cost of all customers at bus *j* is A_j, source G_i takes the share of $\lambda_s\eta_{ij}A_j$. In another word, G_i should pay $\lambda_s\eta_{ij}A_j$ to the equivalent load L_j for the load-loss cost. As for the whole system, the load-loss cost paid by one source is equal to the sum of the cost paid by this source at each bus in this system. So after obtaining the cost that one source pays at each bus, the total cost that this source pays can be determined. We denote by B_i the part of cost G_i is responsible for.

$$B_i = \sum_{h=1}^{m} \lambda_s \eta_{ih} A_h \qquad (5)$$

The total amount of load-loss cost charged by G_i is B_i. Then all this money should compensate to each load according to the LDC.

6. A 5-bus Example

A simple 5-bus power system shown in Figure 2 is employed in this section. An independent

IOP Conf. Series: Earth and Environmental Science **86** (2017) 012038 doi:10.1088/1755-1315/86/1/012038

generator G_1 is at the end of this network, connecting with another source G_2. Between G_1 and G_2 four lines connected in series. All parameters of the system are shown in Figure 2.

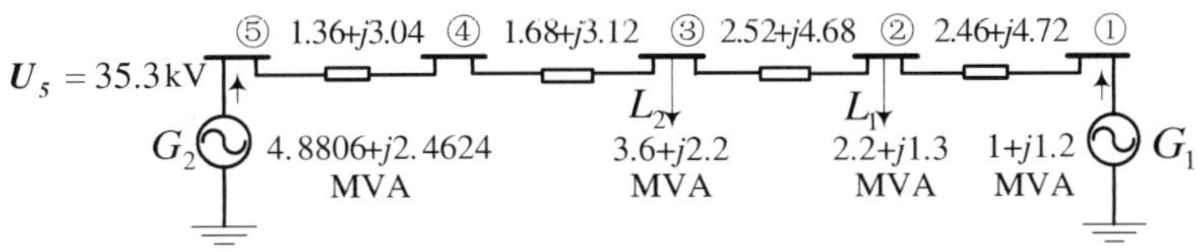

Figure 2. 5-bus distribution grid

Bus 5 in the distribution network is selected as reference. The voltage at this bus is $U_5 = 35.3\angle 0$ (kV). Based on the data in Figure 2, the power flow results are given in Table I. The power delivered to bus 5 and withdrawn from transmission system G_2 is 4.8806+j2.4624 MVA, the total power loss is 0.08105141+ j0.1631660 MVA, and the voltage at each bus is also shown in Table I.

Table.1 Voltage at each bus

bus i	1	2	3	4	5
Voltage V_i(kV)	34.587	34.352	34.456	34.901	35.300
Phase δ_i (rad)	-0.02161	-0.02309	-0.01857	-0.009325	0

The equivalent current through each equivalent impedance of loads and the current withdrawn from each single source are determined based on the initial power flow solution of the system, as shown in Table II.

Table.2 Equivalent currents of sources and loads

	G_1	G_2	L_1	L_2
Current I (kA)	0.0448	0.1565	0.0745	0.1225
Phase (rad)	-0.8977	-0.4636	-0.5571	-0.5671

The equivalent impedance of load L_1 is 394.25+j233.16 Ω and that of L_2 is 244.06+j149.31 Ω. The current through load L_1 and withdrawn from source G_1 is 0.012-j0.012 kA. The current through L_1 and from G_2 is 0.055-j0.024 kA. The active power delivered to load L_1 and withdrawn from source G_1 is 0.458 MVA. The active power to load L_1 and from source G_2 is 1.762 MVA. In a conclusion, η_{11} is 0.21 and η_{21} is 0.79. Similarly, the current through load L_2 from source G_1 is 0.017-j0.022 kA and from G_2 is 0.086-j0.044 kA, the active power delivered to load L_2 from G_1 is 0.787 MVA and from G_2 is 2.783 MVA, and η_{12} is 0.22 and η_{22} is 0.78 respectively.

The load L_1 consists of only one residential customer, L_2 consists of one residential, one commercial, one small and one large industrial customer. Investigation and statistic analysis show that each type of customer's expected demand affected by voltage sags. This demand and the LDC are illustrated in Table III. Thus the load-loss cost is determined, which is also shown in Table III.

Table.3 Customers' load-loss cost

bus	1	2			
Customer type	Resident	Resident	Com-mercialist	Little industr y	Large industr y
LDC($/kW)	18.8	20.4	51.0	33.1	35.4
Peak load (kW)	9.98	12.6	192	236	270
Load-loss cost($)	187.57	257.04	9792	7814.96	9558

In this distribution grid, λ_u, λ_s and λ_l equals 0.5, 0.15 and 0.35 respectively. Consequently, the load-loss cost of load L_1 is 187.57 $ and the cost of L_2 27422 $. At bus 2, the utility owing grids pays 93.79 $ for load-loss cost to customer, all sources pay 28.137 $, and the customer is charged 65.653 $. Furthermore, as for the part of cost which all sources are responsible for, source G_1 and G_2 are charged 5.907 $ and 22.227 $ respectively. Similarly at bus 3, all sources compensate load L_2 4113.3 $ for load-loss cost together, the utility pays 13711 $ and all customers are charged 9597.7 $. Source G_1 pays 904.8 $ and G_2 pays 3208.5 $. As a result, in the total distribution system, G_1 pays 910.707 $ and G_2 pays 3230.727 $. Further at bus 3, the cost allocated to each type customers is λ_l of the total load-loss cost of the same type customers. Thus the residential customers are charged 89.964 $, commercial ones are charged 3427.2 $, little industrial customers are charged 2735.236 $ and large industrial customers are charged 3345.3 $ respectively.

7.Conclusions

For a distribution grid, the transmission grid, the adjacent distribution grid, the independent generator and the load are responsible for voltage sags. As a result, they are responsible for the load-loss cost together from an economical benefit point of view. Obviously, the probable sources of voltage sag are responsible for load-loss cost. Taking example for a simple power system with only residential user connected to the bus, the quantitative evaluation index of load-loss cost is determined. When many sources are in the power system, the share of load-loss cost of all customers at one bus paid by each source is in proportion to the power contribution supplied by corresponding source to equivalent load at the same bus. Further, the compensation from sources is allocated to each type of loads in the direct ratio of load-loss cost. Thus each customer gets its compensation on the basis of its power flow. The load-loss caused by voltage sag is usually analyzed by considering the characterization of voltage change merely. Nevertheless the method proposed in this paper achieves to analyze voltage sag with its physical and financial characteristics. The utility owing grids, generators and loads will do more things to solve the voltage sag problems under financial incentives. So that all sources of voltage sag are responsible for the load-loss cost can improve power system safety.

References

[1] IEEE 1995 Recommended Practice for Monitoring Electric Power Quality. *IEEE Standard 1159-1995.*
[2] Bollen M H J, Sabin D D and Thallam R S 2003 Voltage-sag indices – recent developments in IEEE P1564 Task Force. *CIGRE/IEEE-PES on Quality and Security of Electric Power Delivery Systems* **10** 34-41.
[3] Martinez J A and Martin-Arnedo J 2002 Voltage sag analysis using an electromagnetic transients program. *IEEE Power Engineering Society Winter Meeting* **2** 1135-40
[4] Lamoree J, Mueller D, Vinett P, Jones W, and Samotiy M 1994 Voltage Sag Analysis Case Studies. *IEEE Transaction on Industry Applications* **30** 1083-89
[5] Su C L 2010 Stochastic evaluation of voltages in distribution networks with distributed generation using detailed distribution operation models. *IEEE Transactions on Power Systems* **25** 786-795.
[6] Heine P, Pohjanheimo P, Lehtonen M, and Lakervi E 2002 A method for estimating the frequency and cost of voltage sags. *IEEE Transactions on Power Systems* **17** 290-296.
[7] Tan R H G and Ramachandaramurthy V K 2012 Voltage sag acceptability assessment using multiple magnitude duration function. *IEEE Transactions on Power Delivery* **27** 1984-90
[8] Bollen M H J, Tayjasanant T and Yalcinkaya G 1997 Assessment of the number of voltage sags experienced by a large industrial customer. *IEEE Transactions on Industry Applications* **33** 1465-71.
[9] Park C H, Hong J H and Jang G 2010 Assessment of system voltage sag performance based on the concept of area of severity. *IET Generation, Transmission & Distribution* **4** 683-693.
[10] Lee Geun-Joon, Albu M M and Heydt G T 2004 A power quality index based on equipment sensitivity, cost, and network vulnerability. *IEEE Transactions on Power Delivery* **19** 1504-10.
[11] Peng J C and Jiang H 2002 Contributions of individual generators to complex power losses and flows-part 1: fundamental theory. *IEE Proceedings on Generation, Transmission and Distribution* **149** 182-185.
[12] Peng J C and Jiang H 2002 Contributions of individual generators to complex power losses and flows-part 2: algorithm and simulations. *IEE Proceedings on Generation, Transmission and Distribution* **149** 186-190.

Magnetic Shielding Design for Coupler of Wireless Electric Vehicle Charging Using Finite Element Analysis

W N Zhao [1], X J Yang [2], C Yao[2], D G Ma[2] and H J Tang [2]

[1]Department of Electronic Information and Electrical Engineering, Shang Hai Jiao Tong University, China

Email: wnzhao1992@126.com

Abstract. Inductive power transfer (IPT) is a practical and preferable method for wireless electric vehicle (EV) charging which proved to be safe, convenient and reliable. Due to the air gap between the magnetic coupler, the magnetic field coupling decreases and the magnetic leakage increases significantly compared to traditional transformer, and this may lead to the magnetic flux density around the coupler more than the safety limit for human. So magnetic shielding should be adding to the winding made from litz wire to enhance the magnetic field coupling effect in the working area and reduce magnetic field strength in non-working area. Magnetic shielding can be achieved by adding high-permeability material or high-conductivity material. For high-permeability material its magnetic reluctance is much lower than the surrounding air medium so most of the magnetic line goes through the high-permeability material rather than surrounding air. For high-conductivity material the eddy current in the material can produce reverse magnetic field to achieve magnetic shielding. This paper studies the effect of the two types of shielding material on coupler for wireless EV charging and designs combination shielding made from high-permeability material and high-conductivity material. The investigation of the paper is done with the help of finite element analysis.

1.Introduction

As a new charging way wireless power transmission has attracted attention of all the world. Electric vehicle based on wireless power transmission technology has advantages of security, waterproof, dust-proof and easy to maintain, ensuring safely charging even in cold and damp environment. Inductive power transfer (IPT) is one of the most widely used wireless power transmission method and has an increasing demand in many applications, such as many medical implants, consumer electronics chargers. Compared with plug-in charging, wireless EV charging has several characteristics. In wireless charging system, power is transmitted through magnetic coupler with air gap and the distance of air gap depends on applications. For wireless EV charging, the receive coil pad is usually fixed in the car chassis while the transfer coil pad is on the ground. So the vertical distance of air gap may be 100-400 mm [1]. On the other hand, the output power for EV charging should be several kW depending on the EV types [2]. To achieve this

high output power, wireless EV charging works in high frequency. Due to the air gap between the magnetic coupler, the magnetic field coupling decreases and the magnetic leakage increases significantly compared to traditional transformer, and this may lead to the magnetic flux density around the coil pads more than the safety limit for human. To reduce the operating electromagnetic interference of the wireless charging coupler in the non-working area, magnetic shielding must be considered when designing magnetic coupler.In practice magnetic shielding usually can be achieved by adding high-permeability material or high-conductivity material. For high-permeability material, its magnetic reluctance is much lower than the surrounding air medium because of its high permeability. This will force most of the magnetic line go through the high-permeability material rather than surrounding air. For high-conductivity material, due to the electromagnetic induction effect, the eddy current in the material can produce reverse magnetic field to achieve magnetic shielding.
This paper investigates the characteristics of the two types of shielding material and designs an effective magnetic shielding for coupler of EV charging using finite element analysis. In section II coupler with single kind of material shielding is studied and compared to explore the characteristics of the two kinds of material. In section IV combination shielding composed of high-permeability material and high-conductivity material is designed and investigated.

2. Single kind of material shielding

The coupler design of IPT system is difficult and complex because the magnetic field produced by coupler depends on many factors, such as vertical distance between coupler, geometry, number of turns, shielding material and so on. So the coupler design has to rely on finite element analysis (FEM) software. In this paper, the FEM software JMAG is used to model coupler.
To achieve high transfer power rating and large misalignment tolerance, many different structures of magnetic coupler has been proposed and investigated in [3][4][5]. In this paper square coupler in Fig.1 is chosen as study object to simplify the finite element analysis model. However, the study results and design method also can be applied to other structures.

Fig.1 The structure of square coupler

The coupler in Fig.1 is composed of two coil pad with one layer shielding plate made from high-permeability material or high-conductivity material. In this paper PC47 ferrite manufactured by TDK company and aluminum are selected as representative of high-permeability material and high-conductivity material. The size of shielding plate is 600 mm×600 mm×4 mm and the vertical distance between the magnetic coupler is 200 mm. The distance between shielding plate and winding is 5 mm. The winding in Fig.1 is made from
litz wire whose diameter is 4 mm and the turn of winding is 20. The internal length of winding is 400 mm while the external length of winding is 560 mm. The amplitude of current in the transfer winding is set to 20 A and its frequency is 20 kHz which is consistent with the existing experimental platform.
For wireless EV charging the region between the transfer coil pad and the receive coil pad is the working area contributing to output power and the surrounding air region is non-working area. So it is expected that the magnetic field strength in the working area is strong while in the non-working area magnetic field strength is weak so that high output power and the safety of human standing around the car can be ensured at the same time. To evaluate the magnetic field around the coupler, measurement line along z-axis is selected. The magnetic field strength along the measurement line is shown in Fig.2 and Fig.3. For Fig.2 the simulation model consists of only the transfer coil pad and the measurement line is along z-axis with origin in the center of the transfer coil pad. Fig.3 shows the magnetic

field strength of coupler composed of the transfer coil pad and the receive coil pad. The measurement line for Fig.3 is along z-axis with origin in the center of coupler.

It can be seen that after adding high-permeability or high-conductivity plate back the winding the magnetic field strength in non-working area is reduced to a low level. Furthermore the shielding effect of high-conductivity material is better than high-permeability material. When adding high-permeability shielding to the winding, the magnetic field strength in non-working area is much lower while the magnetic field strength in the working region above is much higher compared to coil pad without any shielding. This means that high-permeability material can help enhance magnetic field strength in the working area and weaken magnetic field strength in non-working area. But after adding high-conductivity shielding the magnetic field strength is very low in both working area and non-working area. This is because the reverse magnetic field produced by the eddy current in the high-conductivity material counteracts the original magnetic field in the whole area.

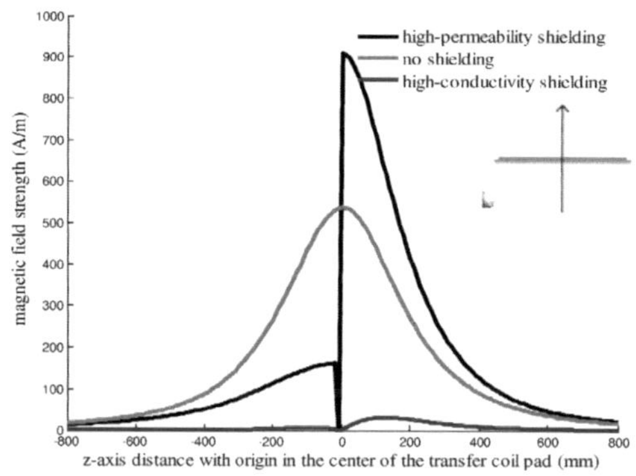

Fig.2 Magnetic field strength along z-axis for only the transfer coil pad

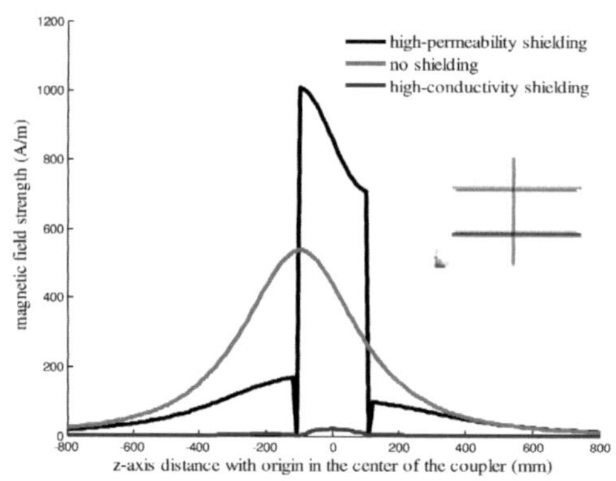

Fig.3 Magnetic field strength along z-axis for coupler

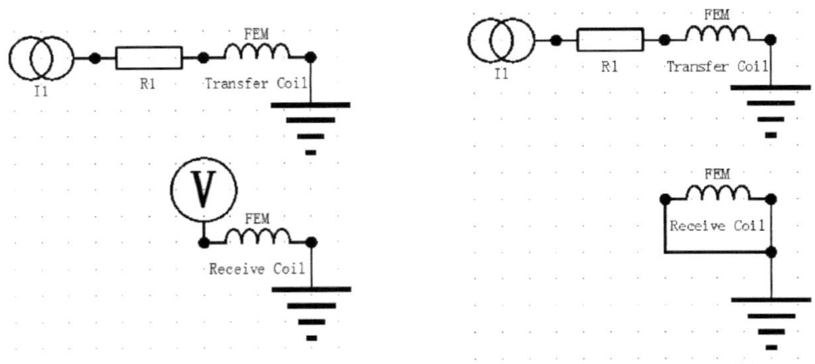

Fig. 4 Circuit calculation of finite element analysis (a) open circuit calculation (b) short circuit calculation

According to [3] power output of wireless EV charging can be calculated by (1)

$$P_{out} = V_{oc} I_{sc} Q_2 = P_{su} Q_2 \qquad (1)$$

Here, V_{oc} is the open-circuit voltage of receive coil. I_{sc} is the short-circuit current in receive coil. P_{su} is the uncompensated power rating and equal to the product of V_{oc} and I_{sc}. Q_2 is defined as quality factor of resonant circuit in receive side.

To calculate uncompensated power rating of coupler, circuit calculation should be added to the coupler model. Fig.4 (a) shows the circuit to calculate open-circuit voltage of the receive coil and Fig.4 (b) shows the circuit to calculate short-circuit current in the receive coil. In Fig.4 (a) AC working current is flowing in the transfer coil while the receive coil is open circuited. In Fig.4 (b) the same current is flowing in the transfer coil while the receive coil is short circuited. In practical application, power supply is usually a voltage source, which can be converted to current source by adding LCL immittance converter [6][7].

In this paper, the receive coil pad is designed to be identical to the transfer coil pad. To ensure strongest coupling effect, the transfer coil pad and the receive coil pad should be center aligned. But in practice application it is very hard for drivers to keep the receive coil pad fixed in the car chassis center aligned to the transfer coil pad on the ground, so the coupler must have enough misalignment tolerance to ensure EV charging under a certain level of misalignment. Fig.5 shows the uncompensated output power rating of coupler under different misalignment. After adding high-conductivity shielding the uncompensated output power rating of coupler increased significantly because the magnetic field strength in the working area is enhanced. The power rating of coupler with high-conductivity shielding is very low. So single kind of high-conductivity shielding plate is inappropriate in wireless EV charging.

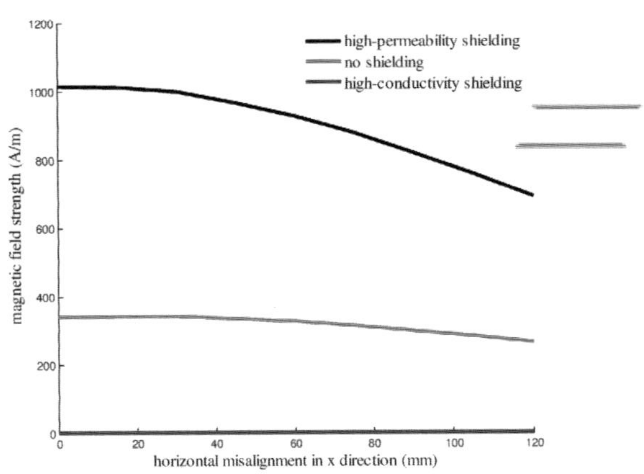

Fig.5 Uncompensated output power rating of coupler under different misalignment

IOP Conf. Series: Earth and Environmental Science **86** (2017) 012039 doi:10.1088/1755-1315/86/1/012039

3.Combination shielding of high permeability material and high conductivity material

In section II high-permeability material shielding and high-conductivity material shielding are compared and the result shows the shielding effectiveness of high-conductivity material is better than high-permeability material. But high-conductivity material will weaken the magnetic field in working area and significantly reduce the output power. So just adding high-conductivity material as shielding is inappropriate in wireless EV charging. The high-permeability material can help enhance magnetic field strength in working area and weaken magnetic field in the non-working area. But its shielding effectiveness is lower than high-conductivity material and in some cases just adding high-permeability material may not guarantee the magnetic field strength in the surroundings meet the safety standard. The International Commission on Non-Ionizing Radiation Protection (ICNIRP) has developed guidelines for different applications to assure the protection of human [8]. The safety limit for general public exposed to flux density is frequency dependent and the spot limit is 27 μT in the frequency of 0.8 to 150 kHz [9].

Considering the shielding characteristics of the two kinds of material, combination of high-permeability material and high-conductivity material may be an optimal design. The position of the two kinds of material is essential and the magnetic field strength along the vertical measurement line under different position of the two kinds of material is shown in Fig.6. It can be seen the high-permeability material should be put close to the winding. That is because high-permeability material acts as shielding by changing the distribution of magnetic line while high-conductivity material acts as shielding by producing reverse magnetic field.

The magnetic field distribution of coupler before and after adding high-conductivity material is shown in Fig.7. The measurement plane is 20 mm above the receive coil pad. It can be seen that the magnetic field strength in non-working area will significantly decrease by adding another shielding layer made from high-conductivity material.

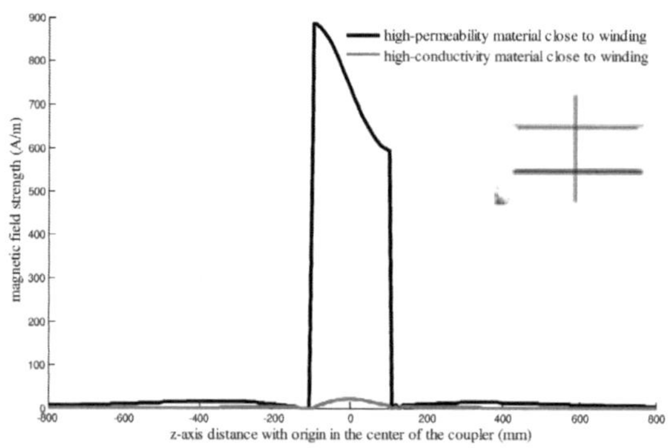

Fig. 6 Magnetic field strength under different position of the two kinds of material

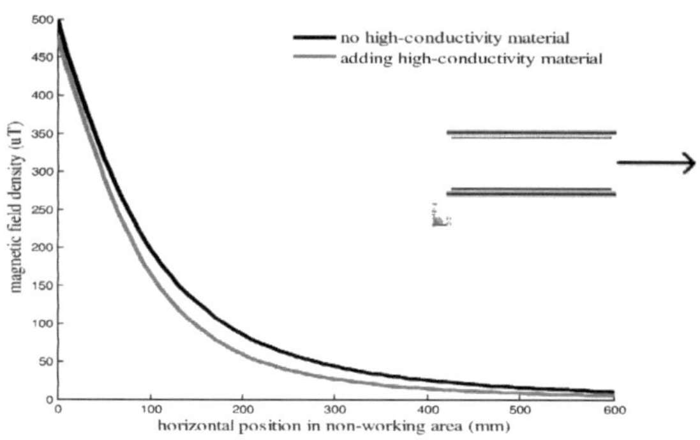

(a)

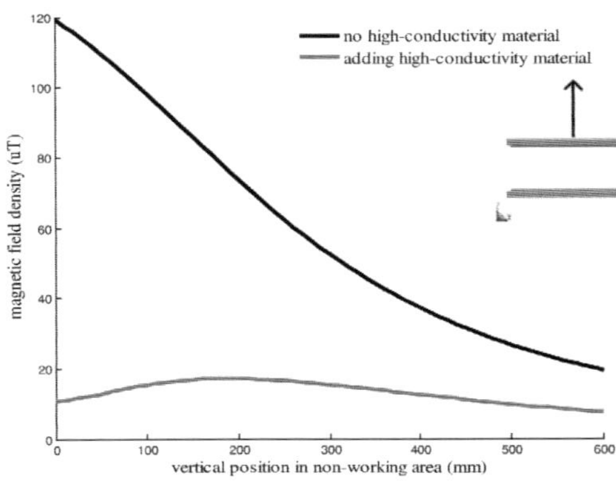

(b)

Fig.7 Magnetic field density in non-working area (a) magnetic field density along horizontal line (b) magnetic field density along vertical line

In IPT system for wireless EV charging, the transfer coil is fixed on the ground and the receive pad is fixed in the vehicle chassis. To evaluate magnetic flux density in non-working area around the magnetic coupler, probe magnetic flux density along x-direction and y-direction when the receive coil and transfer coil is center aligned. Fig.7 (a) shows magnetic flux density along a horizontal line while Fig.7 (b) shows magnetic flux density along a vertical line.

For coupler with high-permeability shielding the magnetic field density in horizontal position is larger than that in vertical position in non-working area. Adding another layer shielding made from high-conductivity material can help reduce magnetic field density in non-working area and magnetic field density in non-working area will drop to a safety level in a short distance which means magnetic radiation has been improved. From Fig.7 it can be seen that the high-conductivity plate back the winding is more effective in improve vertical magnetic radiation. The decrease of magnetic flux density in the horizontal position is very small. In practical high-conductivity shielding ring around the coupler can be added to the coupler to further decrease the magnetic field strength in the horizontal position to ensure human safety.

4.Conclusion

To model and investigate the influence of high-permeability and high-conductivity material shielding coupler of square structure is constructed and simulated using finite element analysis software. This paper compares the shielding effect and analyzes the shielding principle of the two material. The high-permeability material can help enhance magnetic field strength in working area and weaken magnetic field in the non-working area while the high-conductivity material shielding reduces magnetic field strength in both working area and non-working area. So although the shielding effect of high-conductivity material is better than high-permeability material, single kind of high-conductivity shielding is inappropriate in wireless EV charging. To increase the transmission power of the magnetic coupler and further reduce magnetic field strength in the non-working area, combination of the two kinds of material can be used. In practical if the shielding effect is not satisfactory more layer of high-conductivity shielding can be added.

IOP Conf. Series: Earth and Environmental Science **86** (2017) 012039 doi:10.1088/1755-1315/86/1/012039

References

[1] Wang C S, Stielau O H and Covic G A 2005 Design considerations for a contactless electric vehicle battery charger. *IEEE Trans. Industrial Electronics, vol. 52, No. 5* 1308-14.
[2] Nagendra G R, Covic G A. and Boys J T 2014 Determining the Physical Size of Inductive Couplers for IPT EV Systems. *IEEE Journal of Emerging and Selected Topics in Power Electronics, vol. 2, No. 3* 571-583.
[3] Budhia M, Covic G and Boys J Design and Optimisation of Circular Magnetic Structures for Lumped Inductive Power Transfer Systems. *Power Electronics, IEEE Transactions on,9 1-14.*
[4] Stielau O H and Covic G A 2000 Design of loosely coupled inductive power transfer systems. *Proc. Int. Conf. Power Syst. Technol, Perth,Australia* 85–90.
[5] Budhia M, Covic G A, Boys J T and Huang C Y 2011 Development of a single-sided flux magnetic coupler for electric vehicle IPT charging systems. *IEEE Transactions on, Feb.*
[6]Mangesh B, Sunil T and Swama K 2005 Analysis and design of an LCL-T resonant converter as a constant-current *power supply vol. 52, no. 6* 1547-54.
[7]Hao H, Grant A C and John T B 2014 An approximate dynamic model of LCL- T-based inductive power transfer *power supplies vol. 29, no. 10* 5554-66.
[8]International Commission on Non-Ionizing Radiation Protection ICNIRP Guidelines for limiting exposure to time-varying electric and magnetic fields (1Hz to 100 kHz) 2010 Health Phys.99 818–836
[9]Maximum exposure levels to radiofrequency fields – 3 kHz to 300 GHz,Australian Radiation Protection and Nuclear Safety Agency (ARPANSA),2002.

IOP Conf. Series: Earth and Environmental Science **86** (2017) 012040 doi:10.1088/1755-1315/86/1/012040

Optimal Design of Magnetic ComponentsinPlasma Cutting Power Supply

J F Jiang[1],B R Zhu[1], W N Zhao[2], X J Yang[2] and H J Tang[2]

[1]Electric Power Science Research Institute, Shanghai Electric Power Company, State Grid, Shanghai, China.

[2]Department of Electrical Engineering, Shanghai Jiao Tong University, Shanghai, China.

Email: *247721794@qq.com

Abstract. Phase-shifted transformer and DC reactor are usually needed in chopper plasma cutting power supply. Because of high power rate, the loss of magnetic components may reach to several kilowatts, which seriously affects the conversion efficiency. Therefore, it is necessary to research and design low loss magnetic components by means of efficient magnetic materials and optimal design methods. The main task in this paper is to compare the core loss of different magnetic material, to analyze the influence of transformer structure, winding arrangement and wire structure on the characteristics of magnetic component. Then another task is to select suitable magnetic material, structure and wire in order to reduce the loss and volume of magnetic components. Based on the above outcome, the optimization design process of transformer and dc reactor are proposed in chopper plasma cutting power supply with a lot of solutions. These solutions are analyzed and compared before the determination of the optimal solution in order to reduce the volume and power loss of the two magnetic components and improve the conversion efficiency of plasma cutting power supply.

1.Introduction

Cutting and welding has been one of the key technology of many industries such as bridge, construction, machinery, ships, aircraft carriers, railway and power generation equipment industry. The development of modern industry is inseparable from the cutting device and the requirement of cutting quality and cutting efficiency is higher [1]. At present the main metal thermal cutting methods include flame cutting, plasma cutting and laser cutting. Flame cutting has disadvantages of low speed, rough cutting surface and poor cutting accuracy. These problems can be solved through laser cutting and the cutting quality can be significantly improved. But the cost of laser cutting is very high and its cutting thickness is usually limited to 20 mm. However plasma cutting has both advantages of flame cutting and laser cutting, which can achieve faster cutting speed, better cutting quality and low cost at the same time [2-4].

From the beginning of the 1960s, plasma cutting technology continue to progress with the development of power electronic semiconductor devices, from silicon rectifier, thyristor rectifier power supply to chopper type plasma cutting power supply and high power inverter plasma cutting power supply[5]. Among them the chopper type plasma cutting power supply has advantages of simple control circuit, high switching frequency, good control performance and high reliability, so it is commonly used in high power system more than 20kW[6]. In the chopper type plasma cutting power supply, the power transformer and the DC reactor are needed. Due to the high power and high current level the loss of magnetic components may be more than several kW, which will seriously affect the conversion efficiency. For DC reactor the loss is mainly determined by core material and wire type, but when designing transformer its connection mode and winding arrangement should be taken into account.

This paper optimizes the design of phase shifting transformer and the DC reactor based on operation parameters of 45 kW chopper type plasma cutting power supply to decrease volume and power loss of two magnetic components and improve the efficiency of power supply. The second section briefly introduces the topology and the principle of the chopper type plasma cutting power supply. In the third section, the influence of different magnetic material, transformer structure, winding arrangement and wire structure on the characteristics of magnetic components is compared, and the suitable material and structure are selected. In the fourth section, the design procedure of the DC reactor and the power transformer is described, and several design schemes are compared to select optimal design result. The last section is the experimental verification.

2.Principle of Plasma Cutting Power Supply

The topology of chopper type plasma cutting power supply is shown in Fig.1. The chopper type plasma cutting power supply has many advantages, such as simple control circuit, high switching frequency, good control performance and high reliability, so it is the ideal choice in high power plasma cutting power supply [6]. In this paper the power of plasma cutting power supply is set to be 45kW. Because of the high power and large current, a two-channel DC-DC buck converter in parallel operation as shown in Fig.2 is designed. This parallel structure can help partake the system capacity, reduce current stress of switch and improve the system stability. When the output current of plasma cutting machine is less than 130A only one buck DC-DC converter is working and the other one is backup. When the required current is greater than 130A the two converter works at the same time to share the current.

In Fig.2, 1 and 2 represent the two channel Buck DC-DC converter. Their input DC voltage is u_{i1} and u_{i2}. Their DC output current is i_{L1} and i_{L2}. The output current of the power supply is the sum of output current of each channel and it can be adjusted from tens A to 260A continuously according to the type and thickness of the work piece. 3 in Fig.2 represents double secondary winding step-down transformer and 4 means the work piece, cutting torch and arc device.

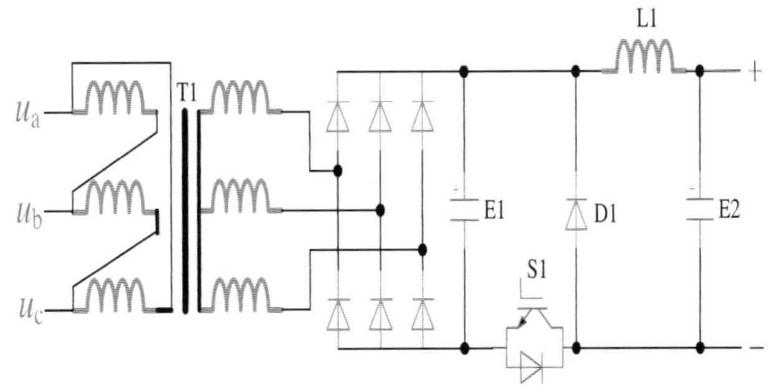

Fig.1 Chopper type plasma cutting power supply

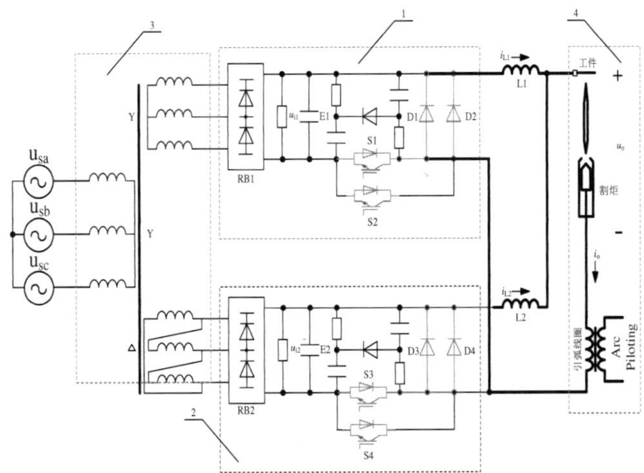

Fig.2 Two channel paralleled plasma cutting power supply

3.Optimal Selection of Magnetic Components

3.1 Magnetic Material. Magnetic component is indispensable key element in high power supply, which can realize energy delivery, storage or work as filter. Its volume and weight generally account for 20% to 30% of the whole power supply system and its loss may account for 30% of total loss. Core loss of magnetic component consists of magnetic hysteresis loss, eddy current loss and residual loss and it can be influenced by magnetic core material, operating frequency, AC magnetic flux density and the waveform. Core loss under sine excitation can be calculated by Steinmetz formula in equation (1):

$$P = K \cdot f_s^{\alpha} \cdot B_{ac}^{\beta} \quad (1)$$

Where,K、 α and $\beta$are determined by the material and the working condition of core and they can be obtained by referring to manual provided by manufacturer or fitting after testing [7]. Core loss under non-sinusoidal excitation can be calculated by correction based on the above formula or Fourier series decomposition [8].

DC reactor in plasma cutting power supply can be divided into two types according to magnetic material of core. One is single magnetic material reactor, the other is mixing magnetic material reactor. For single magnetic material reactor high permeability material with air gap or magnetic powder can be used as core material. In addition, the magnetic core can made from combination

IOP Conf. Series: Earth and Environmental Science **86** (2017) 012040 doi:10.1088/1755-1315/86/1/012040

of high-permeability and low-permeability material or high-permeability and permanent magnet material.The volume, cost, loss and inductance of the above four structure is shown in table 1. Considering these factors select high permeability magnetic material with air gap as magnetic core of DC reactor in this article.

Table 1 Design method comparison of magnet structure

Core material	high-permeability magnetic material with air gap	magnetic powder	combination of high-permeability and low-permeability material	combination ofhighpermeability and permanent magnet material
volume	small	big	medium	small
cost	low	medium	medium	high
loss	medium	medium	low	high
inductance	linear	nonlinear	nonlinear	linear

3.2 Phase-shifted Transformer. Three-phase bridge rectifier is always needed in chopper type plasma cutting power supply. The rectifier is a strong nonlinear and time-varying circuit, which will produce a large amount of harmonic current pollution in the power system and reduce the power factor of the grid. This may lead to distortion of voltage and current and adversely affect other electricity users. Phase-shifting transformer can raise the ladder count of the primary current superposition by increasing number of pulse so as to minimize the harmonic content. Improving power factor by phase-shifted transformer has advantages of easy to realize and high reliability, so it is suitable for high current power supply, high voltage frequency converter and high voltage DC transmission system [9].

Phase-shifting transformer used in plasma cutting power supply in this paper can be divided into two types as shown in Fig.3. One is called 12-pulse phase-shifting transformer, and the other one is 24-pulse phase-shifting transformer. 12-pulse phase-shifting transformer is of simple structure and easy to produce, but the harmonic distortion of current in grid side is 15.22% which means the current distortion is still high. While using 24-pulse phase-shifting transformer the harmonic distortion of current in grid side can be reduced to 7.5%. So the effect of 24-pulse phase-shifting transformer in improving power factor is much better than 12-pulse transformer. But its structure is complex so 24-pulse transformer is fabricated complicatedly. After comprehensive comparison select 12-pulse phase-shifted transformer in this paper.

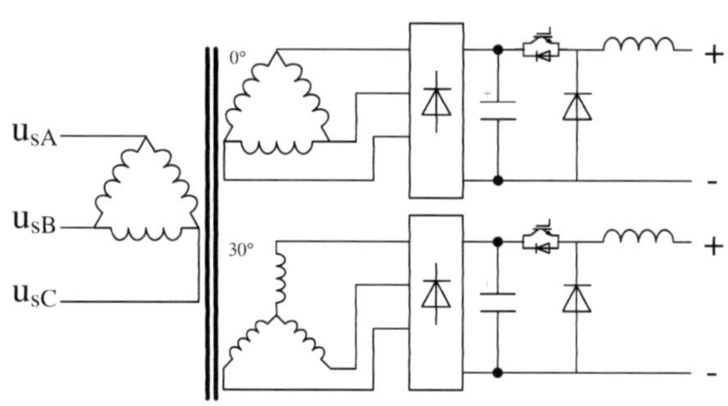

(a) 12-pulse phase-shifting transformer

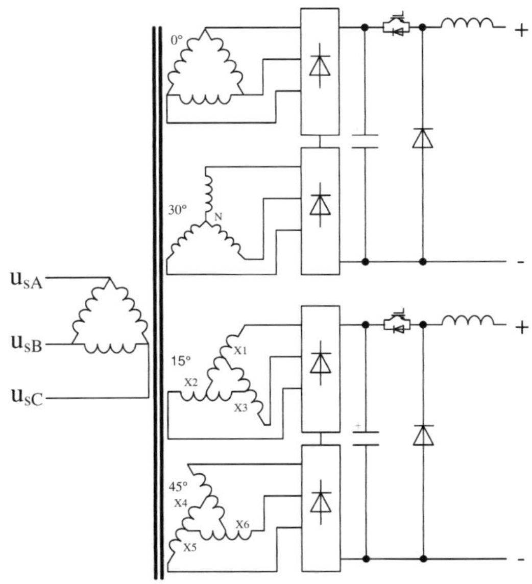

(b) 24-pulse phase-shifting transformer

Fig.3 Design solution of phase shifting transformer

3.3 Winding Arrangement of Transformer. The common winding arrangement of transformer is shown in Fig.4. Here P means primary winding and S means secondary winding. The × in Fig.4 indicates that the current direction is inflow while · means outflow direction. H is the magnetic field strength.

For traditional arrangement, the magnetic field strength H in primary winding increases with the the number of ampere-turns of the winding and reaches maximum in the junction of primary and secondary winding, then H gradually decreases to zero in secondary winding. This traditional arrangement causes great adjacent effect and leakage inductance.

Fig.4 (b) is a sandwich arrangement which can significantly reduce adjacent effect and leakage inductance. The maximum magnetic field strength under sandwich arrangement is about half of that in traditional structure. Fig.4 (c) and Fig.4 (d) are staggered winding and they can help further decrease adjacent effect and leakage inductance. However staggered winding is complex to produce and the parasitic capacitor may increase.

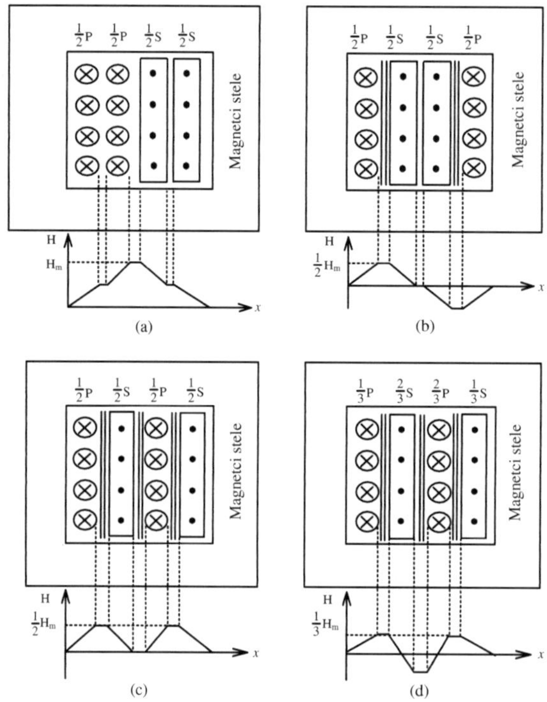

(a) traditional arrangement. (b) sandwich arrangement.
(c) evenly staggered arrangement.(d) unevenly staggered arrangement.

Fig.4 Winding arrangement of transformer

3.4 Structure of Wire. Considering the skin effect and the adjacent effect under high frequency, the current distribution in wire is not average, causing the copper loss increased. The copper loss of wire under high frequency is closely related to the structure of wire so it is necessary to select suitable wire to reduce copper loss. The common used wire in magnetic components includes copper foil, round wire and stranded wire. Stranded wire can realize minimum AC copper loss because its wire diameter is small and stranding techniques can be effective against the skin effect and adjacent effect. But its producing process is complex and the cost is high. The loss of thin copper foil is low in the case of fewer turns, but increasing the number of turns will cause AC loss rise rapidly because of adjacent effect. Round wire is of medium AC loss and has advantages of easy to produce and low cost. But limited by maximum wire diameter and skin effect round wire is not suitable for large current reactor. After comprehensive assessment select copper foil to make DC reactor used in plasma power supply in this paper because it has advantages of moderate cost, small AC loss and good heat dissipation.

4.Design Procedure of Magnetic Components

4.1 Design of DC Reactor. The calculation method of DC reactor is mainly using the AP method, but due to large dc offset the inductance ofDC reactor must be enough high to meet the requirement of the ripple with short-term maximum current operating condition. The design procedure of DC reactor is shown in Fig.5.

When designing DC reactor the input electrical parameters are voltage, the maximum current, rated current, ripple current, frequency and so on. Number of turns, wire diameter, magnetic chip width, thickness and air gap length are selected as design variables. Besides cabinet install size, temperature-rise limit should be added as boundary conditions. The design procedure searches for the scheme which can realize minimum loss or minimum cost. Setting the percentage of loss and cost as design target can make trade-off parameters.

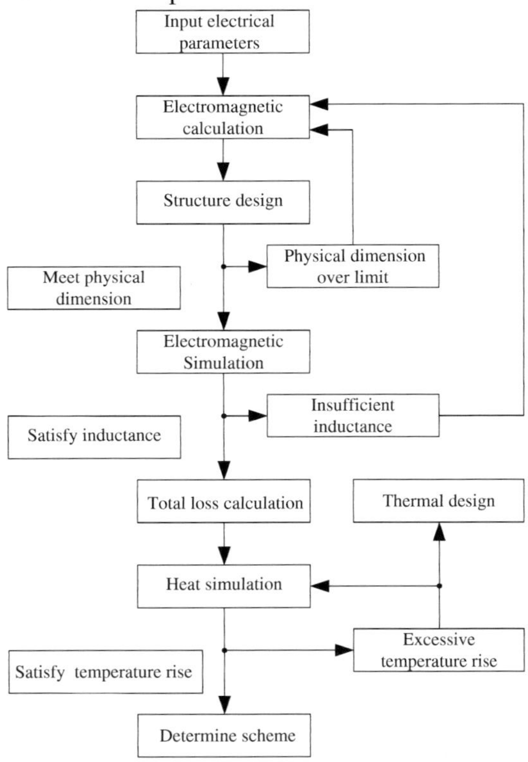

Fig.5Design procedure of DC reactor

The inductance of DC reactor in plasma cutting power supply is determined by ripple current and working frequency. Keeping working frequency constant, the inductance is inversely proportional to ripple current. Assuming the operating frequency of plasma cutting power supply is 10 kHz, the influence of ripple current on weight, loss and overall dimensions of DC reactor is shown in table 2. Increasing ripple current contributes to smaller inductance, lighter DC reactor in weight, smaller loss and smaller dimensions. After comprehensive assessment the current ripple coefficient is set to 5% to ensure stable operation of DC power supply and reduce the weight, volume, loss and cost of DC reactor in plasma cutting power supply.

Similarly when the ripple current is fixed inductance of DC reactor is inversely proportional to working frequency. The current ripple coefficient is set to 5% then change working frequency and analyze weight, loss and dimension of DC reactor, as shown in table 3. Increasing frequency contributes to smaller inductance, lighter DC reactor in weight, smaller loss and smaller dimensions. However due to high power rating of plasma cutting power supply too high working frequency will cause large switching loss. Besides for high power switch IGBT its operating frequency is limited by its switch speed. After comprehensive assessment set switching frequency to be 20kHz. Then the optimal DC reactor design scheme is determined. The inductance of DC reactor is 1 mH and the weight of coil and core are 1.6 kg and 27.1 kg. Its dimension is 156×144×222 mm. The DC coil loss is 163.3 W, while AC coil loss is 1.7 W, and core loss is 32.5 W.

IOP Conf. Series: Earth and Environmental Science **86** (2017) 012040 doi:10.1088/1755-1315/86/1/012040

4.2 Phase-shifted Transformer. The design procedure of phase-shifted transformer is shown in Fig.6. When designing phase-shifted transformer the input electrical parameters are voltage rating, voltage ratio, rated current, leakage inductance and so on. Number of turns, wire diameter, magnetic chip width, airway distance and airway number are selected as design variables. Besides cabinet install size, temperature-rise limit should be added as boundary conditions. The design procedure searches for the scheme which can realize minimum loss or minimum cost. Setting the percentage of loss and cost as design target can make trade-off parameters.

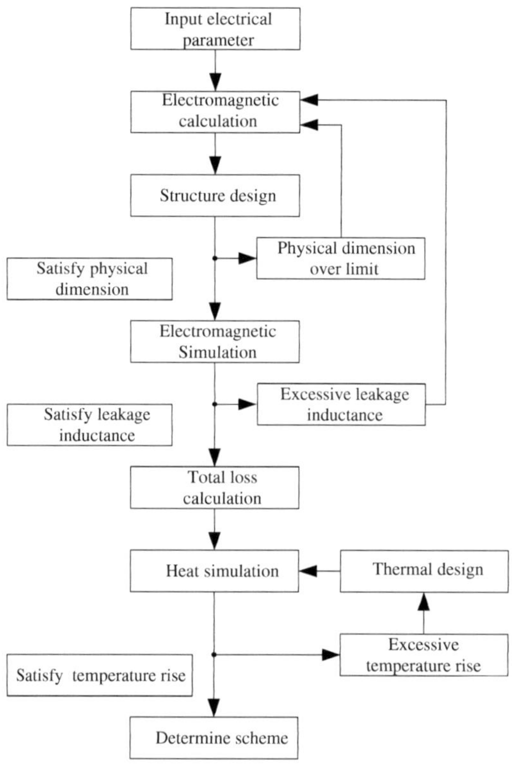

Fig.6 The design procedure of phase-shifted transformer

Table 2 Scheme comparison of fixed frequency

Ripple coefficient	Inductance (mH)	Wire weight (kg)	Core weight (kg)	Total weight (kg)	Wire loss (W)	Core loss (W)	Total loss (W)	Dimension (mm)
0.5%	20	10.4	512.7	523.2	1046.3	0.3	1046.6	416×404×498
1.0%	10	6.5	264.1	270.7	655.7	6.8	662.4	328×342×384
2.0%	5.0	4.3	133.0	137.3	433.2	11.4	444.6	242×302×305
3.0%	3.33	3.8	72.2	76.0	387.1	12.6	399.7	182×288×245
5.0%	2.0	3.0	37.7	40.7	298.9	16.0	314.9	164×196×224

Table 3 Scheme comparison of fixed ripple current

Frequency (kHz)	Inductance (mH)	Wire weight (kg)	Core weight (kg)	Total weight (kg)	Wire loss (W)	Core loss (W)	Total loss (W)	Dimension (mm)
5.0	4.0	4.2	87.7	91.9	422.3	13.1	435.3	204×285×262
7.5	2.67	3.3	57.8	61.1	331.1	25.2	356.3	182×234×242
10.0	2.0	3.0	37.7	40.7	298.9	16.0	314.9	164×196×224
12.5	1.6	2.4	36.8	39.1	238.3	21.9	260.2	160×191×222
15.0	1.33	2.0	35.9	37.8	197.7	28.8	226.6	156×186×222
17.5	1.14	1.8	30.9	32.6	179.0	36.2	215.2	156×162×222
20.0	1.0	1.6	27.1	28.7	165.0	32.5	197.5	156×144×222
22.5	0.89	1.6	21.2	22.8	163.8	30.6	194.4	148×130×222
25.0	0.8	1.5	19.1	20.7	154.4	32.2	186.6	148×119×222

Table 4 Scheme Comparison of aluminium and copper phase shifting transformer

Wire material		Aluminum wire			Copper wire		
Core material		non-oriented silicon steel sheet	orientation silicon steel sheet	amorphous alloy	non-oriented silicon steel sheet	orientation silicon steel sheet	amorphous alloy
Weight (kg)	wire	45	42	48	57	53	85
	core	191	143	169	204	158	154
	total	286	231	264	319	253	286
Wire loss (W)		1422	1340	1535	1397	1361	1618
Core loss (W)		300	143	28	278	146	23
Total loss (W)		1723	1483	1562	1675	1507	1641
Dimension (mm)		580×410×590	550×410×560	610×430×600	550×440×550	520×410×530	570×410×590
Cost		3403	4092	12584	4081	4858	13475

When designing phase-shifted transformer non-oriented silicon steel sheet, orientation silicon steel sheet or amorphous alloy can be selected as transformer core.Aluminum wire or copper wire can be used to make winding. The selection of magnetic material and wire will have an impact on cost and loss. Table 4 compares transformer made from different magnetic material and wire material on weight, size and cost. It can be seen that amorphous alloy has advantage in reducing core loss but its lamination factor and magnetic saturation is lower than silicon steel sheet which leads to large size of transformer and high cost. Compared with non-oriented silicon steel sheet, orientation silicon steel sheet can reduce loss and volume by 10% to 15% even though the cost increases by about 20%. As for the wire, using copper wire can slightly reduce loss and volume, but cause the cost increase by 20%. So after comprehensive analysis and comparison aluminum wire and orientation silicon steel sheet are selected to produce transformer in this paper.

5.Experimental Results

DC reactor and phase-shifted transformer in plasma cutting power supply based on the above design method is shown in Fig.7.

IOP Conf. Series: Earth and Environmental Science **86** (2017) 012040 doi:10.1088/1755-1315/86/1/012040

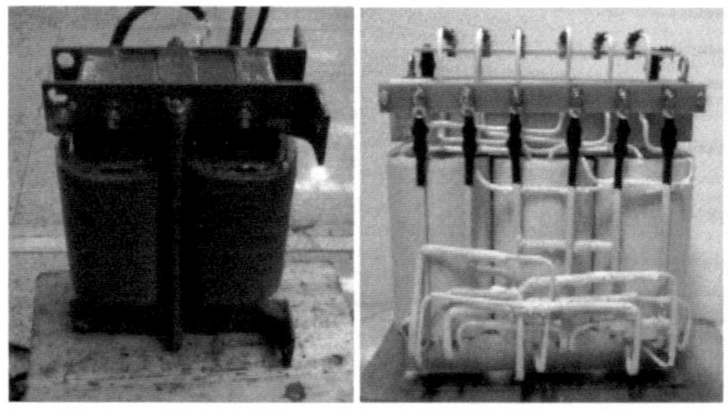

(a) DC reactor (b) phase-shifted transformer

Fig.7 Pictures of DC reactor and phase-shifted transformer

The circuit parameters of plasma cutting power supply with two channel buck DC-DC converter in parallel are as described below. Its input is three-phase AC voltage of 380V then after the transformer the voltage is converted to 220V. Each buck DC-DC converter can realize maximum output current of 130A so the overall maximum output current is 260A. The rated output voltage is 150V. The operating frequency is 20kHz and the ripple current peak-peak value of output current should be less than 10A. According to these parameters the inductance of the DC reactor is designed to be 1.0mH.The waveform under different working stage is shown in the following figures. In these figures the green line represents output voltage and the blue line means output current. Fig.8, Fig.9 and Fig.10 are respectively corresponding to the arc striking stage, start-up stage and normal cutting stage.

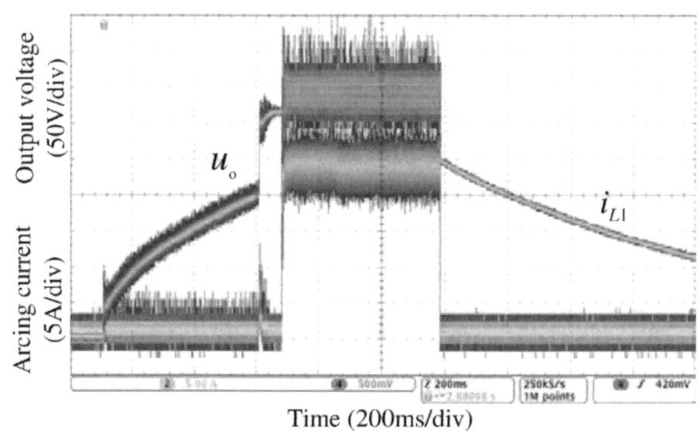

Fig.8 Voltage and current waveforms when arcing

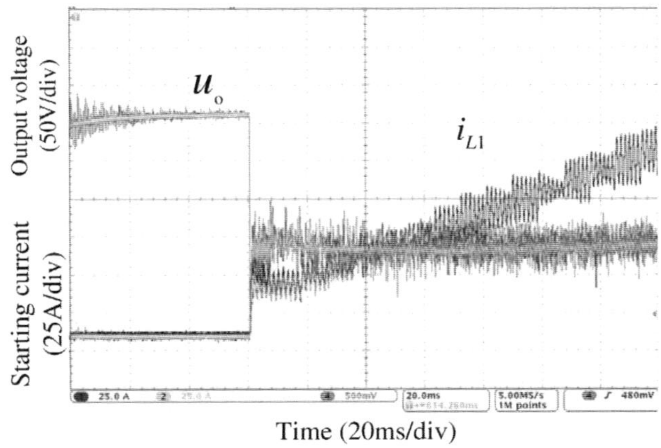

Fig.9 Voltage and current waveforms when starting

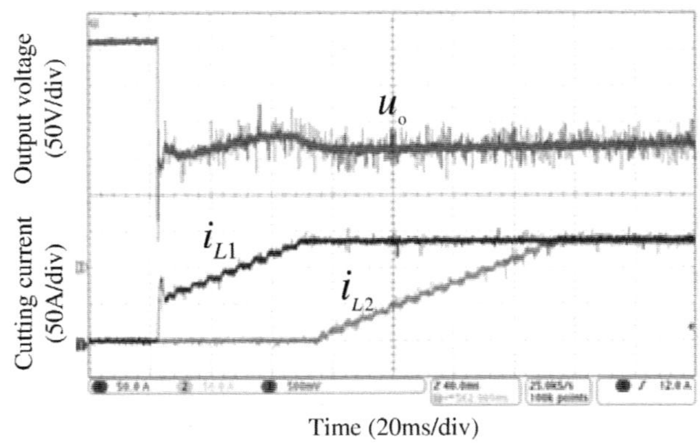

Fig.10 Voltage and current waveforms when cutting

6.Conclusions

This paper optimizes design of DC reactor and phase-shifting transformer in chopper plasma cutting power supply. The paper focuses on the influence of magnetic material, transformer structure, winding arrangement and wire structure on the characteristics of magnetic components. Based on this proposes design procedure of DC reactor and phase-shifting transformer and lists a variety of design schemes. After comparison and analysis optimal solution is determined to ensure lower loss of magnetic components and smaller size.

References

[1] Hao R X, Zheng Q L, You X F, Guo W J and Lin F 2007 Characteristic analysis and experimental research on high-power plasma arc heater power supply. *Trans. of China Electrotechnical Society*V22, Sup.178-82.

[2] Wu J 2010 Research on high-performance plasma cutting power supply system. A thesis in Electrical Engineering for the Degree of Master of Engineering.*Nanjing University of Aeronautics and Astronautics* March.

[3] Chen Y L 2008 IGBT inverted plasma cutting power supply. A thesis in Electrical Engineering for the Degree of Master of Engineering. *Lanzhou University of Technology* May.
[4] Liu B Q, Duan S X and Li X 2012 An improved double closed loop control strategy for air plasma cutting converter. *PrzegladElektrotechniczny***88** 278-282.
[5] Ramakrishnan S, GershenzonM, Polivka F, Kearney T Nand Rogozinski M W 1997 Plasma generation for the plasma cutting process. *IEEE Transactions on Plasma Science, vol. 25, no. 5*937-946.
[6] Yang X J and et al. 2014 Research on two-channel interleaved two-stage paralleled Buck DC-DC Converter for plasma cutting power supply.*2014 International Power Electronics and Application Conference and Exposition, Shanghai* 914-919.
[7] Chas. P S 1984On the law of hysteresis. *Proc. IEEE***72**197-221.
[8] Aguglia D and Neuhaus M 2013 Laminated magnetic materials losses analysis under non-sinusoidal flux waveforms in power electronics systems. *Power Electronics and Applications (EPE)*1-8.
[9] Sikora and Kulesz B 2008 Effectiveness of different designs of 12- and 24-pulse rectifier transformers. *Electrical Machines*1-5.

Institute of Physics
Dirac House, Temple Back
Bristol BS1 6BE UK

ISSN: 1755-1307
ISBN 978-1-5108-5109-2

9 781510 851092

CURRAN ASSOCIATES INC.
proceedings.com